Elements of nuclear physics

Elements of nuclear physics

W. E. Burcham F.R.S.

Longman
London and New York

Longman Group Limited

Longman House,
Burnt Mill, Harlow, Essex, UK

*Published in the United States of America
by Longman Inc., New York*

© Longman Group Limited 1979

First published 1979
Reprinted, with corrections, 1981.

British Library Cataloguing in Publication Data

Burcham, William Ernest
   Elements of nuclear physics.
   1. Nuclear physics
   I. Title
   539.7   QC776

   ISBN 0-582-46027-1

Printed in Singapore by Four Strong Printing Pte Ltd.

Contents

Elements of nuclear physics

Elements of nuclear physics

viii

Preface

This book, like its predecessor *Nuclear Physics, an Introduction* (Longman, 1972) is intended for the undergraduate who is approaching the end of a first degree course in physics. It is not simply an abridged version of the earlier book, but is a new attempt to bring some of the more significant achievements of nuclear physics within the compass of second-and third-year university teaching. Much selection of detail has therefore had to be exercised, and certain topics are expanded by an increased use of examples, for many of which solutions are provided. No formal treatment of elementary-particle physics is offered, but the connection of this subject with nuclear physics is emphasized at many points.

The book is written from the point of view of an experimentalist, but it is assumed that the reader has some acquaintance with atomic physics, special relativity, quantum mechanics and radioactivity. Because it is possible for undergraduates to undertake many useful practical exercises in nuclear physics, actual results obtained in a student laboratory are often given as illustrations, in preference to the more polished data to be found in 'the literature'. In the same spirit, and primarily because of pressure of time in the average course, the number of detailed references has been reduced to a minimum; such references are generally to books or articles in which extensive coverage is given.

The organization of the book follows a simple pattern: Part I deals with the methods and techniques of nuclear physics, Part II surveys nuclear properties through the main present-day models, and Part III is concerned with nuclear reactions. In the last part, a classification into the manifestations of the electromagnetic, weak and strong interactions seemed not inappropriate, although boundaries of this sort cannot always be clearly drawn. As in the earlier book, no account is given of the many important applications of nuclear physics in modern technology, since the concern here is deliberately for the underlying basic discipline.

My thanks for help in the preparation of this book go to many friends and colleagues, both for detailed scrutiny and for informed discussion. Among them are R. J. Blin-Stoyle, G. C. Morrison and D. J. Thouless, together with J. D. Dowell, J. B. A. England, J. Lowe, J. M. Nelson and N. E. Sanderson. A special debt is owed to Dorothy M. Skyrme for her concern for the clarity of parts of Chapter 1 and to Roger New, who prepared the whole of Appendix 5. I am grateful to W. F. Vinen, Head of the Department of Physics of the University of Birmingham, for permission to publish data obtained in the final-year laboratory.

Finally, I acknowledge with gratitude the patience and skill of Catriona Thomas, whose understanding helped vitally in the production of the typescript.

Department of Physics   W. E. Burcham
University of Birmingham
February 1978

I The methods of nuclear physics

1 Introduction

1.1 General survey

It is customary to regard nuclear physics as the field of study that includes the structure of atomic nuclei, the reactions that take place between them, and the techniques, both experimental and theoretical, that shed light on these subjects. Rigid adherence to such limits would, however, exclude much that is both exciting and informative. The nucleus entered physics as a necessary component of the atomic model and nuclear effects in spectroscopy and solid state physics now provide not only elegant methods for determination of nuclear properties but also convincing demonstrations of the powers of quantum mechanics. Equally, those particles sometimes described as elementary or fundamental, although first recognized in the cosmic radiation, soon assumed a role of importance in nuclear problems, especially in the understanding of the forces between neutrons and protons. Advances in the study of particles, or sub-nuclear physics, besides leading to the discovery of new and previously unsuspected physical laws, have frequently stimulated back-reference to complex nuclei for the testing of some new hypothesis or some presumed symmetry. On a less fundamental but socially more significant plane, the exciting problems of astrophysical energy generation and the basic technology of terrestrial power sources relate directly to nuclear structure and nuclear dynamics.

Nuclear physics, therefore, is not a self-sufficient and not an isolated subject. Nor is it strictly a modern subject, since it derives from Becquerel's experiments of 1896, and from Rutherford's wholly classical scattering studies of 1911. Rather, it is an essential ingredient in our understanding of the nature of matter at a level deeper than that of the electronic structure of atoms, molecules and solids but not so deep as that of the particle structure of the proton and neutron.

1.2 Collisions between particles

Collision phenomena provided the original evidence for the hypothesis of the nucleus and have also offered, from the earliest days, powerful methods of studying both nuclear forces and nuclear structure. Although a wave-mechanical treatment is usually necessary for the detailed interpretation of collision experiments, many problems of interest only involve de Broglie wavelengths that are small compared with sizes or geometrical constraints, and classical trajectory calculations can then be made. In all cases it is useful to distinguish between the *kinematic* features of the collision, which are governed by the laws of conservation of energy and of linear and angular momentum, and the *dynamical* features, e.g. the frequency of occurrence of events of a given type, which are determined by the energy available in the collision, by the nature of the forces between the interacting objects and by their internal structure.

1.2.1 Elastic scattering (non-relativistic)

A collision between two particles is *elastic* if the particles of the final state (*f*) are identical with those of the initial state (*i*). 'Identical' means that in both states *i* and *f* the particles are in their ground state, i.e. without internal excitation energy. The principle of conservation of energy then requires that the total kinetic energy is conserved between initial and final states, i.e. $T_i = T_f$.

Figure 1.1 defines kinematic parameters for such a collision in

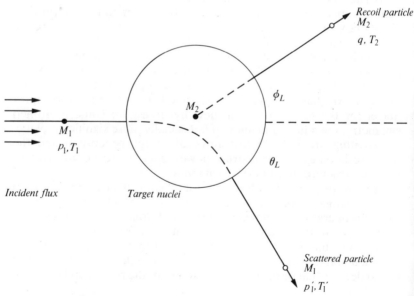

Fig. 1.1 Elastic collision between particles, laboratory coordinates. The circle indicates schematically that within the range of nuclear forces, details of the collision may not be known. The conservation laws for energy and momentum, however, determine the kinematics of the process.

which particles of mass M_1, e.g. from an accelerator, strike particles of mass M_2, equal to nM_1, say. There is a certain region of interaction in which the detailed behaviour of the system is usually not known but when the separation of the particles exceeds the range of their mutual force, some predictions can be made. These follow from the conservation laws for linear momentum and energy, namely

$$\boldsymbol{p}_1 = \boldsymbol{p}_1' + \boldsymbol{q} \tag{1.1}$$

and, for an elastic collision,

$$T_1 = T_1' + T_2 \tag{1.2}$$

together with the non-relativistic relation between momentum and energy

$$p^2 = 2MT \tag{1.3}$$

where p is the absolute magnitude of the vector $\boldsymbol{p}$, i.e. $|\boldsymbol{p}|$. Useful predictions are, referring to Fig. 1.1:

(i) *Angle* $(\theta_L + \phi_L)$ *between particles after collision.* This may be obtained by taking the scalar product of equation (1.1) with itself since the required angle is just that between the two vectors on the right-hand side. The result is

$$p_1^2 = (p_1')^2 + q^2 + 2p_1'q \cos(\theta_L + \phi_L)$$

Substituting for the squared momenta and using $n = M_2/M_1$ we find

$$T_1 = T_1' + nT_2 + p_1'q \cos(\theta_L + \phi_L)/M_1 \tag{1.4}$$

For equal masses, $n = 1$, and from equation (1.2) it follows that $\cos(\theta_L + \phi_L) = 0$, i.e. $\theta_L + \phi_L = \pi/2$. Similarly, for unequal masses with $n > 1$, $(\theta_L + \phi_L) > \pi/2$ while with $n < 1$, $(\theta_L + \phi_L) < \pi/2$. These are the cases of a particle striking a heavier and lighter target respectively.

(ii) *Energy of the scattered particle* T_1' *in terms of the scattering angle* θ_L. In this case we are interested in the angle between $\boldsymbol{p}_1$ and $\boldsymbol{p}_1'$, so equation (1.1) is rearranged before forming a scalar product to read

$$\boldsymbol{q} = \boldsymbol{p}_1 - \boldsymbol{p}_1'$$

whence

$$q^2 = p_1^2 + p_1'^2 - 2p_1p_1' \cos\theta_L$$

Converting to energies

$$nT_2 = T_1 + T_1' - 2(T_1T_1')^{1/2} \cos\theta_L$$

where $T_1^{1/2}$ and $T_1'^{1/2}$ are both positive since p_1 and p_1' are positive. Eliminating T_2 by using the energy equation and rearranging

$$2(T_1'/T_1)^{1/2} \cos\theta_L = (T_1'/T_1)(1+n) - (1-n)$$

whence

$$(T_1'/T_1)^{1/2} = \cos\theta_L[1 \pm (1-(1-n^2)/\cos^2\theta_L)^{1/2}]/(1+n) \quad (1.5)$$

For equal masses $n = 1$ and $T_1' = T_1\cos^2\theta_L$. At $\theta_L = \pi/2$, $\phi_L = 0$, and the incident energy is wholly transferred to the struck particle. The incident particle cannot be scattered backwards $(\theta_L > \pi/2)$. A familiar example in nuclear physics is the scattering of neutrons by protons. For unequal masses, if $n > 1$ the sign $\pm$ in equation (1.5) must be chosen so that $(T_1'/T_1)^{1/2} = p_1'/p_1$ is positive. There is then a *single* value T_1' for each θ_L and any value of θ_L is possible. This is seen in the nuclear scattering of β-particles passing through matter. The maximum kinetic energy transfer to the target nucleus $T_1 - T_1'$ is found from equation (1.5) to be $4T_1/n = 4T_1M_1/M_2$ for large n.

In the case of unequal masses when $n < 1$, as in the collision of an α-particle with an electron, equation (1.5) requires that $\cos^2\theta_L \geqslant 1 - n^2$ so that $\sin\theta_L \leqslant n$, i.e. the scattering angle is always less than $\pi/2$ and tends to zero for $n \ll 1$, so that the heavy particle continues essentially undeviated. In this case the maximum energy conveyed to the target particle is approximately $4nT_1 = 4T_1M_2/M_1$. The alternative signs in equation (1.5) now allow *two* values of T_1' for each allowed θ_L, except $\theta_L = \sin^{-1}n$. This is best understood by reference to the centre-of-mass coordinate system (Sect. 1.2.3).

(iii) *Momentum transfer q to M_2.* In a similar way (see Ex. 1.1) it can be shown that

$$q = |\boldsymbol{q}| = |\boldsymbol{p}_1|(2n/(1+n))\cos\phi_L \quad (1.6)$$

In later chapters, this important quantity will often appear in the form $\boldsymbol{q} = \boldsymbol{q}_i - \boldsymbol{q}_f$ where $\boldsymbol{q}_i$ and $\boldsymbol{q}_f$ are the momentum vectors of a particle before and after scattering.

1.2.2 Inelastic reactions (non-relativistic)

A collision between two particles is *inelastic* if the particles of the final state (f) are different from those of the initial state (i). They may be of a different kind, or in different states of excitation. In such a collision, linear momentum is still conserved but *kinetic* energy is not. The conservation law for energy between states i and f now reads

$$T_i + Q = T_f \quad (1.7)$$

where T_i and T_f are the total kinetic energies in states i, f, and Q is a constant for a given reaction; equation (1.7) defines the Q-*value* as the difference between the total kinetic energies in the final and initial states. For elastic scattering, $Q = 0$.

Typical inelastic *reactions* are the first example of artificial transmutation, discovered by Rutherford in 1919 and later photographed by Blackett in a cloud chamber:

$$^{14}N + {}^4He \rightarrow {}^{17}O + {}^1H \quad (Q = -1\cdot 19\,\text{MeV}) \qquad (1.8a)$$

and the first example of transmutation by accelerated particles (Cockcroft and Walton 1932):

$$^7Li + {}^1H \rightarrow 2\,{}^4He \quad (Q = +17\cdot 35\,\text{MeV}) \qquad (1.8b)$$

These reactions are often abbreviated in the form $^{14}N(\alpha, p)^{17}O$ and $^7Li(p, \alpha)\alpha$ using the special symbols for the light particles.

Inelastic *scattering*, as distinct from an inelastic reaction, does not alter the basic nature of the reacting particles but excites energy levels of one or both, e.g.

$$^{27}Al + p \rightarrow {}^{27}Al^* + p' \quad (Q = -0\cdot 84, -1\cdot 01, \ldots \text{MeV}) \qquad (1.9)$$

This reaction is abbreviated $^{27}Al(p, p')^{27}Al^*$ where the symbol $^{27}Al^*$ indicates an aluminium nucleus in an excited state and p' indicates a proton group with an energy less than the incident energy because of transfer to the particular excited state. Figure 1.2a shows the lowest few levels of the ^{27}Al nucleus and Fig. 1.2b represents the observed energy spectrum of the inelastic proton groups $p_0, p'_1, \ldots, p'_4$, of which p_0 relates to the *ground state* ^{27}Al and is the *elastic* group that is always present. The study of level schemes by inelastic scattering is a basic technique of nuclear spectroscopy. A complementary technique is to measure the energies of the γ-rays emitted when the excited nucleus $^{27}Al^*$ returns to its ground state ^{27}Al (Fig. 1.2c). The discrete energies found may be correlated with the levels as shown in Fig. 1.2a. Together, these two studies present a complete analogy with the Franck–Hertz experiment of atomic physics.

The origin of a finite Q-value in processes (1.8a) and (1.8b) cannot be understood classically, but is immediately clear when relativistic mechanics is used (Sect. 1.2.4) because of the mass–energy relation $E = mc^2$. For the present we simply note that inelastic collisions can be treated in the same way as elastic processes using the equations, in the notation of Fig. 1.3:

$$\left.\begin{array}{r} \boldsymbol{p}_1 = \boldsymbol{p}_3 + \boldsymbol{p}_4 \\ T_1 + Q = T_3 + T_4 \\ p^2 = 2MT \end{array}\right\} \qquad (1.10)$$

The modified energy equation makes calculations more complicated, but the methods are the same as for elastic processes. Thus, we can show that the kinetic energy T_3 of the product particle M_3 is related to the scattering angle θ_{3L} by the equation

$$(M_3 + M_4)T_3 - 2(M_1 M_3 T_1 T_3)^{1/2} \cos\theta_{3L} = (M_4 - M_1)T_1 + M_4 Q \qquad (1.11)$$

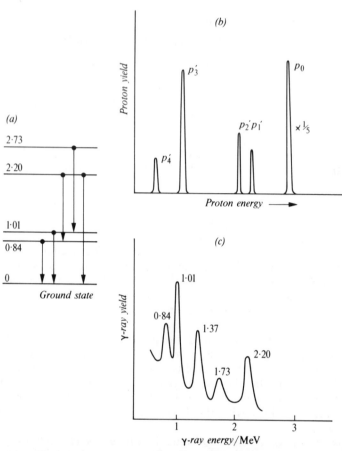

Fig. 1.2 (a) Levels of ^{27}Al nucleus, with excitation indicated in MeV. (b) Spectrum of inelastically scattered protons in the reaction ^{27}Al(p, p')^{27}Al*. (c) Spectrum of de-excitation γ-radiation following inelastic scattering. The radiative transitions are shown on the level diagram (a).

which again predicts a double-valued energy at a given angle under certain conditions (see Ex. 1.4 and Sect. 1.2.3) but in most cases determines a unique energy T_3 at a given θ_{3L}. Observation of the spectrum of M_3 then leads to a series of Q-values corresponding with the excited states of M_4 assuming that M_3 itself is unexcited.

Equations (1.10) show that an exothermic reaction (Q + ve) can occur even when T_1 is zero, assuming that the reacting nuclei can approach sufficiently close. For endothermic reactions (Q − ve), however, the kinetic energy of the incident particle must exceed a certain threshold energy in order to permit momentum conservation. It can be shown (see Ex. 1.5 and Sect. 1.2.3) that the threshold

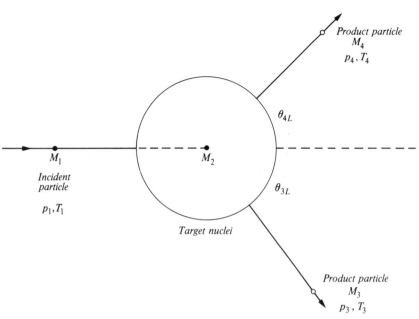

Fig. 1.3   Inelastic collision between particles, laboratory coordinates. The products M_3 and M_4 may be the same as M_1 and M_2 but with some excitation energy; the collision is then described as an *inelastic scattering process*.

energy is $T_1 = |Q|(M_1 + M_2)/M_2$ with the assumption that for kinematic calculations we may write for masses the classical equality $M_1 + M_2 = M_3 + M_4$. The relativistic calculation (eqn. (1.26)) gives a slightly different result because mass is not conserved.

1.2.3   Centre-of-mass system (non-relativistic)

In the classical two-body problem, with no external forces, it is possible to describe the motion by two completely separate equations, one of which defines the motion of the mass centre, while the other gives the *relative* motion of the two particles. *Relative motion* means the variation with time of the distance between the particles. A similar separation may be made when the interaction must be described by Schrödinger's wave equation. The mass centre moves with the same constant velocity before and after the interaction, whatever the nature of the forces between the particles; these forces affect the relative motion only (Ref. 1.2a, Chs. 18 and 19). Thus, the complete motion is the resultant of the centre-of-mass motion and the relative motion, and since collisions are usually studied with the object of learning about the forces between the particles, the first step in comparing observation with theory is to 'subtract-off' the irrelevant motion of the mass centre. That is, one must learn how to transform scattering angles and energies from the laboratory system, which is usually the rest-system of one of the initial-state particles, to those values which would have been obtained had the experiment been carried out in a frame of reference in which the mass centre was stationary. A Galilean transformation is the appropriate one for a non-relativistic treatment.

In the centre-of-mass (c.m.) system the momenta of the particles are equal and opposite both before and after the collision, and for this reason this system is often called the 'zero momentum system'. Only in the special case of a 'storage ring' collision between oppositely moving particles of equal mass and equal energy are the laboratory and c.m. systems identical, and each is of zero momentum. In the more usual case, where the target particle is intially stationary in the laboratory (see Fig. 1.1), the velocity of the centre-of-mass in the laboratory system is given by

$$v_c = (p_1/M_1)M_1/(M_1 + M_2) = p_1/(M_1 + M_2) \qquad (1.12)$$

and the equal and opposite momenta of particles 1 and 2 in the c.m. system are given numerically by

$$p_c = p_1 M_2/(M_1 + M_2) = p_1 n/(1 + n)$$

where $n \; (= M_2/M_1)$ is easily seen to be equal to the ratio of the velocity of M_1 in the c.m. system to v_c, the velocity of the c.m. in the laboratory.

The c.m. momentum p_c is the momentum of a particle of the *reduced mass*

$$\mu = M_1 M_2/(M_1 + M_2) \qquad (1.13)$$

moving with the incident particle velocity (p_1/M_1). The corresponding c.m. wavenumber is

$$k = 1/\lambda = (\mu/\hbar)(p_1/M_1) \qquad (1.14)$$

The c.m. kinetic energies follow from the non-relativistic relation (1.3); thus M_1 has energy $p_c^2/2M_1$, M_2 has $p_c^2/2M_2$ and the total kinetic energy is the sum of these, namely $\frac{1}{2}\mu(p_1/M_1)^2$, which is less than the incident energy $p_1^2/2M_1$ by an amount $\frac{1}{2}(M_1 + M_2)v_c^2$. This is the energy associated with the motion of the mass centre.

The relations obtained so far relate to conditions before the collision. Afterwards, if the collision is *elastic* (Fig. 1.4a) kinetic energies and absolute values of momenta are unchanged in the c.m. system but the momenta are differently directed. Only one scattering angle is needed to describe the collision. To return to the laboratory system, the c.m. velocity must be restored and the corresponding momentum changes are shown in Fig. 1.4b. The

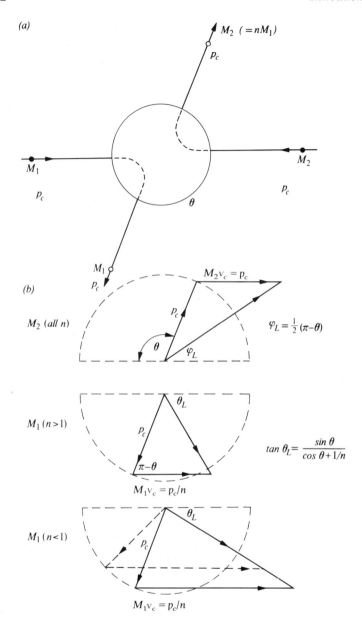

Fig. 1.4 (a) Elastic collision between particles, c.m. system. (b) Momentum diagrams for M_1 and M_2 with $n\ (=M_2/M_1) >$ or < 1. These show the conversion from the c.m. system to the laboratory system. For $n < 1$ the maximum scattering angle is $\sin^{-1} n$.

diagram shows that

$$\phi_L = \tfrac{1}{2}(\pi - \theta) \tag{1.15}$$

$$\tan \theta_L = \sin \theta / (\cos \theta + (1/n)) \tag{1.16}$$

It may also be seen, by comparing the momentum diagrams for M_1, why a double-valued energy occurs for a certain range of θ_L when $n < 1$. This is found when the c.m. velocity *exceeds* that of M_1 in the c.m. system. The two energies coincide at the maximum scattering angle $\theta_L = \sin^{-1} n$ and for larger angles there is no scattering.

If the collision is *inelastic*, then in the c.m. system as well as in the laboratory the final kinetic energy will differ from its initial value by $|Q|$. This energy is distributed between the particles in such a way that their momenta are still equal and opposite. For endothermic collisions the individual c.m. momenta will be zero if the initial c.m. kinetic energy T is equal to $|Q|$. The reaction cannot take place for smaller values of T and $T_0 = |Q|$ is the *c.m. threshold energy*. The corresponding *laboratory threshold energy* T_1 is obtained by putting

$$|Q| = T_0 = \tfrac{1}{2}\mu(p_1/M_1)^2 = (\mu/M_1)T_1 \tag{1.17}$$

and is equal to the value $|Q|(M_1 + M_2)/M_2$ quoted earlier.

The relation between $|Q|$ and the laboratory threshold energy and also the existence of double-valued energies at a given angle for both of the product particles (see Ex. 1.4) are visualized more clearly by use of the c.m. system rather than the laboratory frame of reference.

1.2.4 Relativistic processes

As soon as kinetic energies comparable with rest energies are considered and in all cases in which calculations to the highest accuracy are necessary, relativistic kinematics must be used. In the theory of beta decay, for instance, the electron rest-energy is only $0 \cdot 511$ MeV and typical decay energies are about 1 MeV.

The fundamental difference between classical and relativistic kinematics is that classical definitions of linear momentum and energy lead to conservation laws that are invariant under Galilean but not Lorentz transformations. The theory of special relativity (see, for instance, Ref. 1.2*a*, Ch. 15) shows that for a particle of mass M moving with velocity v, if the linear momentum is defined as $\gamma M v$ and the energy as $\gamma M c^2$, where the Lorentz factor $\gamma = (1 - \beta^2)^{-1/2}$, $\beta = v/c$, then the corresponding conservation laws are indeed Lorentz invariant. The classical definition of momentum is the low-velocity limit of $\gamma M v$; for energy, a stationary particle has rest energy $M c^2$ and a moving particle a total energy $\gamma M c^2$ so that the kinetic energy is $T = Mc^2(\gamma - 1)$. The low-velocity limit of T of course has the classical form $\frac{1}{2}Mv^2$.

It is now possible to see the physical meaning of the Q-value for an inelastic process $1 + 2 \rightarrow 3 + 4$. The energy conservation law is

$$E_1 + E_2 = E_3 + E_4 \qquad (1.18)$$

and in terms of rest energies and kinetic energies this is

$$T_i + (M_1c^2 + M_2c^2) = T_f + (M_3c^2 + M_4c^2) \qquad (1.19)$$

It follows by equation (1.7) that

$$[(M_1 + M_2) - (M_3 + M_4)]c^2 = T_f - T_i = Q \qquad (1.20)$$

so that the Q-value is simply the difference between the initial and final masses. In nuclear reactions such as (1.8a,b) the resulting difference of kinetic energies is generally easily detected, in contrast with the *mass* difference which is small compared with either $(M_1 + M_2)$ or $(M_3 + M_4)$. In particle physics, however, energy changes are often comparable with rest energies.

In equation (1.20) M_1, M_2, M_3 and M_4 should be the masses of bare nuclei in kilograms, and Q is then in joules; conversion to the more useful MeV is straightforward. Since in reactions such as (1.8) the inclusion of the same number of electrons in the initial and final states leads to *neutral atoms* it is permissible to use atomic instead of nuclear masses in Q-value calculations. Equation (1.20) is then written

$$(M_1 + M_2) - (M_3 + M_4) = Q \qquad (1.21)$$

with M_1, M_2, M_3 and M_4 in atomic mass units (a.m.u. or u) as defined in Section 6.1. These masses are tabulated, and again conversion to $\text{MeV}/c^2(\propto \text{MeV})$ or kilograms is direct. In nuclear reactions or decay processes leading to the emission of *positive* electrons, the Q-values calculated from atomic masses must be reduced by an amount $2m_ec^2$ (Sect. 6.5).

In particle physics, absolute masses are written $m_p, m_e, m_\pi, \ldots$ and are specified in kg or MeV/c^2, while momenta $p_p, p_e, p_\pi, \ldots$ are given in MeV/c. In natural units (Ref. 1.1a, p. 22) $c = 1$, and masses, energies and momenta are quoted in MeV or GeV. The momentum energy mass relation is then $E^2 = p^2 + m^2$.

Relativistic collision problems can be solved by the methods of Sections 1.2.1 and 1.2.2 but it is better to use the fact that energy and momentum together form a *four-vector* $p_\mu = (E, \boldsymbol{p})$. Four-vectors (Ref. 1.2b, Ch. 25) are quantities whose magnitude or norm is invariant under a Lorentz transformation; they have a scalar component, e.g. E, and an ordinary vector component, e.g. $\boldsymbol{p}$. Two four-vectors $A_\mu = (A_0, \mathbf{A})$ and $B_\mu = (B_0, \mathbf{B})$ have the following properties, in which a particular sign convention is adopted:

$$\left.\begin{aligned}
A_\mu B_\mu &= A_0 B_0 - \mathbf{A} \cdot \mathbf{B} \\
A_\mu A_\mu &= A_0^2 - A^2 \\
A_\mu + B_\mu &= (A_0 + B_0, \mathbf{A} + \mathbf{B}) \\
(A_\mu + B_\mu) \cdot (A_\mu + B_\mu) &= (A_0 + B_0)^2 - (\mathbf{A} \cdot \mathbf{A} + \mathbf{B} \cdot \mathbf{B} + 2\mathbf{A} \cdot \mathbf{B}) \\
&= A_\mu^2 + B_\mu^2 + 2A_\mu B_\mu
\end{aligned}\right\} \quad (1.22)$$

For the energy momentum four-vector the Lorentz invariant is $E^2 - p^2 = M^2$, the particle mass squared. Energy-momentum conservation implies conservation of p_μ.

For the collision problem we now have

$$p_\mu^1 + p_\mu^2 = p_\mu^3 + p_\mu^4 \tag{1.23}$$

and problems are solved by taking suitable scalar products (see Exs. 1.21–1.24). Thus, if particle 2 is at rest in the laboratory system $p_\mu^2 = (M_2, 0)$ and the (norm)^2 for the initial system by equation (1.22) is

$$M_1^2 + M_2^2 + 2M_2 E_1 = M_i^2 + 2M_2 T_1 \tag{1.24}$$

where $M_i = M_1 + M_2$ is the total mass in the initial state. At a reaction threshold, all the final-state particles are at rest in the c.m. system and in this frame the four-momentum is $p_\mu^c = (M_f, 0)$ where $M_f = M_3 + M_4$. The (norm)^2 for the c.m. system at threshold is M_f^2 and from equation (1.24) we find that the laboratory threshold energy is given by

$$M_i^2 + 2M_2 T_1 = M_f^2 \tag{1.25}$$

or

$$T_1 = (M_f^2 - M_i^2)/2M_2 = (M_f - M_i)(M_f + M_i)/2M_2 = |Q|(M_f + M_i)/2M_2 \tag{1.26}$$

If we put $M_f \approx M_i$ we obtain the value already noted in Section 1.2.3, i.e.

$$T_1 = |Q|(M_1 + M_2)/M_2$$

Equation (1.25) also shows that at very high energies, where $T_1 \gg M_i$, the useful energy M_f only increases as $T_1^{1/2}$, so that colliding beams with a total energy $\approx 2T_1$ offer an attractive alternative to fixed-target accelerators.

It is often necessary to transform four-vectors from one inertial frame of reference to another and the Lorentz transformation for the components of $p_\mu = (E, \boldsymbol{p})$ is

$$\left.\begin{array}{l} E' = \gamma(E - \beta p_z) \\ p'_x = p_x \\ p'_y = p_y \\ p'_z = \gamma(p_z - \beta E) \end{array}\right\} \qquad (1.27)$$

for a velocity β of the accented system relative to the unaccented system in the positive z-direction; c is taken equal to 1. Since the transformation equations are linear they apply to the total energy $E = E_1 + E_2$ of two colliding particles and to their total momentum $\boldsymbol{p} = \boldsymbol{p}_1 + \boldsymbol{p}_2$. It then follows from equations (1.27) that the velocity of the c.m. in the laboratory system is

$$\beta_c = \boldsymbol{p}/E \qquad (1.28)$$

and the Lorentz factor for the c.m. system is then

$$\gamma_c = E/M \qquad (1.29)$$

where M is the total energy in the c.m. system.

In concluding this brief review of the high-energy collision it should perhaps be remarked that the choice of sign for the energy-momentum invariant is arbitrary and values of both $+M^2$ (as here) and $-M^2$ will be encountered. It should also be noted that the connection between mass and energy does not imply the creation or destruction of particles such as protons or neutrons, to which conservation laws apply, unless antiparticles are also involved. Energy changes in reactions relate to the energy of binding of these particles in more complex systems such as nuclei, or to the energy of formation of resonant states (Sect. 2.2).

1.2.5 Cross-sections

The basic probability of a nuclear reaction process is measured by finding the yield of reaction products under well-defined geometrical conditions for a known incident flux of particles or of radiation. Experimental data are reduced to *cross-sections* in the centre-of-mass system for comparison with theoretical calculations.

Consider a two-body process of the type shown in Fig. 1.1 or Fig. 1.3, namely, an elastic scattering of incident particles a (M_1) by nuclei X (M_2)

$$X + a \rightarrow X + a \qquad (1.30)$$

or an inelastic process producing particles b (M_3) and Y (M_4)

$$X + a \rightarrow Y + b \qquad (1.31)$$

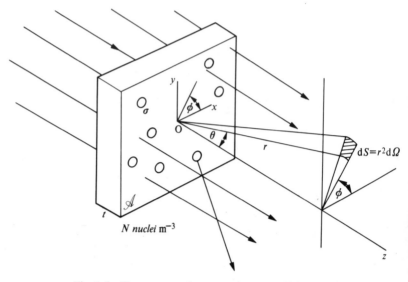

Beam n_i s^{-1}
Flux $= n_i/\mathscr{A}$ m^{-2} s^{-1}

N nuclei m^{-3}

$dS = r^2 d\Omega$

Fig. 1.5 The concept of cross-section, or collision area.

Let such a reaction take place in a target foil of thickness t and area $\mathscr{A}$ containing N target nuclei X per unit volume (Fig. 1.5) and suppose that n_i incident particles a strike the foil per second in a parallel beam, so that the incident flux Φ_i is $n_i/\mathscr{A}$. Suppose also that each target nucleus acts independently of all the others and that the beam is not significantly attenuated by the reaction processes in passing through the foil. The number of such processes that take place per second is then proportional to the number of particles incident (n_i) and to the number of target nuclei $(N\mathscr{A}t)$ and may be written

$$\mathscr{Y} = n_i \cdot (N\mathscr{A}t) \cdot \sigma/\mathscr{A} \tag{1.32}$$

This equation defines the *cross-section* σ as an area. It is not necessarily the area of a target nucleus X, but is rather the area of the incident beam associated with each nucleus through which an incident particle has to pass if it is to cause the specified interaction. It is thus a property of both particles involved in the collision and may depend on the size of each, e.g. on the sum of their radii if the de Broglie wavelength of relative motion is small. Nuclear cross-sections are conveniently measured in *barns* (1 b $= 10^{-28}$ m^2).

Equation (1.32) may be rewritten in terms of the incident flux as

$$\mathscr{Y} = (n_i/\mathscr{A})(N\mathscr{A}t)\sigma = \Phi_i \cdot (N\mathscr{A}t)\sigma \tag{1.33}$$

so that setting $N_s \mathcal{A} t = 1$, the cross-section may be defined as the yield per unit time for one target nucleus placed in unit incident flux. This particular definition is useful for theoretical comparisons because the quantity $\Phi_i \sigma$ is the probability per unit time of the process (1.30) or (1.31), and for this a general formula is provided by quantum mechanics. The expression (1.33) is also useful in practice for computing yields of radioactive elements produced by neutron activation, e.g. in a beam from a reactor; it is only necessary to know the flux, the total number of target atoms subject to this flux, and the cross-section. The yield may also be expressed as

$$\mathcal{Y} = n_i \,.\, (Nt) \,.\, \sigma \tag{1.34}$$

in which (Nt) is the number of target nuclei per unit area of the target, equal for an element to $N_A \rho_s / A$ where N_A is Avogadro's number, A is the atomic weight and ρ_s is the mass per unit area.

If only one process occurs in the target, with total cross-section σ, the actual *attenuation* of the incident beam is easily obtained by letting t become a small thickness dz and writing

$$- dn = d\mathcal{Y} = n(N\,dz)\sigma$$

or

$$n = n_i \exp(-\mu z) \tag{1.35}$$

where n_i is the unattenuated number of particles, $\mu = N\sigma$ is the *linear attenuation coefficient*, sometimes also known as the *macroscopic cross-section* and z is a length. If (1.35) is rewritten

$$n = n_i \exp(-\mu z'/\rho) = n_i \exp(-\mu_m z') \tag{1.35a}$$

then $\mu_m = \mu/\rho = N_A \sigma/A$ (m^2 kg^{-1}) is the *mass attenuation coefficient* which is independent of the physical state of the absorber, and $z' = \rho z = \rho_s$ is a mass per unit area (kg m^{-2}). The attenuation in a thin sample is

$$(n_i - n_z)/n_i = 1 - e^{-\mu z} \approx \mu z \quad \text{or} \quad \mu_m z' \tag{1.36}$$

Consider now the angular dependence of the yield of reaction products a, or b. If $d\mathcal{Y}$ particles are observed per unit time in the solid angle $d\Omega$ defined by a detecting area $dS = r^2\,d\Omega$ (Figs. 1.5 and 1.9), then for one target nucleus, equation (1.33) gives

$$d\mathcal{Y} = \Phi_i\,d\sigma \tag{1.37}$$

and this yield constitutes a flux Φ_s of particles passing through the area dS at distance r from the target nucleus. We may therefore also write

$$d\mathcal{Y} = \Phi_s\,dS = \Phi_s r^2\,d\Omega \tag{1.38}$$

and comparing equations (1.37) and (1.38)

$$d\sigma/d\Omega = r^2 \Phi_s/\Phi_i \qquad (1.39)$$

This quantity is conventionally called the *differential cross-section* and gives the angular distribution of the yield of reaction products with respect to the beam direction. It is numerically equal to the ratio of scattered to incident flux for one target nucleus and unit detector distance.

The differential cross-section is generally a function of the polar angles θ, ϕ (shown in Fig. 1.5) that define the detector axis and is often written $\sigma(\theta, \phi)$. The angle θ relates to the *scattering plane* defined by the direction of incidence and the direction of scattering. The angle ϕ (not that used in Sect. 1.2.1) is the azimuthal angle, between the scattering plane and a fixed reference plane, normally the zOx plane of Cartesian axes centred at O. In nuclear physics, differential cross-sections are measured in millibarns (or some other submultiple of the barn) per steradian, mb sr^{-1}.

The total cross-section for the reaction is obtained by integration of dσ, where

$$d\sigma = \sigma(\theta, \phi)\, d\Omega = \sigma(\theta, \phi) \sin \theta\, d\theta\, d\phi \qquad (1.40)$$

In the case that there is no dependence of yield on the azimuthal angle ϕ, the ϕ integration gives 2π and

$$\sigma = 2\pi \int_0^\pi \sigma(\theta) \sin \theta\, d\theta \qquad (1.41)$$

This cross-section is the same in the laboratory and c.m. system, since the total number of processes per second is the same. The differential cross-sections, however, are related by the equation

$$\sigma(\theta) \sin \theta\, d\theta = \sigma_L(\theta_L) \sin \theta_L\, d\theta_L \qquad (1.42)$$

the azimuthal angle ϕ being the same in each system for coplanar events. Observed cross-sections are transformed to c.m. values by using equation (1.16) with n as in Example 1.10.

The concept of cross-section is easily applied to an incident *wave system* because in practice it is necessary to collimate the incident beam and to place detectors at a distance of many wavelengths from the target nuclei. Interference between incident and scattered waves may then be neglected and asymptotic forms for radial waves may be used.

For *electromagnetic radiation*, with an incident flux of I_i joules m^{-2} s^{-1} the energy loss per second from the beam due to the target foil is written

$$I_i \cdot (N\mathscr{A}t)\sigma \qquad (1.43)$$

in direct analogy with equation (1.33). Alternatively, in analogy with equation (1.34) the loss can be expressed as

$$(I_i \mathscr{A}) . (Nt)\sigma \qquad (1.44)$$

in which $(I_i \mathscr{A})$ is the total energy incident per second on the scattering foil. If both the energy loss and the incident energy are divided by the photon energy $h\nu$, the flux and energy loss then relate to a number of particles as in the earlier discussion. The attenuation of a beam of gamma radiation, for instance, is then given by equation (1.35).

For the *de Broglie waves* that describe particles there is a quantum-mechanical formula for flux

$$\Phi = (\hbar/2im)(\Psi^* \text{ grad } \Psi - \Psi \text{ grad } \Psi^*) \text{ m}^{-2}\text{ s}^{-1} \qquad (1.45)$$

where Ψ is the wavefunction of a particle of mass m. For plane waves representing particles of total energy E and momentum p

$$\Psi_i = A_i \exp i(kz - \omega t) \qquad (1.46)$$

where k is the wavenumber, related to the particle velocity v_i by the equation

$$k = 2\pi/\lambda = 2\pi m v_i/h = p/\hbar \qquad (1.47)$$

and ω is the angular frequency, given by

$$\omega = E/\hbar \qquad (1.48)$$

we find

$$\Phi = (k\hbar/m)|\Psi|^2 = v_i|A_i|^2 \qquad (1.49)$$

For a scattering process equation (1.39) then gives

$$v_s|A_s|^2 = v_i|A_i|^2 . (1/r^2) . d\sigma/d\Omega \qquad (1.50)$$

In the particular case of elastic scattering in the c.m. system we have $v_s = v_i = k\hbar/m$ at *all* angles.

1.2.6 Angular momentum

The angular distribution of a nuclear reaction process about the direction of the incident beam is intimately connected with the angular momentum transfer taking place in the process. Figure 1.6 shows a particle a approaching a fixed target nucleus X along a path which, if continued without deflection, defines an *impact parameter* b. If the linear momentum of particle a is p at a large distance from X and if b is the position vector corresponding to b, then the angular momentum of a about X is

$$L = b \times p \qquad (1.51)$$

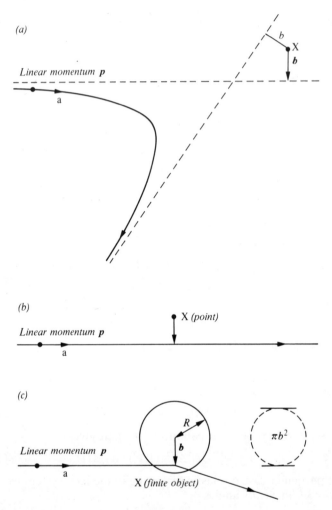

Fig. 1.6 Impact parameter **b** in collision of a particle a with a nucleus X. (*a*) Long-range force. (*b*) Short-range force, point nucleus. (*c*) Short-range force, finite nucleus, showing projected area πb^2 corresponding to collision with impact parameters 0 to *b*.

In the absence of external couples, and of any change of internal angular momentum, this quantity is strictly conserved in the collision so that if an elastic scattering takes place along the trajectory shown in the figure, the final direction of the scattered particle also defines a perpendicular distance *b* from the target X.

Kinematically **b** and therefore **L** can have any magnitude, but for an interaction to take place, a must approach close enough to X for the influence of the interparticle force to be felt. For the long-range forces of electromagnetism and gravity this can be a very large distance and in such processes (e.g. the scattering of α-particles by nuclei as illustrated in Fig. 1.6a, or the passage of comets near the Sun) very large angular momenta may enter the equations of motion as conserved quantities. In processes such as neutron scattering, however, in which the main force is a short-range nuclear attraction (range $\approx 10^{-15}$ m) no effect would take place for impact parameters greater than this if the target nucleus were a point object (Fig. 1.6b). For a finite nucleus of radius R, in which R includes the range of nuclear forces, interaction may take place for impact parameters b up to the value R, i.e. for angular momenta up to

$$L = R \times p \tag{1.52}$$

but not for larger values (Fig. 1.6c).

If the incident particles are described by de Broglie waves, with wavenumber and frequency as defined in equations (1.47) and (1.48), the angular momentum about the centre X is quantized. The quantum values for the total angular momentum and for its component along one of the axes are given by the equations

$$\left.\begin{array}{l} |L|^2 = l(l+1)\hbar^2 \\ L_z = m\hbar, |m| \leqslant l \end{array}\right\} \tag{1.53}$$

where l is the angular momentum quantum number. Frequently, a particle with L^2 and L_z given by (1.53) is said to have angular momentum $l(\hbar)$, although this quantity is strictly the maximum observable component of the vector.

If the z-axis is the axis of incidence then there is no orbital angular momentum component along it, i.e. $L_z = 0$, $m = 0$. The condition (1.52) for interaction is then

$$|L| \leqslant |R \times p|$$

or, using (1.53) and substituting for p in terms of the wavenumber k

$$\sqrt{l(l+1)}(= kb_l) \leqslant kR \tag{1.54}$$

This equation, which is often given in the form $l \leqslant kR$, limits the angular momentum transfer possible from an incident beam with given wavenumber and defines a maximum l-value $l_{\max} \approx kR$. It will be seen in Section 1.2.7 that this limitation is reflected in the degree of complexity of the angular distribution of the reaction process.

Figure 1.6c also shows that the cross-section for an interaction produced by incident particles with classical impact parameters uncertain by the equivalent of one unit of angular momentum cannot exceed

$$\pi(b_{l+1/2}^2 - b_{l-1/2}^2) = (\pi/k^2)[(l+\tfrac{1}{2})(l+\tfrac{3}{2}) - (l-\tfrac{1}{2})(l+\tfrac{1}{2})]$$
$$= (\pi/k^2)(2l+1) \quad \text{or} \quad \pi\lambda^2(2l+1) \quad (1.55)$$

This semi-classical result provides a useful limit on cross-sections.

When the target nucleus X is free to move the linear momentum of each particle with respect to the centre-of-mass is $k\hbar = \mu v_a$ where μ is the reduced mass. The total angular momentum relative to the centre-of-mass is $\mu v_a b = k\hbar b$. These quantities may be used in equations (1.51–1.55).

Angular momentum, like linear momentum and energy, is normally specified for nuclear problems in the centre-of-mass system, since the relation between external couples and rate of change of angular momentum

$$G = dL/dt \qquad (1.56)$$

is independent of the motion of the c.m.s. If $G = 0$ then L is a constant of the motion. Such cases arise when a particle interacts with another through a purely *central* force (directed along the line of centres), and also when a charged particle moves in a plane perpendicular to the lines of force of a uniform magnetic field B (Fig. 1.7). In the latter case, for a particle of mass M and charge e, the angular momentum about an axis parallel to the lines of force and through a point O is

$$L = r \times Mv \qquad (1.57)$$

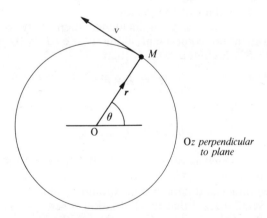

Oz *perpendicular to plane*

Fig. 1.7 Circular motion of a charged particle in a plane. This may be due to (*a*) a uniform magnetic field of induction B perpendicular to the plane, or (*b*) an attractive, radial force due to another particle at the centre O. In both cases the angular momentum is a constant of the motion.

and the force due to the magnetic field is

$$\boldsymbol{F} = e \cdot \boldsymbol{v} \times \boldsymbol{B} \qquad (1.58)$$

This force produces an acceleration which can be equated to the acceleration of the particle moving in a circle of centre O, and radius r with a uniform angular velocity $\omega(= v/r)$ given by

$$evB/M = \omega^2 r = v^2/r$$

or

$$\omega = eB/M \qquad (1.59)$$

the so-called *cyclotron frequency* in angular measure (see Sect. 4.3.1). The angular momentum $\boldsymbol{L}$ is then also constant, as required. This is in fact an example of an axially symmetric situation, in which the strict requirement is that the *component* L_z of angular momentum along the direction of the magnetic field should be constant.

Particles and nuclei may possess an intrinsic angular momentum or *spin* $\boldsymbol{S}$ whose maximum observable z-component is $S\hbar$ where the quantum number S is integral or half-integral. For a single particle moving with respect to a centre of force with orbital angular momentum $\boldsymbol{L}$, the total angular momentum is then

$$\boldsymbol{J} = \boldsymbol{L} + \boldsymbol{S} \qquad (1.60)$$

Conventionally, these symbols are used for the total electronic momenta of an atom; for a single electron within the atomic structure, or for a single nucleon in the nuclear shell model (Ch. 7), lower-case letters are used, i.e.

$$\boldsymbol{j} = \boldsymbol{l} + \boldsymbol{s} \qquad (1.60a)$$

It is also conventional to use $\boldsymbol{J}$ for the spin of an elementary particle, and $\boldsymbol{I}$ for the spin of a complex nucleus, rather than $\boldsymbol{S}$. In problems in which both nuclear and electronic momenta are involved, such as hyperfine structure (Sect. 3.4), the total mechanical angular momentum of an atom is written $\boldsymbol{F}$ and

$$\boldsymbol{F} = \boldsymbol{I} + \boldsymbol{J} \qquad (1.60b)$$

For a charged particle there is a magnetic moment $\boldsymbol{\mu}_l$ associated with any orbital angular momentum $\boldsymbol{L}$ because this creates a circular current, and for many particles with mass there is an intrinsic magnetic moment $\boldsymbol{\mu}_s$ associated with the spin $\boldsymbol{S}$ if this is non-zero.

If a nucleus with spin $\boldsymbol{I}$ (and $\boldsymbol{L} = 0$) is subject to a magnetic field of induction $\boldsymbol{B}$ along the z-axis (Fig. 1.8), a couple $\boldsymbol{G} = \boldsymbol{\mu}_I \times \boldsymbol{B}$ arises tending to alter the angle β between the nuclear spin axis and O. Because of conservation of angular momentum along Oz, this cannot happen, but as in the case of the electrons of an atom in the

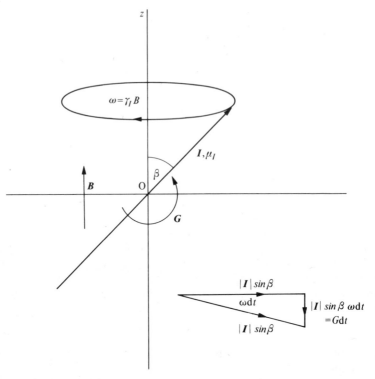

Fig. 1.8 Precession of nuclear spin **I** in a magnetic field **B**. The couple is due to the interaction of **B** with the (positive) nuclear magnetic moment μ_I. The angular momentum created along the axis of the couple **G** by the precession in time dt is represented in the vector triangle.

Zeeman effect, a *precessional motion* occurs in which the z-component of **I** stays constant but the **I**-vector rotates round Oz. The couple is balanced because this rotation ensures that a changing component of **I** continually appears along the axis of the couple. The motion is such that

$$I_z = |\mathbf{I}| \cos \beta = \text{constant}$$
$$\mathbf{G} = \boldsymbol{\mu}_I \times \mathbf{B} = \text{d}\mathbf{I}/\text{d}t = \boldsymbol{\omega} \times \mathbf{I} \tag{1.61}$$

where ω is the angular velocity of the precession. This is immediately obtained as

$$\omega = (\boldsymbol{\mu}_I / \mathbf{I})B = \gamma_I B \tag{1.62}$$

where $\gamma_I = \boldsymbol{\mu}_I / \mathbf{I}$ is the *gyromagnetic ratio* for the nucleus; it is a ratio of magnetic moment to mechanical angular momentum. This precessional frequency occurs in all nuclear problems involving interactions with both static and time-varying magnetic fields.

The total energy of a particle of mass M moving about a centre of force under radial, conservative forces may be written to show the effect of conservation of angular momentum explicitly. Thus, using coordinates r and θ in the plane of motion (Fig. 1.7):

$$E = T + V(r)$$
$$= \tfrac{1}{2}M(\dot{r}^2 + r^2\dot{\theta}^2) + V(r)$$
$$= \tfrac{1}{2}M\dot{r}^2 + [L^2/2Mr^2 + V(r)] \tag{1.63}$$

where $L = Mr^2\dot{\theta}$ (writing $v = r\dot{\theta}$ in equation (1.57)). If then E is expressed as

$$E = \tfrac{1}{2}M\dot{r}^2 + V_{eff} \tag{1.64}$$

the new quantity V_{eff} is an effective potential energy arising partly from the centrifugal effect; it tends to prevent the close approach of the orbiting particle to the centre of force. In the quantum theory of the motion of a particle in a potential well the orbital momentum-dependent potential has an effect on the allowed energy levels of the system which varies from zero in the case of the hydrogen atom to a displacement comparable with that due to a change of principal quantum number for a nucleon in a nucleus (see Sect. 7.2).

1.2.7 Partial-wave theory

Wave theory provides a description of collisions in both nuclear and particle physics which pays particular attention to the role of the conserved quantity angular momentum. The partial-wave formalism, originated by Lord Rayleigh, is especially suitable when scattering is due to a localized spherically symmetrical potential of an extent R comparable with the reduced de Broglie wavelength λ ($= 1/k$) of the incident particle.

The formalism does not explain *why* scattering occurs; that information is contained in the interaction potential for the problem. It does, however, *describe* the scattering in terms of phase shifts which are readily determined from experimental differential cross-sections and which may be predicted according to hypotheses about the potential. The phase shifts indicate which angular momentum states are important in the scattering and their variation with c.m. energy may help to locate favoured states or resonances in the interaction, e.g. states of the compound system formed by the incident particle plus target.

The procedure is to set up the problem as a superposition of spherical wave solutions of the Schrödinger equation in the coordinates (r, θ, ϕ). If the interaction potential involves only a central force (depending only on the particle–target distance r), such solutions are eigenfunctions of the angular momentum operator and describe particles with definite $\mathbf{L}^2$ and L_z with respect to, say, the target as origin and a selected quantization axis Oz. For most problems the time variation of the wavefunction will disappear when particle fluxes are calculated, and it will be omitted from the beginning, so that essentially an energy-conserving steady state is

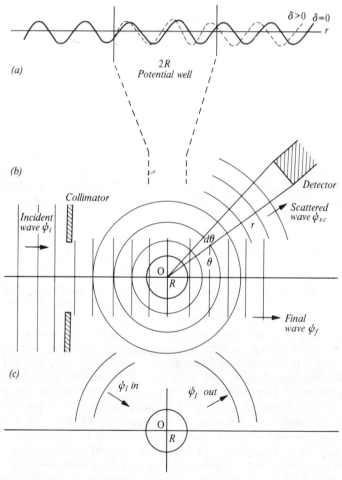

Fig. 1.9 (*a*) Radial de Broglie wave amplitude $r\psi$ for a free particle as a function of r showing the phase shift $\delta > 0$ produced by an attractive potential over a distance $2R$. No amplitude change is shown. For a repulsive potential a similar diagram shows that $\delta < 0$. (*b*) Scattering diagram (schematic) showing collimation of incident plane wave and scattered wave (not necessarily isotropic in amplitude) arising from the target nucleus. The final wave system is no longer an undisturbed plane wave. The direction Oz and the detector axis define the scattering plane, which makes an angle ϕ, not shown, with a fixed reference plane xOz. (*c*) Partial waves; for $l > 0$ the waves are non-isotropic.

under consideration. A particular de Broglie spherical wave amplitude as a function of r is shown in Fig. 1.9a; at a potential boundary there may be a change in amplitude (η) and/or a change in phase (δ). If only a phase change occurs then elastic scattering is described; an amplitude change means that inelastic processes occur.

The partial-wave formalism will be illustrated for the scattering of uncharged, spinless particles a (M_1) incident upon a spinless target nucleus X (M_2) at the origin O (Fig. 1.9b). The incident particles are represented by a plane de Broglie wave whose direction of propagation defines the z-axis:

$$\Psi_i = A_i \exp ikz \tag{1.65}$$

where the wavenumber k is given by equation (1.47) and the time factor $\exp(-i\omega t)$ is omitted. Strictly k and ω should refer to the c.m. system, in which

$$k = \mu v_1/\hbar \quad \text{(see eqn (1.14))} \tag{1.66}$$

where μ is the reduced mass and v_1 the laboratory velocity of the incident particles. For simplicity M_2 will be assumed infinite so that the target nucleus is fixed and the laboratory and c.m. systems are identical. The target interacts with the incident particle through a spherically symmetrical potential which may be assumed to exist within the volume of radius R shown in Fig. 1.9b.

The wavefunction (1.65) is an eigenfunction of the *linear* momentum operator $-i\hbar\partial/\partial z$ with eigenvalue $\hbar k = M_1 v_1$ and the flux of incident particles is given by equation (1.49) as

$$\Phi_i = (k\hbar/M_1)|A_i|^2 = v_1|A_i|^2 \tag{1.67}$$

To convert to a spherical wave representation it is first noted that the particles have no angular momentum component about the z-axis, i.e. $L_z = 0$. The angular part of the wavefunction, therefore, only involves spherical harmonics of the type $Y_l^0(\theta, \phi)$ and these may be expressed in terms of Legendre polynomials using the relation

$$Y_l^0(\theta, \phi) = [(2l+1)/4\pi]^{1/2}P_l^0(\cos\theta) = [(2l+1)/4\pi]^{1/2}P_l(\cos\theta) \tag{1.68}$$

The incident wavefunction is then written, using (r, θ) coordinates and a known expansion

$$\Psi_i = A_i \exp(ikr\cos\theta)$$

$$= A_i \sum_{l=0}^{\infty} (2l+1)i^l j_l(kr)P_l(\cos\theta) \tag{1.69}$$

where $j_l(kr)$ is a *spherical Bessel function* and l is the quantum number for the orbital angular momentum about the origin, such that

$$|\boldsymbol{L}|^2 = l(l+1)\hbar^2 \qquad (1.70)$$

(and $L_z = 0$ by choice of axis).

The spherical Bessel functions have a simple form in the asymptotic limit, when observations are made at a great many wavelengths distant from the scattering centre. This is, in fact, the practical case, Fig. 1.9b, in which the incident beam is collimated so that at the detector there is no interference with the scattered wave. The asymptotic form (valid for all kr for $l = 0$) is

$$j_l(kr) \to \sin(kr - l\pi/2)/kr \qquad \qquad .$$
$$= [\exp i(kr - l\pi/2) - \exp -i(kr - l\pi/2)]/2ikr. \qquad (1.71)$$

in which we see two terms, corresponding when the time factor is inserted with an *outgoing and incoming partial wave* indexed by l, the angular momentum number (Fig. 1.9c).

If no target nucleus is present there is no scattering and the incoming and outgoing fluxes for large r obtained from equation (1.45) are identical.

If the target nucleus is present there will be some effect on the outgoing partial wave (but not on the incoming partial because of the causality principle) and the final asymptotic wave system takes the form

$$\Psi_f = A_i \sum_{l=0}^{\infty} (2l+1)i^l P_l(\cos\theta)[\eta_l \exp 2i\delta_l \exp i(kr - l\pi/2)$$
$$- \exp -i(kr - l\pi/2)]/2ikr \qquad (1.72)$$

in which η_l is the amplitude change (a real number between 0 and 1) and $2\delta_l$ is the phase difference between the outgoing partial wave with and without the scatterer present. The formula (1.45) now shows that the incoming and outgoing fluxes in the lth partial wave are not identical and the difference may be written

$$(\text{Incoming} - \text{outgoing flux}, l\text{th wave}) \propto (k\hbar/M_1)[1 - |\eta_l \exp 2i\delta_l|^2]$$
$$\propto (k\hbar/M_1)(1 - \eta_l^2) \qquad (1.73)$$

The wavefunction ψ_f of (1.72) describes just the elastically scattered particles, whether or not inelastic processes take place. It differs from the wavefunction ψ_i without the scatterer because the target nucleus originates a physical scattered wave of amplitude

$$\Psi_{sc} = \Psi_f - \Psi_i$$
$$= A_i \sum_{l=0}^{\infty} (2l+1)i^l P_l(\cos\theta)[(\eta_l \exp 2i\delta_l - 1)\exp i(kr - l\pi/2)]/2ikr$$

by direct subtraction of equations (1.72) and (1.69), using (1.71)

$$= A_i \sum_{l=0}^{\infty} (2l+1)P_l(\cos \theta)[\eta_l \exp 2i\delta_l - 1] \exp ikr/2ikr \quad (1.74)$$

(using the identity $i^l = \exp il\pi/2$)

$$= A_i f(\theta) \exp (ikr)/r \quad (1.75)$$

where $f(\theta)$ is a length known as the *scattering amplitude*. The existence of a scattered wave implies removal of particles from the direct beam, and from the definition (1.39) we may immediately obtain the differential cross-section $d\sigma_{el}(\theta)/d\Omega$. The necessary quantities are

$$\text{incident flux} = k\hbar/M_1 . |A_i|^2$$
$$\text{scattered flux} = k\hbar/M_1 . |A_i|^2 |f(\theta)|^2/r^2 \quad (1.76)$$

whence

$$d\sigma_{el}/d\Omega = |f(\theta)|^2$$

Often, and especially in particle physics, it is useful to write

$$f(\theta) = 1/k \sum_{l=0}^{\infty} (2l+1)P_l(\cos \theta)f_l \quad (1.77)$$

where

$$f_l = (\eta_l \exp (2i\delta_l) - 1)/2i \quad (1.78)$$

is a complex quantity known as the *partial-wave amplitude*. The integrated cross-section for elastic scattering is then

$$\sigma_{el} = \int d\sigma_{el} = \int 2\pi |f(\theta)|^2 \sin \theta \, d\theta$$

$$= (1/k^2) \sum (2l+1)^2 . 4\pi/(2l+1) . |f_l|^2$$

using the orthogonality and conventional normalization of the Legendre polynomials. It follows that

$$\sigma_{el} = \sum \sigma_{el}^l = (4\pi/k^2) \sum (2l+1) |f_l|^2$$

$$= (\pi/k^2) \sum (2l+1)[1 + \eta_l^2 - 2\eta_l \cos 2\delta_l] \quad (1.79)$$

where σ_{el}^l is the integrated cross-section for the lth partial wave.

We now consider the particular cases of zero and finite inelastic scattering, and some general results.

(i) *No inelastic processes* (elastic scattering only).   If elastic scattering is the only process that takes place in the interaction, there is no absorption and $\eta_l = 1$ for all partial waves. From (1.79) the cross-section then takes the simpler form

$$\sigma_{el} = (4\pi/k^2) \sum_{l=0}^{\infty} (2l+1) \sin^2 \delta_l \qquad (1.80)$$

Substitution of $\eta_l = 1$ into the asymptotic forms of the final wave (1.72) and scattered wave (1.74) shows that δ_l is the phase shift of the lth component of these waves, whereas the phase shift of the outgoing *partial* wave is $2\delta_l$ (eqn (1.72)).

(ii) *Finite inelastic scattering.*   When non-elastic processes occur, the corresponding outgoing particles have a different velocity and perhaps a different nature from the incident beam and ψ_{sc} cannot be obtained in the form (1.74). The integrated cross-section for all such processes, however, may be obtained from equation (1.73) and definition (1.39). Integrating over the whole solid angle using the normalization of $P_l(\cos \theta)$ and summing the partial-wave series we obtain

$$\sigma_{inel} = (\pi/k^2) \sum (2l+1)(1-\eta_l^2) \qquad (1.81)$$

In this case, both elastic and inelastic processes occur together and the total removal of particles from the beam is represented by

$$\sigma_{total} = \sigma_{el} + \sigma_{inel}$$

$$= (\pi/k^2) \sum (2l+1)[2-2\eta_l \cos 2\delta_l] \qquad (1.82)$$

(iii) *Limiting cross-sections.*   The cross-sections (1.79), (1.81) and (1.82) are expressed as the incoherent sum of partial cross-sections corresponding to particles incident upon the scattering centre with angular momentum component $l\hbar$, i.e. with a classical impact parameter $l\hbar/p_1 = l/k$. Such particles will interact with the target nucleus if $l/k < R$, the nuclear dimension. If they so interact, the maximum cross-sections that may be observed are seen to be

$$\left. \begin{array}{l} \sigma_{el}^l = (4\pi/k^2)(2l+1) \\ \sigma_{inel}^l = (\pi/k^2)(2l+1) \\ \sigma_{total}^l = (4\pi/k^2)(2l+1) \end{array} \right\} \qquad (1.83)$$

The elastic cross-section reaches its maximum (resonance) value when $\eta_l = 1$ and $\delta_l = \pi/2$ and the inelastic cross-section then vanishes. The inelastic cross-section reaches its maximum value, which is exactly that given in equation (1.55) as a result of semi-classical considerations, when $\eta_l = 0$, corresponding to complete absorption of the outgoing partial wave; the elastic cross-section then has the same value and the total cross-section is $(2\pi/k^2)(2l+1)$. It is also clear from equations (1.79) and (1.81) that although elastic scattering may take place without absorption ($\eta_l = 1$) the converse is not true. An absorption process in which particles of orbital angular momentum $l\hbar$ are removed creates an elastic scattering, known as *shadow scattering* with an angular distribution characteristic of l.

The angular distribution of elastic scattering, in the present case in which Coulomb forces are neglected, is given by equation (1.76). From the formulae for $P_l (\cos \theta)$ (Appendix 1) it can be seen that if the maximum l-value entering into the scattering for spinless particles is L then the angular distribution contains powers of $\cos \theta$ up to $\cos^{2L} \theta$ only.

(iv) *Phase shifts and the scattering potential.* Within the radial extent of the force between the incident particle and target nucleus, an interaction potential must be added to the Schrödinger equation describing the elastic scattering. For an attractive potential the kinetic energy within the range of the force is increased and the deBroglie wavelength decreases. Figure 1.9a shows that this means a positive phase shift δ_l in the asymptotic wave. Conversely, a repulsive potential gives a negative phase shift.

The partial-wave analysis so far outlined must be modified if a Coulomb potential is involved in the scattering, since this is of long range and affects waves with all l-values. The modified analysis shows that a calculable term α_l representing the Coulomb phase shift appears in the elastic scattering amplitude together with the nuclear phase shift δ_l.

(v) *An example of phase-shift analysis.* A practical case is shown in the work of Clark *et al.* (Fig. 1.10) on the elastic scattering of

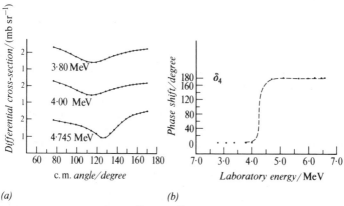

Fig. 1.10 Partial-wave analysis, $^{12}C(\alpha, \alpha)^{12}C$. (a) Differential cross-sections near 4·0 MeV bombarding energy. (b) The $l = 4$ phase-shift energy variation. (From Clark, G. J. *et al.*, Nuclear Physics, **A110**, 481, 1968.)

31

α-particles by ^{12}C:

$$^{12}C + \alpha \rightarrow {}^{16}O^* \rightarrow {}^{12}C + \alpha \qquad (1.84)$$

Differential cross-sections were obtained over the c.m. angular range 70–170° for laboratory energies 2·8 MeV to 6·6 MeV and were analysed by the use of equations (1.76)–(1.78) with $\eta_l = 1$ and with the inclusion of Coulomb scattering terms to take account of the nuclear charge. Figure 1.10a shows the differential scattering cross-section for energies near to 4·00 MeV and Fig. 1.10b gives the variation of the $l = 4$ phase shift with energy. This phase shift passes through 90° at an energy of about 4·3 MeV, and indicates a level of spin 4 and even parity at an excitation energy of 10·36 MeV in the intermediate nucleus $^{16}O^*$ in process (1.84).

Another well known and simple example of phase shift analysis is a study of the $\Delta(1236)\pi^+$-proton scattering resonance, as described by Perkins (Ref. 1.1a, p. 272) and shown in Fig. 2.10. In this and the preceding case of $^{12}C(\alpha, \alpha)^{12}C$, one partial wave only is dominant and inelastic processes are negligible. The partial-wave method is generally useful in sorting out complex situations in which overlapping resonances are present, and measurements of both elastic and inelastic (or total) cross-sections as a function of energy are necessary.

1.3 Conservation laws and constants of motion

In the preceding sections, and indeed in the whole of classical physics, the principles of conservation of total energy, of linear and of angular momentum have been assumed for macroscopic systems not acted on by external forces or couples. In atomic and nuclear physics we assume that the same principles apply, albeit that they should be expressed in quantum mechanical language. In these domains, however, other plausibly conserved quantities such as parity and isobaric spin have appeared. The discovery of the non-conservation of these quantities in the weak interaction has led to a deeper examination of the meaning of conservation laws in physics and their relation to the basic forces of nature.

In classical mechanics it is possible to formulate Newton's laws of motion in terms of the Hamiltonian function H giving the total energy ($H = T + V$) of a system of n interacting particles, and the position coordinates q_i and momentum values p_i of individual particles. The equations of motion are then

$$\left. \begin{array}{l} \dot{q}_i = \partial H / \partial p_i \\ \dot{p}_i = \partial H / \partial q_i \end{array} \right\} \quad \text{for} \quad i = 1 \text{ to } n \qquad (1.85)$$

Suppose now that a small and equal change is made to all the position coordinates so that q_i becomes $q_i + \Delta q$. The change in the total energy is then

$$\Delta H = \Delta q(\partial H/\partial q_1 + \partial H/\partial q_2 + \cdots)$$
$$= \Delta q(\dot{p}_1 + \dot{p}_2 + \cdots) \tag{1.86}$$

If, however, linear momentum is a constant of the motion the sum $\sum \dot{p}_i$ vanishes and $\Delta H = 0$ so that the Hamiltonian is unchanged by, or *invariant* to, the spatial displacement Δq. Conversely, one may postulate that if the Hamiltonian is invariant to a particular transformation (e.g. Δq) then the associated (or conjugate) mechanical quantity (e.g. p) is a *constant of the motion*.

This is so far simply a re-statement of what is already understood as a consequence of Newton's third law of motion. But the idea of invariance under transformations is a new and important one because it underlies the existence of symmetry in natural phenomena. A simple geometrical example is given in Fig. 1.11; the rotation of an equilateral triangle through 120° about a central axis perpendicular to its plane reproduces the original figure. This shape is invariant to the 120° rotation, or in other words, there is three-fold symmetry in the pattern.

These ideas can be immediately transferred to a system which must be described by quantum mechanics. There will then be a state wave function Ψ which is in general a linear combination of normalized eigenfunctions of the Hamiltonian operator H. The expectation (or average) values of observable quantities D are obtained by applying suitable operators to Ψ and integrating over configuration space, e.g.

$$\langle D \rangle = \int \Psi^* D \Psi \, \mathrm{d}^3 r \tag{1.87}$$

gives the expectation value of quantities such as position, kinetic energy, momentum.

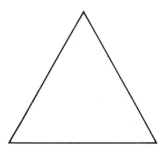

Fig. 1.11 The equilateral triangle has a three-fold axis of symmetry.

The effect of a transformation may be assessed by enquiring whether or not the expectation value of such an operator depends on time, i.e. whether or not it is a constant of the motion. To do this it is convenient to use the Heisenberg, rather than the Schrödinger formulation of quantum mechanics, because this uses time-independent state functions and gives an explicit formula for the time dependence of an operator D, namely

$$\mathrm{d}D/\mathrm{d}t = i/\hbar[HD - DH] = i/\hbar[H, D] \qquad (1.88)$$

where the bracketed quantity is called the *commutator* of the operator D with the Hamiltonian or total energy operator H. If D is a constant of the motion then the commutator vanishes, i.e. $DH = HD$, which means as an operator equation, that the Hamiltonian is unaffected by, or invariant to, the D operation.

If now D represents space translation, as in the foregoing classical example, then for an infinitesimal displacement

$$D = 1 + \Delta x \, . \, \partial/\partial x \qquad (1.89)$$

But quantum mechanics relates the gradient operator directly to the x-component of linear momentum by the equation

$$p_x = -i\hbar \frac{\partial}{\partial x} \qquad (1.90)$$

and conservation of linear momentum follows. A conservation law is thus connected with the invariance of the Hamiltonian under the appropriate transformation or, in equivalent terms, to the fact that the operator for this transformation commutes with the Hamiltonian.

In Table 1.1a the transformations corresponding with the classically familiar conserved quantities are listed. These quantities are as

TABLE 1.1

Conserved quantity	Transformation
(a) Linear momentum p	Space translation Δx
Angular momentum L	Space rotation $\Delta\phi$
Total energy E	Time translation Δt
Mass M^2	Lorentz transformation $(\Delta r, \Delta t)$
(b) Parity π	Space inversion P
Isobaric spin T	Rotation in isospin space $T_x + iT_y$
Charge parity π_c	Particle–antiparticle conversion C
(no quantum number)	Time reversal T
(c) Charge Q ⎫	
Baryon number B ⎪	
Lepton number L ⎬	Gauge transformations
Strangeness S ⎪	
Charm C ⎭	

far as one knows at present rigorously conserved in quantum mechanical systems and for all interactions, subject only to the relaxation permitted by the uncertainty relations:

$$\Delta p_x \,.\, \Delta x \approx \hbar, \qquad \Delta L_\phi \,.\, \Delta \phi \approx \hbar, \qquad \Delta E \,.\, \Delta t \approx \hbar \qquad (1.91)$$

In Table 1.1*b* are listed some of the quantities used in nuclear physics that are not universally conserved, but may nevertheless provide useful quantum numbers for certain classes of process, especially those governed by the strong nuclear interaction. It will be noted that parity and charge parity relate to discontinuous transformations of the nature of mirror-reflection processes. Isobaric spin is discussed in Section 5.6.

Finally, in Table 1.1*c* are noted the important particle numbers, including charge, which can be brought into the general framework of the present discussion in terms of a special type of transformation which just adds a phase factor to the corresponding particle wave-functions. Baryons and leptons are defined in Section 2.3; for strangeness and charm see Sections 2.2.2/3.

The conservation laws discussed in this section lead directly to the *selection rules* that determine which quantum states of a nuclear system may be connected by a specific operation, e.g. the emission of a quantum of electric dipole radiation or of a particle carrying total angular momentum *J*.

1.4 The spherically symmetrical potential well

Angular momentum plays an essential role, not only in the dynamics of collision processes in which both attractive and repulsive forces may appear, but also in determining the order of the energy levels of a particle bound within an attractive potential well. This is really a special case of the general formalism already discussed for scattering problems, namely that the system 'particle plus target' has a negative total energy. The zero of the energy scale is taken to be the state of the system when the particle is removed to an infinite distance from the centre of the potential well. As it moves in under an attractive force, its kinetic energy increases and its potential energy diminishes, but the total energy remains zero. Finally, in the process of binding some energy is released, usually as radiation, and the total energy becomes negative. Both the scattering and the bound-state problems are described by the Schrödinger equation and solutions for the bound states in simple wells are given in Reference 1.3. Only a very brief treatment will be presented here.

Let the interaction between particles M_1 and M_2 be described by a spherically symmetrical potential $V(r)$ and suppose that M_2 is very large so that c.m. motion may be disregarded. In such a field (Ref. 1.3 and Sect. 1.2.7) solutions of the wave equation may be sought that are simultaneous eigenfunctions of the Hamiltonian or total energy operator and of the operators L^2 and L_z for angular momentum. These eigenfunctions have the form

$$\Psi_{nlm} = R_{nlm}(r) Y_l^m(\theta, \phi)$$

where $Y_l^m(\theta, \phi)$ is a spherical harmonic and $R_{nlm}(r)$ is a radial function; the index n will be defined later and the polar coordinates used are shown in Fig. 1.12. The eigenvalues for the angular momentum operators are given, as in Section 1.2.6, by the equations

$$\left. \begin{array}{ll} L^2 \Psi_{nlm} = l(l+1)\hbar^2 \Psi_{nlm}, & l = 0, 1, 2, \ldots \\ L_z \Psi_{nlm} = m\hbar \Psi_{nlm}, & |m| \leqslant l \end{array} \right\} \quad (1.92)$$

where l is the orbital angular momentum quantum number and m is the magnetic quantum number. Following spectroscopic convention, states with $l = 0, 1, 2, 3, 4, \ldots$ are known as s-, p-, d-, f-, g-, \ldots states.

The orbital motion, as noted in Section 1.2.6, adds a repulsiye centrifugal potential to the radial wave equation so that the motion takes place in an effective potential

$$V_{eff} = V(r) + l(l+1)\hbar^2/2Mr^2 \quad (1.93)$$

The one-dimensional radial wave equation, which remains after removal of the angular part of the Schrödinger equation, is then

$$\left[-\frac{\hbar^2}{2M} \frac{1}{r^2} \frac{d}{dr} \left(r^2 \frac{d}{dr} \right) + l(l+1) \frac{\hbar^2}{2Mr^2} + V(r) \right] R(r) = ER(r) \quad (1.94)$$

where E is the total energy of the particle of mass $M\,(=M_1)$. This

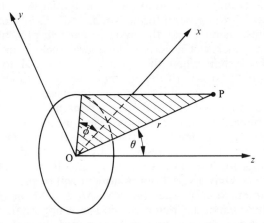

Fig. 1.12 Spherical polar coordinates, oriented to correspond with the scattering problem.

equation is simplified by the substitution

$$u(r) = rR(r) \tag{1.95}$$

and then becomes

$$\frac{d^2u}{dr^2} + \frac{2M}{\hbar^2}\left[E - V(r) - l(l+1)\frac{\hbar^2}{2Mr^2}\right]u = 0 \tag{1.96}$$

Solution of this equation for discrete eigenfunctions u_{nl} and energy eigenvalues E_{nl} is equivalent to fitting appropriate waves to the effective potential V_{eff}, remembering that when the coefficient of u_{nl} becomes negative the wavenumber becomes imaginary. The energy E_{nl} in an eigenstate is determined by the depth of the effective potential well and by the number of wavelengths within it, so that the states are characterized by two suffixes, l and n. Although there are $(2l+1)$ eigenfunctions ψ_{nlm} for each l, corresponding with the permitted m-values, these are degenerate, i.e. they all have the same energy because in the absence of external fields the direction of the z-axis is arbitrary. The *orbital angular momentum* quantum number l determines not only the effective well shape for a given $V(r)$ but also the *parity* of the orbital motion as $(-1)^l$ where $+1$ means even and -1 odd parity. It counts the number of angular nodes in the wavefunction Ψ_{nlm} as θ varies from 0 to π.

The suffix n could be the number of radial nodes, or changes in sign, in the function $u(r)$ or $R(r)$ as r varied from 0 to infinity. It would then be the *radial quantum number* n_r (≥ 0), and this usage is appropriate to the short-range potentials of nuclear physics. More generally, however, n is the *principal quantum number*, equal to $n_r + l + 1$ (≥ 1), which is the total number of nodes or changes of sign in both radial and angular eigenfunctions plus 1; for a given principal quantum number n we have $l \leq (n-1)$. This definition of n is useful for long-range potentials.

Further progress requires specification of the form of the potential $V(r)$ and it will be useful to consider the Coulomb potential and the oscillator potential, representing respectively forces of a long-range and a short-range character.

1.4.1 The Coulomb potential

If

$$V(r) = -Ze^2/4\pi\varepsilon_0 r \tag{1.97}$$

and if the particle is an electron $(M = m_e)$, we have the case of a hydrogen-like atom which is fully discussed in Reference 1.3 in which diagrams of the effective potential and radial eigenfunctions (known as associated Laguerre polynomials) are to be found. The energy eigenvalues are

$$E_{nl} = -\tfrac{1}{2} \cdot 1/(n_r + l + 1)^2 \cdot Z^2 e^2/4\pi\varepsilon_0 a_0 = -(1/2n^2)Z^2 e^2/4\pi\varepsilon_0 a_0$$

(1.98)

where a_0 is the Bohr radius $4\pi\varepsilon_0 \hbar^2/m_e e^2 = 5\cdot 29 \times 10^{-11}$m and $n = 1, 2, 3, \ldots$.

For a given principal quantum number the energy does not depend on l, a degeneracy that arises from the special shape of the Coulomb potential. The sequence of hydrogen-like levels shows a decreasing spacing as the dissociation energy $E = 0$ is approached.

In terms of the quantum numbers n_r, l the levels of the electron in this field are

$$(n_r, l) \equiv (0, 0); \qquad (1, 0), (0, 1); \qquad (2, 0), (1, 1), (0, 2)$$

or in terms of n, l

$$(n, l) \equiv (1, 0); \qquad (2, 0), (2, 1); \qquad (3, 0), (3, 1), (3, 2)$$

i.e.   1s ;   2s , 2p ;   3s , 3p , 3d

The hydrogen-like degeneracy of these atomic levels with respect to l is removed when the central field departs from a Coulomb shape. For the *screened* Coulomb shape, in which the potential decreases towards $r = 0$ relatively more rapidly than in the pure Coulomb field, the wavefunctions with low l-values, which tend to exist closer to the central and deeper part of the potential, are bound relatively more tightly than those with larger l for the same principal quantum number. The sequence of levels then becomes, in the (n, l) scheme

1s;   2s;   2p;   3s;   3p;   3d; ...

but a general dominance of the central charge still persists in the sense that levels of a given principal quantum number remain close together and the main energy displacement occurs with change of n.

In a hydrogen-like atom these states may be occupied by the single electron of the atom and they are connected by radiative transitions. In the structure of more complex atoms the hydrogen-like states provide basic levels for occupation and are filled by electrons in accordance with the Pauli principle which permits two electrons (with opposite spin) for each quantum state (n, l, m). It follows that for a principal quantum number (n) the corresponding *shell* contains $2n^2$ electrons.

1.4.2 The oscillator potential

If

$$V(r) = -U + \tfrac{1}{2}M\omega^2 r^2$$

(1.99)

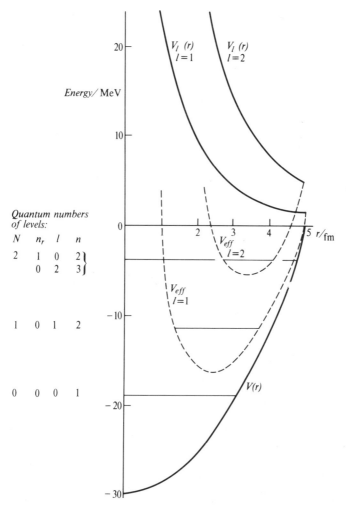

Fig. 1.13   An oscillator potential drawn with $V(r) = (-30 + 1 \cdot 2r^2)$ MeV for $0 < r < R$ where $R = 5$ fm and with $V(r) = 0$ for $r > R$. Centrifugal potentials $V_l(r)$ are drawn for the case $V_l = 20 \cdot 5l(l+1)/r^2$ MeV corresponding with a proton moving in the well. Levels of the effective potential (dotted) are shown, with quantum numbers.

the motion of the particle of mass M is that of a three-dimensional harmonic oscillator of angular frequency ω.

The potential well is shown in Fig. 1.13 together with centrifugal potentials $V_l(r)$ for $l = 1$ and 2. The well may be regarded as a short-range potential for nuclear physics if the value of r corresponding with $V(r) = 0$ is comparable with the size of an average nucleus. The radial eigenfunctions for the oscillator are also associated Laguerre polynomials, or Hermite polynomials if Cartesian coordinates are used. For a one-dimensional oscillator it is well known that the energy levels, measured from the bottom of the well, are at excitations

$$E'_n = (n + \tfrac{1}{2})\hbar\omega \qquad n = 0, 1, 2 \cdots \qquad (1.100)$$

and in the three-dimensional case

$$E'_{n_1 n_2 n_3} = (n_1 + n_2 + n_3 + \tfrac{3}{2})\hbar\omega = (N + \tfrac{3}{2})\hbar\omega \qquad (1.101)$$

where $N = n_1 + n_2 + n_3$ ($\geqslant 0$) is the oscillator quantum number. This gives the sequence of equally spaced levels shown in Fig. 1.13. In contrast with the case of the Coulomb potential, the oscillator levels do not close up as the dissociation energy is approached.

For each value N a range of values of the radial quantum number n_r and of the orbital momentum number l is possible, such that

$$N = 2n_r + l \qquad (1.102)$$

There is thus a degeneracy with respect to l since, for example, for $N = 2$ both s-states (with $n_r = 1$) and d-states (with $n_r = 0$) have the same energy. This degeneracy may be removed if the well is distorted from the pure simple-harmonic shape. The level order is then

$$(n_r, l) \equiv (0, 0); (0, 1); (1, 0); (0, 2); (1, 1); (0, 3)$$

or in terms of (n, l)

$$(n, l) \equiv (1, 0); (2, 1); (2, 0); (3, 2); (3, 1); (4, 3)$$

$$\text{i.e.} \quad 1s \quad ; 2p \quad ; 2s \quad ; 3d \quad ; 3p \quad ; 4f$$

It will be noted from Fig. 1.13 that, in further contrast with the levels of the long-range potential, the effect of changes of n_r and of l on the energy is similar for the oscillator. Moreover, the effect of the well shape on the level order is opposite to that of the screened Coulomb potential. For the same principal quantum number ($n = n_r + l + 1$), states of high l-value are found to be more strongly bound than those of lower l. This is because the hydrogen-like degeneracy is removed, but the well has no singularity at the origin, so that the s-states are less strongly disturbed.

The number of fermions, e.g. electrons, protons or neutrons, that may be accommodated in the levels corresponding with oscillator quantum number N is $(N+1)(N+2)$, allowing for two possible spin directions for each spatial state (n, l, m). The oscillator states, re-ordered by the addition of spin-orbit coupling, provide a set of basic levels for nuclear shell model calculations (Ch. 7).

1.5 Summary

This introductory chapter reviews some of the kinematic relations used in the description of nuclear collisions and sets up the formal definition of cross-section, which is one of the main experimental observables for such collisions. The role of angular momentum in the wave mechanical theory of scattering is examined, and shown to provide a link with the sequence of energy levels for a particle moving in a central potential well.

Examples 1

Sections 1.2.1–1.2.3

1.1* Prove the formula (1.6) given in the text.

1.2 For the collision of Fig. 1.1, show that conservation of momentum leads to the relation

$$p_1/q = \sin(\theta_L + \phi_L)/\sin\theta_L$$

and hence, using equation (1.6) obtain the relation

$$M_1/M_2 = \sin(2\phi_L + \theta_L)/\sin\theta_L$$

(This expression was used in early track-chamber studies for identifying the masses of colliding particles).

1.3* Deduce equation (1.11).

1.4* By inspection of equation (1.11) show that the condition for observing a single group of particles M_3 at any given angle θ_{3L} is that

$$Q + T_1 \geqslant M_1 T_1/M_4$$

Show also that when this condition is not fulfilled there is a maximum value $\theta_{3\max}$ at which particles of mass M_3 can be seen, given by

$$\cos^2\theta_{3\max} = [(M_3 + M_4)/M_1 M_3][(M_1 - M_4) - M_4 Q/T_1]$$

and that for all $\theta_3 < \theta_{3\max}$ the energy is double-valued. Show further that these two expressions include the conditions already stated for the case of elastic scattering ($Q = 0$).

1.5 From the last expression in Example 1.4, deduce that when Q is negative, the threshold energy for production of particles $M_3 + M_4$ is

$$T_1 = |Q|(M_1 + M_2)/M_2$$

(Assume that $(M_3 + M_4) = (M_1 + M_2)$. Note that at threshold the product particles are at rest in the c.m. system).

1.6 Show that in elastic scattering with $n > 1$ the minimum energy T_1' of the scattered particle is $T_1[(M_1 - M_2)/(M_1 + M_2)]^2$. Show also that when $n < 1$ the energy of the scattered particle at the greatest possible scattering angle is $T_1(M_1 - M_2)/(M_1 + M_2)$.

1.7 An α-particle ($M_1 = 4$) is scattered through an angle of 56° by collision with a heavier nucleus, which is observed to move off at an angle of 54° with the original direction of the α-particle. Find the probable mass number of the struck nucleus. [≈ 12]

1.8 By use of Fig. 1.4 or otherwise, verify that for elastic scattering the momentum transfer to M_2 is the same in both laboratory and c.m. systems.

1.9 The scattering of neutrons by protons may be regarded as a collision between equal particles each of mass 1 (in a.m.u.). Show that the laboratory angle of scattering of the neutron θ_L is half the corresponding c.m. angle θ. Compute the angle through which deuterons (mass 2 a.m.u.) are scattered by protons for c.m. angles of 30°, 45°, 90° and 135°.

Elements of nuclear physics

1.10 For the inelastic collision of Fig. 1.3 show that for M_3 the angle θ_{3L} is determined by equation (1.16) but with n given by

$$n^2 = (T+Q)/T \cdot M_2 M_4/M_1 M_3 \quad \text{where} \quad T = \tfrac{1}{2}\mu(p_1/m_1)^2$$

(Note that n is the ratio of the velocity of M_3 in the c.m.s. to the velocity of the c.m.s. in the laboratory).

1.11* Draw a momentum diagram for particle M_3 in the inelastic collision of Fig. 1.3. If the angle and energy of this particle in the laboratory system are T_L, θ_L and in the c.m.s. system T and θ, show that

$$T = T_L + a - 2\sqrt{T_L a}\cos\theta_L$$

and

$$T_L = T + a + 2\sqrt{Ta}\cos\theta$$

where

$$a = \tfrac{1}{2}M_3 v_c^2$$

Show also that $\sin\theta_L/\sin\theta = (T/T_L)^{1/2}$.

1.12 The reaction $^{38}\text{Ar}(p, n)^{38}\text{K}^m$ has a Q-value of $-6\cdot824$ MeV. Find the energy of the bombarding protons at the threshold for the production of $^{38}\text{K}^m$ using integral masses ($\text{Ar} \equiv \text{K} = 38$, $n = p = 1$). What is the c.m. energy of the neutrons for a proton energy of 10 MeV? For this proton energy write a programme that tabulates θ_L and T_L for θ from 0 to 135° in 5° intervals. Use the formulae given in Example 1.11 or equation (1.11). [$2\cdot84$ MeV]

Section 1.2.4

1.13* A particle of mass m and velocity β collides with a stationary particle of mass M. Show that the velocity of the c.m. system is

$$\beta_c = \beta\gamma m/(M + \gamma m)$$

and that the Lorentz factor of this system is

$$\gamma_c = (M + \gamma m)/E_c$$

where E_c is the total energy of the two particles in the c.m. system.

Evaluate E_c in terms of M, m and γ and show that when $M = m$, $\gamma_c = [(1+\gamma)/2]^{1/2}$.

1.14 Two particles of mass M each travelling with speed β approach head-on. Show that the total energy of one in the rest frame of the other is $M(1+\beta^2)/(1-\beta^2)$.

1.15 Neutral mesons ($\pi°$) are produced by the reaction $pp \rightarrow pp\pi°$. Show that the threshold proton kinetic energy is $(m_\pi^2/2m_p + 2m_\pi)$.

1.16 By using energy and momentum transformations show that if a particle moves at an angle θ_L with the z-axis in the laboratory frame, the corresponding angle θ' in a system moving with speed β in the positive z-direction is given by

$$\tan\theta_L = \sin\theta'/\gamma(\cos\theta' + \beta/\beta')$$

where β' gives the velocity of the particle in the second system. Compare this result with the non-relativistic equation (1.16).

1.17 For an energy above threshold in the reaction $1+2 \rightarrow 3+4$ show that the energy of particle 3 in the c.m. system is

$$E_{3c} = (E^2 + M_3^2 - M_4^2)/2E$$

(Use $E = E_{3c} + E_{4c}$ and $p_{3c} = p_{4c}$.)

1.18 Calculate the total energy and momentum of an electron which has been accelerated through a potential difference of 0.511 MV. What are its Lorentz factor and its β-value? $[2, \sqrt{\frac{3}{4}}]$

1.19 What is the laboratory kinetic energy of the incident pion that just forms the baryon resonance $\Delta(1236 \text{ MeV})$ when incident on a proton? (Take $m_\pi = 139$ MeV, $m_p = 938$ MeV.) [196 MeV]

1.20* Find the momentum transfer $|\boldsymbol{q}|$ in the following processes, expressing your result in units of $\hbar$, i.e. as an inverse length:

(*a*) the scattering of electrons of energy 100 MeV through a laboratory angle of 25° by a heavy nucleus. $[0.22 \text{ fm}^{-1}]$

(*b*) the scattering of 65 MeV α-particles by an ^{16}O nucleus if the recoil nucleus is projected at a laboratory angle of 60°. $[2.84 \text{ fm}^{-1}]$

1.21* A particle of mass M collides elastically with a stationary particle of equal mass. After the collision the angle between the trajectories of the two particles is ψ, and their kinetic energies are T_1' and T_2. Show that:

$$\cos \psi = [(1 + 2M/T_1')(1 + 2M/T_2)]^{-1/2}$$

Comment on the important differences between this relation and the classical one and show that it reduces to the latter when T_1' and T_2 are both much less than M.

Calculate ψ when the particles are nucleons, the kinetic energy of each one after the collision being 2 GeV. [60°]

1.22 In the collision described in the previous example, the projectile particle is scattered through an angle θ. Show that its kinetic energy, T_1', after the collision is:

$$T_1' = T_1 \cos^2 \theta / [1 + T_1 \sin^2 \theta / 2M]$$

where T_1 is its initial kinetic energy. Compare this result with the classical one.

1.23 A Λ-hyperon decays at rest into the two particles proton (p) and negative pion (π). If the energy release in the decay is Q, show that the ratio of the kinetic energies of the two decay particles is

$$(Q + 2m_p)/(Q + 2m_\pi)$$

Find the individual kinetic energies, given that $m_\Lambda = 1116$ MeV, $m_p = 938$ MeV, $m_\pi = 139$ MeV. $[T_p = 5.5 \text{ MeV}, T_\pi = 33.5 \text{ MeV}]$

1.24 If the Λ particle in Example 1.23 decayed while it was moving with a velocity $\beta = 0.5$, what would be the laboratory angles of the decay products for a centre-of-mass angle of 90°? (Use the result given in Ex. 1.16) (p, $10.5°$; $\pi, 45°$]

1.25* The reaction ^{34}S(p, n)^{34}Cl has a threshold at a proton energy of 6.45 MeV. Calculate (non-relativistically) the threshold for the production of ^{34}Cl by bombarding a hydrogenous target with ^{34}S ions. For the (p, n) reaction, use equation (1.26) to investigate the nature of the relativistic correction necessary in deducing an accurate Q-value. [219.3 MeV]

Section 1.2.5

1.26 Calculate the de Broglie wavelength of

(*a*) an electron of energy 1 MeV; $[8.8 \times 10^{-13} \text{m}]$

(*b*) a proton of energy 1000 MeV; $[7.3 \times 10^{-16} \text{m}]$

(*c*) a nitrogen nucleus of energy 140 MeV. $[6.5 \times 10^{-16} \text{m}]$

1.27 Show that the differential cross-section $\sigma_L(\theta_L)$ for the scattering of protons by protons in the laboratory system is related to the corresponding quantity in the c.m. system by $\sigma_L(\theta_L) = 4 \cos \theta_L \sigma(\theta)$.

Elements of nuclear physics

1.28 The cross-section for the reaction $^{10}B+n \rightarrow {}^{7}Li+{}^{4}He$ is 4×10^3 barn at a certain energy and it is the only reaction that takes place. Calculate the fraction of a ^{10}B layer that disappears in a year in a flux of neutrons of 10^{15} m^{-2} s^{-1}. [0·013]

1.29 A finely collimated beam of α-particles having a current of 1 μA bombards a metal foil (thickness 2 g m^{-2}) of a material of atomic weight 107·8. A detector of area 10^{-4} m^2 is placed at a distance of 0·1 m from the foil so that the α-particles strike it normally on the average and counts at a rate of 6×10^4 particles s^{-1}. Calculate the differential cross-section for the scattering process at the angle of observation, neglecting c.m. corrections. [1·72 b sr^{-1}]

Section 1.2.6

1.30 Show that the relativistic form of equation (1.59) for the cyclotron frequency is

$$\omega = eB/\gamma M \quad \text{where} \quad \gamma = (1-\beta^2)^{-1/2}, \beta = v/c.$$

1.31 Calculate the perpendicular distance of the line of flight of a 10-MeV proton from the centre of a nucleus when it has an orbital momentum, with respect to this point, of $3\hbar$. Calculate also the distance for a photon of energy 5 MeV and angular momentum $\hbar$. [$4 \cdot 3 \times 10^{-15}$ m, $3 \cdot 9 \times 10^{-14}$ m]

1.32 The impact parameter for Rutherford scattering by a fixed charge Z_2e is given by

$$b = (Z_1 Z_2 e^2 / 4\pi\varepsilon_0 M_1 v_1^2) \cot \theta/2$$

where the symbols have their usual meaning. Calculate the angular momentum quantum number l of a 1 MeV proton with respect to the scattering centre ($Z_2 = 90$) for an angular deflection of 90°. [14]

1.33* Particles of mass m_1 (charge $-e$) and mass m_2 (charge $+e$) describe circular orbits around their centre of mass with angular velocity ω under their mutual interaction. Calculate the total angular momentum and magnetic moment of the system. If the angular momentum is quantized, show how the magnetic moment is related to the Bohr magneton μ_B.

Section 1.2.7

1.34* Show that the energy distribution of protons projected from a thin hydrogenous target by a homogeneous beam of neutrons is uniform up to the maximum available energy T_0 providing that the interaction involves only s-waves.

1.35 By comparing the expressions (1.77) and (1.82) for $f(\theta)$ and σ_{total} prove the relation:
Imaginary part of $f(\theta)$ for $\theta = 0° = k\sigma_{\text{total}}/4\pi$. This is known as the *optical theorem*, because it relates the scattered wave to the total disturbance of the incident wave as in the case of optical refraction.

1.36* Show that if a scattering centre is a perfect absorber for all particles with a classical impact parameter up to R (black disc) then

$$\sigma_{\text{el}} = \sigma_{\text{inel}} = \pi(R+\lambda)^2$$

1.37* At the surface of an impenetrable sphere of radius R the de Broglie wave amplitude must vanish. Use equation (1.72) to deduce the corresponding s-wave phase shift, and calculate the differential and integrated elastic cross-section in the case $\lambda \gg R$.

Compare the latter with the result obtained classically by integrating over all impact parameters for the elastic scattering of a point object by a hard sphere.

'Hard-sphere' scattering is often a good approximation to represent background scattering effects in nuclear processes.

1.38 Write a programme that tabulates σ_{el}, σ_{inel} and σ_{total} as a function of η_l for intervals of η_l of 0·05 between the values 0 and 1. Display the results graphically and check the maximum values noted in equation (1.83).

1.39 Calculate the s-wave phase shift for the elastic scattering of a particle of mass M and kinetic energy T scattered by a square well of depth U_0 and radius R. [$\tan \delta = (k/k') \tan k'R - \tan kR$]

1.40 The measured integrated elastic scattering cross-section for a beam of 1-MeV neutrons incident on a certain heavy target is 2×10^{-28} m². On the assumption that there are no inelastic processes calculate the corresponding phase shifts: (*a*) for pure s-wave, (*b*) for pure p-wave, and (*c*) for pure d-wave. How could these cases be distinguished? [63°, 31°, 23°]

Section 1.4

1.41* The radial eigenfunctions for a particle of mass M moving in a spherically symmetrical square well potential of radius R and infinite depth are of the form $(1/\sqrt{Kr})J_{l+1/2}(Kr)$ where l is the angular momentum quantum number and K is the internal wavenumber.

Assuming that $J_{1/2}(x) \propto \sin x$, and $J_{3/2}(x) \propto \sin x/x - \cos x$, determine the relative order of the first three s- and p-states (use a computer programme for $J_{3/2}$).

1.42* By using equation (1.102) and summing over m-values, prove the statement that $(N+1)(N+2)$ nucleons may be accommodated in the levels with oscillator number N.

1.43* The energy levels for the oscillator potential cannot be found by the method of Example 1.41 because the potential does not suddenly become infinite. They are obtained by solving the wave equation in a series, as for the hydrogen atom.

Determine the minimum value of the effective potential for the motion in which the angular momentum quantum number is l.

1.44* Consider a square-well potential of depth V and radius R, such that the $3p$ level is just bound. Using the eigenvalues for an infinite well (see Example 1.41) show that this level is also just bound in other square-well potentials with the same value of VR^2. If the $3p$ level is observed as an (unbound) scattering resonance in the interaction of protons of energy 6·2 MeV with a target of ^{117}Sn, estimate the well depth and also the energy at which a similar resonance might be seen for a target of ^{124}Sn. Assume that the nuclear radius is $R = 1·25A^{1/3}$ fm.

2 Radioactivity, radiations, particles and interactions

It is the business of the nuclear physicist, having identified the nucleus as the central core of the atom, to enquire what are its component parts. At the first approximation, an answer is given in terms of neutrons and protons, but these, the nucleons, have also come under scrutiny. Moreover, the binding together of nucleons by the internucleon force seems plausibly, since the advent of Yukawa's theory in 1935, to involve mesons at least and the binding of constituents of the nucleon is one of the current problems of the subject. The second approximation for the nuclear physicist therefore establishes a link between nuclear structure and particle physics. This is not even wholly confined to the so-called 'non-strange' particles because 'strange' particle physics has been found to reflect quantitatively on that most historical of nuclear processes, the phenomenon of β-decay.

The particles of nuclear physics and their properties are listed in Appendix 2. A surprising number of these particles were first characterized without the use of high-energy accelerators as a result of experiments on the cathode and canal rays of the gaseous discharge, as a result of the discovery of radioactivity, and in the world-wide attack on the nature of cosmic radiation. The main events in the chronicle of particle discoveries are given in Table 2.1 and will be reviewed in part in this chapter. As far as radioactivity and cosmic rays are concerned, the particles to which they gave birth have amply repaid their debt, by leading to a new understanding of the nature of the primary phenomena involved.

2.1 From electron to muon

2.1.1 Electrons

In 1881 Helmholtz remarked, 'If we assume atoms of chemical elements, we cannot escape from drawing the further inference that electricity, too, positive as well as negative, is divided into definite

TABLE 2.1 Main advances in the growth of nuclear physics

Advance	Date	Physicists
Periodic system of the elements	1868	Mendeléev
Discovery of X-rays	1895	Röntgen
Discovery of radioactivity	1896	Becquerel
Discovery of electron	1897	J. J. Thomson[a]
The quantum hypothesis	1900	Planck
Mass–energy relation	1905	Einstein
The expansion chamber	1911	Wilson
Isotopes suggested	1911	Soddy
Nuclear hypothesis	1911	Rutherford
Nuclear atom model	1913	Bohr
Atomic numbers from X-ray spectra	1913	Moseley
Positive ray parabolas for neon isotopes	1913	J. J. Thomson
Transmutation of nitrogen by α-particles	1919	Rutherford
Mass spectrograph	1919	Aston
Wavelength of material particles	1924	de Broglie
The wave equation	1926	Schrödinger
Diffraction of electrons	1927	Davisson and Germer; G. P. Thomson
Uncertainty principle	1927	Heisenberg
Wave-mechanical barrier penetration	1928	Gamow, Condon, Gurney
The cyclotron	1930	Lawrence
The electrostatic generator	1931	Van de Graaff
Discovery of deuterium	1932	Urey
Discovery of the neutron	1932	Chadwick
Transmutation of lithium by artifically accelerated protons	1932	Cockcroft and Walton
Discovery of the positron	1932	Anderson
Hypothesis of the neutrino	1933	Pauli
Neutrino theory of beta decay	1934	Fermi
Discovery of artificial radioactivity	1934	Curie and Joliot
Neutron-induced activity	1934	Fermi
Hypothesis of heavy quanta (mesons)	1935	Yukawa
Discovery of the μ-meson (muon)	1936	Anderson and Neddermeyer
Magnetic resonance principle	1938	Rabi
Discovery of fission	1939	Hahn and Strassmann
The principle of phase-stable accelerators	1945	McMillan, Veksler
Discovery of π-meson	1947	Powell
Discovery of strange particles	1947	Rochester and Butler
Use of space-time diagrams	1949	Feynman
Production of π^0-mesons	1950	Bjorklund *et al.*
Hypothesis of associated production	1952	Pais
Discovery of hyperfragments	1953	Danysz and Pniewski
Strangeness	1953	Gell-Mann; Nakano and Nishijima
Hypothesis of $\overline{K^0}$ and K_1^0, K_2^0 particles	1955	Gell-Mann and Pais
Discovery of antiproton	1955	Chamberlain *et al.*
Non-conservation of parity	1956	Lee and Yang; Wu; Garwin
Observation of antineutrino	1956	Reines and Cowan

TABLE 2.1 (continued)

Advance	Date	Physicists
Prediction of heavy meson	1957	Nambu
Helicity of neutrino	1958	Goldhaber, Grodzins and Sunyar
Hypothesis of conserved vector current	1958	Feynman and Gell-Mann
Discovery of ω-meson	1961	Maglic *et al.*
Unitary symmetry	1961	Gell-Mann; Ne'eman
Muon neutrino	1962	Danby *et al.*
Discovery of Ω^-	1964	Barnes *et al.*
Non-conservation of CP	1964	Cronin, Fitch and Turlay
Hypothesis of charm	1964– 1974	Bjorken, Glashow and others
Evidence for point-like objects in proton	1968	SLAC (Stanford)
Discovery of neutral currents	1973	CERN (Geneva)
Discovery of J/ψ particles	1974	Richter, Ting
Discovery of explicit charm	1976	SLAC (Stanford)

[a] See *Physics Today* July 1966, p. 12.

elementary quanta that behave like atoms of electricity'. This conclusion from the laws of electrolysis related to positive and negative ions; the former appeared always to be associated with matter, but negative charges could be detached from matter and could appear in the free state. They were most convincingly seen as rays emerging from the cathode of a low-pressure discharge tube and experiments with such tubes in magnetic fields led to a determination of the charge to mass ratio e/m_e or *specific charge* of the atom of electricity or *electron* by application of the formula

$$e\boldsymbol{v} \times \boldsymbol{B} = m_e v^2/r \qquad (2.1)$$

already discussed in Section 1.2.6 (eqn (1.59)). In relativistic form this equation must be written

$$e\boldsymbol{v} \times \boldsymbol{B} = \gamma m_e v^2/r \quad \text{where} \quad \gamma = (1-\beta^2)^{-1/2}, \quad \beta = v/c \qquad (2.2)$$

Knowledge of the value of the specific charge led to the identification of the *beta rays* emitted by radioactive substances as electrons. These particles, which were originally characterized by their negative charge and their penetrating power (≈ 1 mm of lead) were soon found to have kinetic energies considerably in excess of the electron energies that had been produced in the gaseous discharge. It was possible even in the early experiments to verify the presence of the relativistic factor γ in equation (2.2). Beta particles (Ch. 10) arise in a particular form of nuclear decay; in addition, the radiations from radioactive substances include *internal conversion electrons* which originate from the interaction of nuclear excitations with the surrounding atomic structure.

Electrons behave in all respects as structureless point charges. At energies of a few hundred MeV, when their reduced de Broglie wavelength is about 0.5×10^{-15} m, they are important probes of the nuclear charge distribution.

2.1.2 Protons and alpha particles

The simplest positive ion was identified in the 'canal' rays that pass through a hole in the cathode of a low-pressure hydrogen discharge. Its charge-to-mass ratio e/m_p was about $\frac{1}{2000}$ of that for the electron and was very close to the ratio of the Faraday F $(=N_A e)$ to the atomic weight of hydrogen $(=N_A m_H)$. This particle is the *proton*, the nucleus of the hydrogen atom.

The proton is the only stable member of the strongly interacting family of fundamental particles, or *hadrons* (Sect. 2.3). It is, therefore, expected as a constituent of complex nuclei. The spontaneous emission of protons from a nucleus however, although not entirely unknown, is a rare phenomenon because alternative processes usually compete successfully. The radiations from radioactive substances do include a heavy-particle component, namely the *alpha rays*, characterized by positive charge and low penetrating power (≈ 0.02 mm of lead) and these were found to have a specific charge about half that of the proton. Experiments such as that of Rutherford and Royds, in which helium gas was found to appear in a tube into which α-rays were being emitted, disposed of the suggestion that the particle concerned was a hydrogen molecule and confirmed its identity with the helium nucleus, with charge $+2e$ and mass approximately $4m_p$.

The proton, unlike the electron, does not behave as a point charge. The reduced de Broglie wavelength of a proton of energy 1 GeV $(=10^9$ eV$)$ is 0.12×10^{-15} m and the size of the proton should, therefore, be studied at energies considerably higher than this. In fact, elastic electron–proton scattering experiments at an energy of about 1 GeV (Stanford) first demonstrated the finite size of the proton. Later, when elastic proton–proton collisions were observed in the CERN storage rings for a c.m. energy of 62 GeV the process was found to exhibit a diffraction-like angular distribution indicating a proton radius of about 10^{-15} m. Further electron–proton experiments examining *inelastic* processes at energies up to 20 GeV gave clear evidence for some internal structure of the proton (Sect. 2.2.3).

2.1.3 The displacement laws; isotopes

The physical nature of the α- and β-particles leads to the laws formulated by Soddy, Russell and Fajans in 1913 to describe the production of different chemical elements as a result of radioactive processes:

(*a*) the loss of an α-particle displaces an element two places to the left in the periodic table and lowers its mass by four units;

(*b*) the loss of a β-particle displaces an element one place to the right in the periodic table but does not essentially alter the atomic mass.

If these rules are applied to the decay of thorium

$$^{232}_{90}\text{Th} \xrightarrow{\ \alpha\ } {}^{228}_{88}\text{MsThI} \xrightarrow{\ \beta\ } {}^{228}_{89}\text{MsThII} \xrightarrow{\ \beta\ } {}^{228}_{90}\text{RdTh} \xrightarrow{\ \alpha\ } {}^{224}_{88}\text{ThX} \longrightarrow$$
$$(2.3)$$

it is clear that ^{232}Th and ^{228}RdTh have the same atomic number (and therefore chemical nature), but a different mass. They are in fact *isotopes* of the element thorium; many elements occur with two or more stable isotopes, and all elements have isotopes if unstable nuclei are counted. Figure 2.1 is a section of a chart on which isotopic constitution of naturally-occurring elements is displayed by plotting charge number Z against neutron number $N = A - Z$, where A is the mass number, equal to the total number of neutrons and protons.

Isotopes were demonstrated objectively by J. J. Thomson (1913) and Aston (1919) using electromagnetic techniques which are now the basis of methods of separating them.

2.1.4 Photons

The theory of Maxwell is admirably successful in describing the propagation of electromagnetic waves, and accounts for their diffraction and interference. Such phenomena appear over the frequency spectrum from the long waves of radio to the shortest-wavelength radiations from accelerators. The spectrum includes the third and most penetrating of the radioactive radiations, the *gamma rays*, capable of traversing about 10 mm of lead, as well as the X-rays characteristic of atomic inner-shell transitions.

Just as the electron and proton exhibit a duality of nature, in that each may participate in particle-like and wave-like phenomena, so does the duality extend to the electromagnetic spectrum. In the production and absorption of radiation by interaction with matter, quantum theory asserts that the frequency involved is connected with an exchange of energy between the radiation field and initial and final energy levels of the material system (e.g. an atom or a nucleus) (Fig. 2.2) according to the equation

$$h\nu = E_i - E_f \tag{2.4}$$

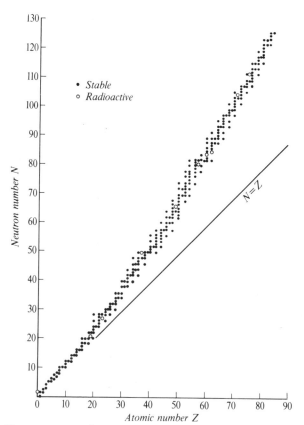

Fig. 2.1 Neutron–proton diagram, sometimes known as a Segrè chart, for the naturally occurring nuclei with $Z < 84$ (Ref. 2.1).

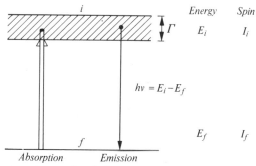

Fig. 2.2 Quantum picture of radiative processes between levels of energy E_i, E_f. Angular momenta I_i, I_f are also indicated. The upper level is assigned an energy width Γ, which is connected with the lifetime τ of the level against spontaneous emission by the relation

$$\Gamma\tau = \hbar = 6 \cdot 6 \times 10^{-16} \text{ eV s}.$$

51

This replaces the classical picture of radiation interacting with an oscillating charge distribution. The quantum of radiation emitted is known as a *photon* and may be ascribed particle-like properties, although a real photon has zero rest mass and of course travels with the velocity of light. Its momentum is

$$p = h\nu/c \qquad (2.5)$$

The concept of the photon simplifies the discussion of many radiative processes. In the case of the Compton scattering of radiation by a free electron for instance (Fig. 2.3), the change of wavelength or of wavenumber k in the scattering may easily be calculated by treating the photon as a particle of four-momentum $p_\mu^1 = (E, \boldsymbol{p}) = (\hbar k c, \hbar \boldsymbol{k}) = (k, \boldsymbol{k})$ putting $\hbar = c = 1$. Conservation of four-momentum gives

$$p_\mu^1 + p_\mu^2 = p_\mu^3 + p_\mu^4 \qquad (2.6)$$

where index 2 refers to the struck electron, assumed to be initially at rest, index 3 to the scattered photon of wave vector k', and index 4 to the recoil electron.

To obtain the dependence of k' on k and the scattering angle θ, equation (2.6) is written to permit a suitable scalar product to be taken, i.e.

$$p_\mu^1 + p_\mu^2 - p_\mu^3 = p_\mu^4$$

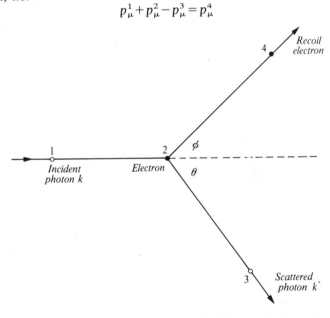

Fig. 2.3 Compton effect. The circular wavenumber k' of the scattered photon is ≤ the incident wavenumber k for an electron at rest. The electron is assumed to be free.

and then

$$(p_\mu^1 + p_\mu^2 - p_\mu^3) \cdot (p_\mu^1 + p_\mu^2 - p_\mu^3) = (p_\mu^4)^2$$

or using equation (1.22)

$$[(E_1 + E_2 - E_3)^2 - (\boldsymbol{p}_1 + \boldsymbol{p}_2 - \boldsymbol{p}_3)^2] = (p_\mu^4)^2 = m_e^2$$

or, in terms of k, m_e

$$(k + m_e - k')^2 - (k^2 + k'^2 - 2kk' \cos \theta) = m_e^2$$

whence

$$1/k' - 1/k = (1 - \cos \theta)/m_e$$

Inserting $\hbar$ and c to yield the correct dimensions

$$1/k' - 1/k = (\hbar/m_e c)(1 - \cos \theta)$$

or

$$\lambda' - \lambda = (h/m_e c)(1 - \cos \theta) \tag{2.7}$$

The verification of this energy–angle relationship and of the associated expression of the energy of the recoil electron lends support to the photon hypothesis, as of course also do Planck's theory of blackbody radiation and Einstein's theory of the photoelectric effect.

2.1.5 The positron

Cosmic radiation is a flux of high-energy particles, largely protons, continuously incident on the earth from outer space, together with secondary radiations deriving from interactions of the primary protons in the atmosphere. The radiation at sea level was shown by absorption experiments to contain an easily absorbed or soft component and a penetrating or hard component with an absorption coefficient less than the least expected for gamma radiation; Fig. 2.4

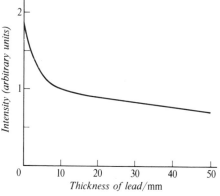

Fig. 2.4　Absorption in lead of cosmic radiation at sea level, showing soft and hard components (based on Auger, P. et al., J. de Phys., **7**, 58, 1936).

gives a typical absorption curve. Cloud chamber photographs taken at random by Anderson (1932) and under conditions of counter control by Blackett and Occhialini (1933) showed that the soft component contained, in addition to electrons, particles of electronic mass but positive charge. This was established by measurements of curvature in a magnetic field, of range and of density of ionization along the track. Figure 2.5 shows the trajectory of one of the positive particles.

Anderson had, in fact, discovered the anti-electron or *positron*, the first example of the fact that particles may have antiparticles of equal mass but opposite charge. The origin of these particles in the cosmic radiation was not fully understood until Bhabha and Heitler in 1937 pointed out that electrons deriving from the primary radiation could give rise to energetic 'bremsstrahlung' quanta as a result of deflections in the electric field of a nucleus (Sect. 3.1.3) and

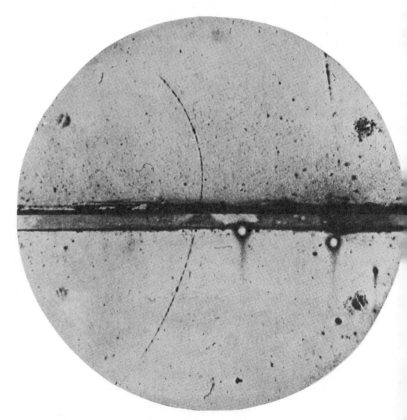

Fig. 2.5   A positron of energy 63 MeV passes through a lead plate and emerges with an energy of 23 MeV (Anderson, C. D., *Phys. Rev.*, **43**, 491, 1933).

that these radiations could subsequently produce electron–positron pairs (Sect. 3.1.4) in further electromagnetic interactions:

$$\gamma \rightarrow e^+ + e^- \tag{2.8}$$

These processes would build up as the radiations passed through the atmosphere into cascade showers, of which Anderson's particle was probably a component.

The possibility of the existence of antiparticles had already been considered theoretically by Dirac who noted that the expression $E^2 = p^2c^2 + m^2c^4$ for the square of the total energy of a particle of mass m led to the conclusion that

$$E = \pm(p^2c^2 + m^2c^4)^{1/2} \tag{2.9}$$

suggesting that electrons might occupy states of negative energy, extending from $-m_ec^2$ to $-\infty$. The difficulty that ordinary electrons should all make transitions to such states was circumvented by the suggestion that normally the states were all occupied, so that the restrictions offered by the Pauli principle would apply. If an electron in a negative energy state were raised by an electromagnetic process to a positive energy level, with an expenditure of energy greater than the minimum of $2m_ec^2$ required, the remaining 'hole' would behave as a normal particle of opposite charge and pair production would have been achieved, i.e. process (2.8).

The positron, as shown in Appendix 2, is a stable particle like the electron, but it may disappear by annihilation with an electron

$$e^- + e^+ \rightarrow 2\gamma \tag{2.10}$$

If the electron and positron are both essentially at rest, the two oppositely-directed annihilation quanta have each an energy of $m_ec^2 = 511$ keV. Fast-moving positrons may both radiate and annihilate in flight in a Coulomb field, yielding energetic quanta which contribute to the build-up of the cosmic-ray showers.

2.1.6 The neutron and the neutrino

In 1919 Rutherford detected the emission of protons from nitrogen bombarded by α-particles. This was the discovery of artificial transmutation, the actual process being the $^{14}N(\alpha, p)^{17}O$ reaction already set out in equation (1.8a). Figure 2.6 reproduces the celebrated cloud chamber picture by Blackett and Lees which gave visual confirmation of this disintegration.

The (α, p) reactions for light nuclei were studied thoroughly in the years following Rutherford's discovery and Rutherford himself was certainly aware of the possibility that some of the products of such reactions might be radioactive and that some might be neutral

Fig. 2.6 Expansion chamber photograph showing ejection of a proton from nitrogen nucleus by an α-particle (Blackett, P. M. S., and Lees, D. S., *Proc. Roy. Soc* **A136,** 325, 1932).

(though here he envisaged a close combination of a proton and electron). The neutral particle or *neutron* was, however, not identified until in 1932 Chadwick established the occurrence of the reaction

$$^9\text{Be} + \alpha \rightarrow {}^{12}\text{C} + \text{n} \tag{2.11}$$

(Q-value now known to be 5·7 MeV).

Reaction (2.11) is similar in nature to the (α, p) reactions. It was known for some years before 1932, especially from the work of Bothe and Becker, that bombardment of light elements with α-particles could produce a penetrating radiation, assumed then to be gamma radiation connected with the production of excited states of the residual nuclei, e.g. ^{17}O, formed in the reactions. For beryllium particularly, the penetrating power of the radiation seemed exceptionally great ($\mu_m = 2 \times 10^{-3}$ m^2 kg^{-1} for lead). It is now clear that μ_m was impossibly low if the radiation were electromagnetic but at the time the effect of pair production in increasing absorption coefficients (see Sect. 3.1.4) was not realized. More significant, however, was the observation by Mme Curie-Joliot and M. Joliot that the radiation from beryllium was able to eject energetic *protons* from hydrogenous material. This they ascribed to a Compton scattering of the supposed electromagnetic radiation by the target protons and from the observed range of the protons the energy of the radiation was calculated to be 35–50 MeV.

This energy seemed larger than might be expected to originate in an α-particle reaction with a light element, and the Compton scattering hypothesis was therefore further studied by Chadwick (1932). He found, using a simple ionization chamber and amplifier, that the beryllium radiation could produce recoil ions of many light elements as well as of hydrogen. This was most simply explained by the supposition that the radiation was not electromagnetic, but was a stream of neutral particles (*neutrons*) of mass approximately equal to that of the proton.

The apparatus of Chadwick is shown in Fig. 2.7. In the presence of the source assembly the pulse counting rate in the air-filled ionization chamber increased by a factor of about forty and by a further factor when a sheet of paraffin wax was placed near the entrance to the chamber. The former increase was interpreted as due to the production of recoil nitrogen ions and the latter as due to the detection of recoil protons from the paraffin wax. The initial velocity u_p of the protons (3×10^7 m s^{-1}) was deduced from their measured range, and the initial velocity of the nitrogen recoils u_N was found to be $4 \cdot 7 \times 10^6$ m s^{-1} from a separate expansion chamber study of recoil ranges made by Feather.

If the unknown radiation is assumed to consist of particles of mass M_n and velocity u_n and if M_p and M_N are the masses respectively of

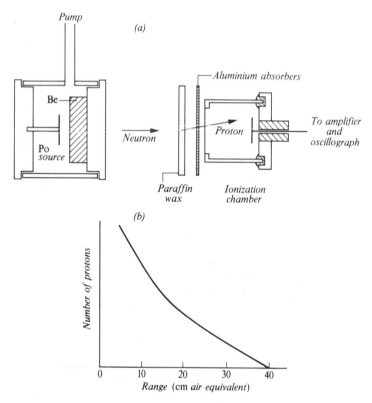

Fig. 2.7 Discovery of the neutron. (*a*) Apparatus, in which neutrons were produced by the (α, n) reaction in a block of beryllium and then ejected protons from a sheet of paraffin wax. (*b*) Number–range curve for protons (Chadwick, J., *Proc. Roy. Soc.* **A136**, 692, 1932).

the proton and the nitrogen nucleus, then it is easy to show that

$$u_p = 2M_n u_n/(M_n + M_p) \quad \text{and} \quad u_N = 2M_n u_n/(M_n + M_N) \quad (2.12)$$

so that

$$u_p/u_N = 3 \cdot 3 \times 10^7/4 \cdot 7 \times 10^6 = (M_n + M_N)/(M_n + M_p)$$

Substitution of $M_N = 14$ mass units and $M_p = 1$ mass unit gave $M_n = 1 \cdot 16$ mass unit with an error of some 10 per cent. This value was soon refined, by observations on other neutron-producing reactions, to $1 \cdot 005 < M_n < 1 \cdot 008$ atomic mass units. This completed the proof of the existence of a neutral nuclear particle of mass closely similar to that of the proton.

The neutron and proton are known as the *nucleons*, and are essential constituents of a complex nucleus. The masses of the nucleons are close enough for them to be recognizable as an *isobaric doublet* but the neutron is, in fact, slightly the heavier and decays into the proton with the emission of an electron and an (anti-) neutrino

$$n \rightarrow p + e^- + \bar{\nu} \qquad (2.13)$$

The neutron was the first unstable elementary particle to be discovered; the probability of such decay is measured by a decay constant λ_n and the disappearance of neutrons from an initial assembly by decay with *halflife* $t_{1/2}$ or *mean life* $\tau_n = 1/\lambda_n$ is given by the radioactive-decay law

$$N_t = N_0 \exp -\lambda_n t; \qquad \lambda_n = 1/\tau_n = 0.693/t_{1/2} \qquad (2.14)$$

The reaction shown in (2.13) is a beta-decay process and is typical of similar processes observed with unstable nuclei in all mass ranges. For reasons that will be discussed later (Ch. 10) Pauli (1931–33) suggested that the further neutral particle shown in (2.13), now called the (anti-) neutrino, should be created in the act of decay. Because of its weak interaction with matter this particle has proved extremely elusive, but in 1959 Reines and Cowan (Sect. 10.4.2) gave objective proof of its existence by observing the inverse of reaction (2.13), namely

$$\bar{\nu} + p + e^- \rightarrow n \qquad (2.15)$$

in the form

$$\bar{\nu} + p \rightarrow n + e^+ \qquad (2.16)$$

which is equivalent to (2.15) in terms of Dirac's theory of holes.

2.1.7 The muon

The penetrating component of the cosmic radiation at sea level (Fig. 2.4) was shown by expansion chamber photographs to consist of singly charged particles which could pass through lead plates without producing a shower. In 1936–37 Anderson and Neddermeyer presented evidence based on range, momentum and ionization measurements that some of these particles had electronic charge (of both signs) and a mass of about 200 times that of the electron. The high mass accounted for the absence of the radiative effects known for electrons. If, further, it was assumed that the new particles were unstable, disappearing from a flux by decay as well as by scattering, then an anomaly in the absorption coefficient of the penetrating radiation could be explained. Subsequently, decay electrons from the new particles were seen and the mean lifetime was determined by coincidence counter methods to be 2.16×10^{-6} s.

This new particle is now known as the muon (though originally as the μ-meson) and decays by what is essentially a beta-decay process into an electron, a neutrino and an antineutrino

$$\mu^{\pm} \rightarrow e^{\pm} + \nu + \bar{\nu} \tag{2.17}$$

An example of muon decay is shown in Fig. 2.9.

The observed lifetime (τ_0) of the muon in its rest-frame only permits the particle to traverse large thicknesses of the atmosphere if it is moving with a relativistic velocity. The number of muons from an initial number N_0 after traversing a path l (disregarding processes other than decay) is obtained directly from equation (2.14) as

$$N_l = N_0 \exp(-t/\gamma\tau_0) = N_0 \exp(-l/c\gamma\tau_0) \tag{2.18}$$

where as usual $\gamma = (1 - \beta^2)^{-1/2}$ and $\gamma\tau_0$ is the Lorentz-transformed lifetime seen by an observer on Earth. For this observer the probability of decay in an atmospheric path dl is

$$p = dN_l/N_l = -dl/c\gamma\tau_0 \tag{2.19}$$

An observer travelling with the muon measures the same probability, but now the mean life is indeed τ_0. The difference is that the atmospheric path is Lorentz contracted so that

$$p = -(dl/\gamma)(1/c\tau_0)$$

as before.

The muon has turned out to behave literally as a heavy electron, and measurement of the anomalous part of its magnetic moment, by observations on the decay of muons circulating in a storage ring at CERN, has confirmed some of the most refined calculations of quantum electrodynamics.

2.2 From pion to parton

2.2.1 The pions and the deltas

The discovery of the muon was at first acclaimed as a triumph for the theory of nuclear forces announced by Yukawa in 1935. The force that binds together the components of a complex nucleus is not apparent at distances much greater than the nuclear radius, as was clear from Rutherford's α-particle scattering experiments which indicated only Coulomb forces down to a distance of 3×10^{-14} m from the centre of a nucleus of gold. To understand an attractive force of an even shorter range Yukawa proposed a potential function

$$U \propto -\exp(-Kr)/r \tag{2.20}$$

with K an adjustable parameter such that K^{-1} is the range of the force. If this length is set equal to the reduced Compton wavelength

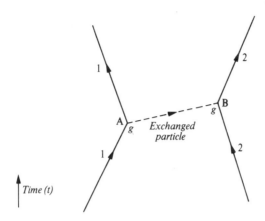

Fig. 2.8 Space–time (Feynman) diagram representing an interaction between two particles (e.g. protons) resulting from exchange of a third particle (e.g. a meson). This is essentially the scattering process shown kinematically in Fig. 1.1 with now an indication of the dynamical process involved. The quantity g^2 is a coupling constant (Sect. 2.3).

of a material particle

$$1/K = \hbar/\mu c \tag{2.21}$$

then for $K^{-1} = 2 \times 10^{-15}$ m, $\mu = 200m_e$, i.e. about equal to the observed mass of the muon. The physical picture of the intervention of this particle is illustrated in the space–time diagram of Fig. 2.8 in which the new particle is emitted spontaneously from a proton (say) at point A and is absorbed a short time later by a second proton (say) at point B. For such a process to happen an energy fluctuation of μc^2 must occur in the system A and B and according to the Uncertainty Principle $\Delta E \cdot \Delta t \approx \hbar$, this must be restored within a time $\hbar/\mu c^2$. In such a time the exchanged particle could travel a distance $c \times \hbar/\mu c^2$ at velocity c and this should be of the order of the range of the force, in agreement with (2.21). Alternatively, one may retain conservation of momentum and energy at the vertices but the exchanged particle must then have a non-physical mass. It is then said to be *off the mass shell*.

Unfortunately, the particle required by the Yukawa theory must interact strongly with nuclei, with a cross-section of the order of the nuclear area, whereas muons are penetrating particles, sometimes traversing large thicknesses of the atmosphere without loss by nuclear interaction and continuing even through many metres of earth. Finally, slow negative muons can form muonic atoms, in which a muon replaces an electron, and in such an atom the muon can stay in effective contact with the nucleus for a time long enough to permit its normal decay. They must, therefore, interact only weakly with nuclei and cannot be the Yukawa particle.

It was soon apparent that if the primary cosmic radiation (or the first results of its interaction with atmospheric nuclei) did contain a strongly interacting particle, such particles would reach sea level only in small numbers. Mountain-top exposures of the conveniently small nuclear photographic emulsions were therefore made and in 1947 Lattes, Muirhead, Occhialini and Powell found tracks of charged particles estimated by grain counts and scattering observations to have a mass of 200–300 electron masses. Some of these particles emitted a secondary particle of constant range and mass also about $200m_e$, at the end of their track. Subsequent work with more sensitive emulsions revealed that this secondary particle itself disintegrated at rest, emitting an electron in the way already known for the muon; an example of the whole sequence is shown in Fig. 2.9, and the discoveries are reviewed in detail by Powell (Ref. 2.2).

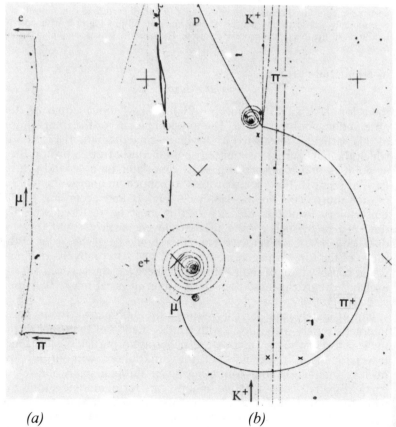

(a)   *(b)*

Fig. 2.9   π–μ–e decay seen in (*a*) nuclear emulsion (Powell, C. F., Ref. 2.2) and (*b*) a bubble chamber in a magnetic field (Colley, D. C.).

This remarkable and beautiful series of observations made with equipment of extreme simplicity immediately confirmed and illuminated the emerging theoretical hypothesis that there must be two particles of intermediate mass (mesons) in the cosmic radiation, one being the muon and the other a strongly interacting parent. This latter particle is now known as the π-meson or pion and it is produced by interactions of the primary cosmic radiation (mainly protons) with nuclei in the atmosphere. The charged pion is now known to have a rest-frame mean lifetime of $2 \cdot 6 \times 10^{-8}$ s and to decay both in flight and at rest (as shown in Fig. 2.9) into a muon and neutrino

$$\pi \rightarrow \mu + \nu \quad (\text{or } \bar{\nu}) \tag{2.22}$$

the neutrino being invoked because momentum must be conserved in the decay and no other identifiable particle or photon appears to be emitted. The muon may subsequently itself decay

$$\mu \rightarrow e + \nu + \bar{\nu} \tag{2.23}$$

and the π–μ–e process shown in Fig. 2.9 is complete.

The pion exists in three charge states π^+, π^-, π^0, with the masses shown in Appendix 2, and each of these may be envisaged as the exchanged particle in a nucleon–nucleon interaction as shown in Fig. 2.8. Its appearance in this role, as well as reaction (2.22), suggests that unlike the electron and muon it is not subject to a number-conservation law.

In matter, charged pions slow down to low velocities just as do other particles. Slow π^+ particles are kept away from nuclei by Coulomb repulsion and finally decay in accordance with process (2.22). Negative pions interact strongly with complex nuclei producing spectacular disintegration 'stars'. Neutral pions decay electromagnetically and very rapidly by the process

$$\pi^0 \rightarrow \gamma + \gamma \tag{2.24}$$

with the occasional alternative of an (e^+e^-) pair instead of one or both of the photons.

The *interaction* of fast charged pions with nucleons has a special significance for nuclear forces and for nuclear structure because of its connection with the Yukawa process. When pion beams from accelerators became available it was possible to determine the energy variation of the total scattering cross-section for the pion–proton collision; the results for positive pions are shown in Fig. 2.10. The outstanding feature of these results is the peak at a pion laboratory momentum of 300 MeV/c, representing a preferred association, or *resonance* of the pion and nucleon at the corresponding centre-of-mass energy. This comes out to be 1236 MeV and although the energy width (120 MeV) of the state shows that it can

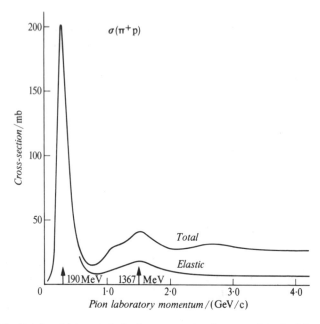

Fig. 2.10   Total and integrated elastic cross-section for scattering of π^+ by protons.

only have a transitory existence ($\tau \approx 5 \times 10^{-24}$ s using $\Delta E \cdot \Delta t \approx \hbar$), there is virtue in regarding it as an unstable particle, or excited nucleon; it is known as the delta particle, $\Delta(1236)$. The delta occurs in four charge states Δ^{++}, Δ^+, Δ^0, Δ^- according to the way in which it is formed from pions ($\pm$) and nucleons ($+$ and 0 charge). If pions play a part in nuclear structure, then so should the delta.

2.2.2 *The kaons and the hyperons*

Even before the elucidation of the π–μ–e decay has been completed, isolated tracks deriving from new types of particle had been seen in cloud chambers exposed to the cosmic radiation at sea level. Rochester and Butler, using a counter-controlled cloud chamber in a magnetic field of 1·4 tesla, concentrated on the components of the showers of penetrating particles that were produced occasionally near the chamber by protons or neutrons. Figure 2.11 shows one of the pictures obtained in 1946–47 in which characteristic 'vee' tracks were found. These were boldly, and correctly, diagnosed as due to the *decay* of a new type of neutral particle produced by a shower-component particle interacting (in the case of Fig. 2.11) with a nucleus in the lead plate mounted in the chamber. From the distances travelled in the chamber by the neutral particle before

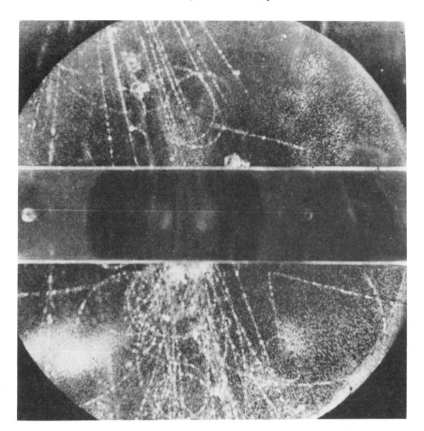

Fig. 2.11 Discovery of strange particles. A K°-meson produced by a cosmic-ray interaction in a lead plate in an expansion chamber decays at bottom right into $\pi^+ + \pi^-$ (Rochester, G. D., and Butler, C. C., *Nature*, **160**, 855, 1947).

decay, a lifetime of 10^{-10} s was assessed; from simple assumptions about the decay product a mass $\approx 1000 m_e$ was indicated.

From these results and from similar work at mountain altitudes and finally from accelerator experiments, the neutral 'vee' particles were grouped into two main classes now known as kaons and hyperons. For these the main decay schemes are

$$(kaon)\quad K^0 \rightarrow \pi^+ + \pi^- \tag{2.25}$$

and

$$(hyperon)\quad \Lambda^0 \rightarrow p + \pi^- \tag{2.26}$$

Masses and mean lifetimes are shown in Appendix 2.

The table in Appendix 2 includes other particles discovered in later experiments, namely the charged kaons $K^\pm$ and the heavier hyperons Σ, Ξ and Ω, all of which, like the Λ^0, contain a nucleon.

The *strangeness* of these particles resides in the fact that although their lifetimes are long $(10^{-8}-10^{-10}\,\text{s})$ on a nuclear time scale, e.g. in comparison with that of the Δ, they are nevertheless produced in accelerators and by cosmic radiation, with a cross-section that is a few per cent of the geometrical area of the target nucleus. If, then, reaction (2.26) is considered in reverse

$$\pi^- + p \to \Lambda^0 \to \pi^- + p \tag{2.27}$$

one might expect a lifetime comparable with that of the delta.

This difficulty was overcome by Pais who proposed that whereas particles or resonances like the Δ are indeed formed in analogy with (2.27) the new particles have quantum numbers which require their production in *association*, e.g.

$$\pi^- + p \to \Lambda^0 + K^0 \tag{2.28}$$

The familiar quantum numbers are not enough to require this process rather than (2.27), but if the kaon and hyperon are ascribed *strangeness* numbers -1 and $+1$ respectively and if this quantity strangeness is conserved in the reaction (2.28), then the production of single strange particles from the non-strange $\pi^- p$ system is forbidden. On the other hand, the decay process (2.26) does violate strangeness conservation, and this means that it cannot proceed via an interaction of the fast type allowed for (2.28).

It might be thought that strange particles have little to do with nuclear structure, since nuclei are non-strange. Nuclear beta decay is, however, quantitatively affected by the existence of strange particle decays in a way that will be discussed in Chapter 10. Furthermore, a nuclear neutron may be replaced by a hyperon with consequences of some significance to nuclear structure calculations (Sect. 2.4).

2.2.3 *The quark, the parton and the* J/ψ

The particles so far described are all objective entities in the sense that they can be produced and characterized and in many cases used to induce reactions. Quarks (Table 2.2) have not yet acquired this status. Their role is a powerful but descriptive one; by their postulated existence as substructures of pions, kaons, nucleons and hyperons they coordinate the grouping of these particles and their associated resonances into multiplets (Ref. 1.1).

The quark hypothesis in its present form is due to Gell-Mann and to Zweig, who postulated the three objects labelled u, d, s in Table 2.2. The u-quark has isobaric spin component (see Sect. 5.6) $T_z = +\frac{1}{2}$ (u = up), the d-quark has $T_z = -\frac{1}{2}$ (d = down) and the s-quark carries strangeness; each has the singular property of non-integral charge. The non-strange quark mass is now thought to be of the order of $\frac{1}{3}$ of the nucleon mass but the quark components of the proton (u + u + d) or of the neutron (u + d + d) are very tightly bound. The binding is supposed, in extension of the quark hypothesis, to result from the exchange of *gluons* between the quarks, just as pions and other particles are exchanged between nucleons to create the internucleon force. No conclusive evidence has yet been found in nature for the existence of fractional charges in a free state, although quark searches continue.

There is, from an entirely different set of phenomena, good evidence for substructure in the proton. It is well known that when Rutherford probed the structure of atoms with α-particles, his conclusions from work with particles of energy much greater than the electron binding energy was that they interacted with a point charge Ze at the centre of the atom. When similar experiments were made on the proton with electrons of several GeV energy from the Stanford linear accelerator both elastic and inelastic scattering involving pion production were observed. The elastic scattering (Sect. 2.1.2) conveyed useful, but expected information on the extent of the proton charge distribution. The inelastic scattering, however, when examined as a function of four-momentum transfer, behaved just as if the scattering was taking place from a point charge or point charges within the proton, in analogy with the Rutherford experiment on the atom. Similar evidence for such localization of interacting centres came from the observation of the production of particles of high transverse momentum (cf. Rutherford large-angle scattering) from the collision of 30 GeV protons in the CERN storage rings. The point charges inferred in both cases are known as *partons*.

It is almost certain that partons are to be identified with the (u, d, s) quarks proposed on the basis of particle systematics. There is, for instance, now sufficient data on neutrino–nucleon scattering

TABLE 2.2 Quark properties

Name	Q	B	T_z	S	Y	C
u (up)	$\frac{2}{3}$	$\frac{1}{3}$	$\frac{1}{2}$	0	$\frac{1}{3}$	0
d (down)	$-\frac{1}{3}$	$\frac{1}{3}$	$-\frac{1}{2}$	0	$\frac{1}{3}$	0
s (strange, or sideways)	$-\frac{1}{3}$	$\frac{1}{3}$	0	-1	$-\frac{2}{3}$	0
c (charmed)	$\frac{2}{3}$	$\frac{1}{3}$	0	0	$\frac{4}{3}$	1

Q = charge/e
B = baryon number
T_z = third component of isobaric spin (Sect. 5.6)
S = strangeness number
Y = hypercharge = $2(Q - T_z) = S + B + C$
C = charm number
To each particle shown corresponds an antiparticle with opposite quantum numbers. The quark spin is assumed to be $J = \frac{1}{2}$.

to make a quantitative comparison with inelastic electron–nucleon scattering. The difference between the two should be due to the quark charge and charge numbers $\frac{1}{3}$ and $\frac{2}{3}$ agree with the data. Future experiments on both neutrino and muon scattering by nucleons are expected to confirm this conclusion and to reveal further details of the quark–gluon interaction.

Other properties, beyond those listed in Table 2.2, may be necessary to complete the quark hypothesis. A difficulty relating to the statistics obeyed by quarks (Ref. 1.1) has suggested a new quantum number known as *colour*, which may be carried from quark to quark by the gluons. But the major development in the quark theory has been the introduction, rendered highly plausible since 1974 by the discovery of the J/ψ particles, of a *charmed* quark c. Such a quark had already been discussed for some time by theoreticians because it offered the possibility of explaining, by some appropriate cancellations, the absence of certain anticipated decay processes, especially $K^0 \rightarrow \mu^+ + \mu^-$, which could be effected by a *neutral* current. When the existence of neutral currents was finally verified by the observation in a heavy liquid bubble chamber at CERN of the production of events by muon-neutrinos (ν_μ) *without* the appearance of a muon, the charm hypothesis received much support (Ref. 2.3). Then in 1974, Ting at Brookhaven and Richter at Stanford, using quite different reaction processes, each reported the production of particles of mass 3·1 GeV which could be formed by, or could decay into, an electron and positron, $e^+ + e^-$. Decay into muon pairs and into pions, kaons and other strongly interacting particles was also established.

These new particles, called J by one group and ψ by the other, were only formed over a relatively sharply defined energy range, so that their lifetime against breakup into pions was about 10^{-20} s, at least 1000 times longer than expected for a strongly decaying object of comparable mass, e.g. the $\Delta(1236)$. The excitement at this discovery was intense, because, as with the discovery of strangeness, it meant that a new kind of matter with entirely unsuspected properties had been discovered. Both theoretical and experimental progress became extremely rapid, and it is now believed that the J/ψ particles have a cc̄ constitution, and that they form a group of states with strong analogies to the ortho- and para-systems of two more mundane spin-$\frac{1}{2}$ particles, e.g. the electron and positron that together form positronium. Many of the predicted states have now been found (Ref. 2.4).

After the J/ψ discovery, effort intensified to observe reactions in which a single charmed quark c would make its presence felt as a constituent of a charmed particle. Charm would then be explicit rather than concealed, as in the J/ψ. In 1976 a narrow state of the e^+e^- system at 1876 MeV was found at Stanford to decay in the way expected for a charmed particle and at about the same time neutrino-induced events with appropriate properties were seen both in nuclear emulsions and in the CERN heavy-liquid bubble chamber. A typical process is

$$\nu_\mu + \text{nucleon} \rightarrow \mu^- + C^+ + \text{hadrons}$$
$$C^+ \rightarrow e^+ + \Lambda^0 \quad (\text{or } K^\circ) + \nu_e \tag{2.29}$$

The signature of the presence of charm in the particle C^+ is the production of a strange particle in a leptonic process. (Leptons and hadrons are defined in Sect. 2.3.)

2.3 Particle classification and interactions

The general classification of particles, roughly according to mass, is as follows:

(*a*) The photon, i.e. the quantum of radiation, with zero mass.

(*b*) The leptons (light particles), i.e. the electrons, the muons and the neutrinos, although the latter almost certainly have zero mass. Leptons behave as point particles.

(*c*) The mesons (intermediate-mass particles), i.e. particles such as the pion and the kaon but excluding the muon that have masses between that of the electron (m_e) and that of the proton ($1836m_e$). Some resonant states that are clearly excited mesons may exceed the latter limit in mass.

(*d*) The baryons (heavy particles), which are subdivided into: (i) nucleons, the neutron and proton; (ii) hyperons, the particles of mass greater than that of the neutron ($1839m_e$) but which behave as if they contain just one nucleon.

The main types of force known to exist between a pair of particles are compared in Table 2.3*a*. Of these, the gravitational and electromagnetic interactions were familiar in classical physics. Their early recognition was due to the long range of the inverse-square-law force, which ensures that the interaction is sensed by objects of familiar size. Gravitation also has the special feature that it provides a cumulative force, because negative masses do not exist.

The weak interaction describes the production and behaviour of leptons and the decay processes of those strange particles that cannot decay electromagnetically. It is of short range, though how short is not yet certain except that it is very much less than nuclear dimensions. It was not known until the discovery of radioactivity.

The strong interaction, which provides the specifically nuclear forces between nucleons, is also of short range ($\approx 10^{-15}$ m) and was not known until the nucleus was discovered. It affects the behaviour, not only of nucleons, but also of mesons and hyperons, so that it is useful to group all these particles into a strongly interacting class called the *hadrons*. Hadrons, unlike leptons, have an extended structure.

All particles with mass feel gravity and all, if charged, feel electric and magnetic forces. Strongly interacting particles, or complex particles such as nuclei that are held together by the strong interaction, may exhibit small effects due to the weak interaction. The relative strengths of the interactions may be expressed by a rough calculation of the absolute magnitude of the force between two protons at rest, separated by a distance of say 10^{-15} m, i.e. within the range of the strong nuclear force; the orders of magnitude are:

gravitational force (mass)	10^{-34}	N
electromagnetic force (charge)	50	N
strong nuclear force	10^5	N

These figures are based on experimentally determined *coupling constants*, which relate to a Yukawa-like description of the forces. Returning to Fig. 2.8, and applying it to the strong interaction between two protons, the square root (g) of the coupling constant gives at each vertex A, B the probability amplitude for the emission or absorption of the exchanged particle. The corresponding potential energy, from which the force may be derived, contains the coupling constant g^2 itself. Particles that may be exchanged in the interactions are shown in Table 2.3 together with the coupling constants reduced to a suitable dimensionless form.

The important physical quantities to which both strict and conditional conservation laws apply were listed in Table 1.1, Section 1.3. The operation of these laws in processes governed by the interactions of concern to nuclear physics are shown in Table 2.3*b*, the conservations of energy and momentum being omitted because of their universal validity.

TABLE 2.3*a* Interactions between particles

Interaction	Range	Dimensionless coupling constant	Exchanged particle
Gravitational	long	10^{-39} [a]	'graviton'
Weak	short ($<10^{-15}$ m)	$1 \cdot 2 \, 10^{-5}$	'vector boson'
Electromagnetic	long	1/137	photon
Strong	short ($\approx 10^{-15}$ m)	15	meson

[a] Using the proton mass as unit.

TABLE 2.3*b* Conservation laws for nuclear interactions

Conserved quantity or operation	Interaction		
	Weak	*Electromagnetic*	*Strong*
Parity	No	Yes	Yes
Isobaric spin	No	No	Yes
Charge parity	No	Yes	Yes
(Time reversal)	Yes[a]	Yes	Yes
Charge Q	Yes	Yes	Yes
Baryon number B	Yes	Yes	Yes
Lepton number L	Yes	Yes	Yes
Strangeness S	No	Yes	Yes
Charm C	No	Yes	Yes

[a] There is a small violation in K° decay.

Although the interactions are discussed in this section as separate and distinct fields of force it is likely that a more unified picture will gradually emerge. Already, the behaviour of the weak and electromagnetic interactions shows much more than coincidental similarity (Ch. 10) and theories now exist that identify them, at least at high momentum transfers.

2.4 The nucleus and the hypernucleus

A complex nucleus is a many-body system in which attention must be paid to the mutual interaction of a number of particles; the number is, in general, neither very small nor very large, but because of its intermediate nature involves individual properties as well as statistical considerations. From the particles mentioned in Sections 2.1 and 2.2 one may build up:

(*a*) The plasma, a gas of ions and electrons of importance for thermonuclear energy generation. In laboratory plasma experiments the interactions holding the system together are electromagnetic.

(*b*) The neutron star, a condensed degenerate system held together by the gravitational interaction.

(*c*) The typical nucleus, in which protons and neutrons but not electrons exist. The nucleus is too small for gravitational forces to have any major significance, and stability is due to strong interactions between the constituent nucleons, with a disruptive influence due to the Coulomb repulsion of the nuclear protons. The balance between Coulomb and nuclear forces is crucial in determining the stability of a heavy nucleus against deformations and hence the possibility of fission. Because of the nature of the internucleon force it may be claimed that at any given time the nucleus contains some pions and perhaps heavier mesons and, therefore, probably also delta particles.

The distribution of all nuclear constituents, excepting any obeying the Bose–Einstein statistics, between available quantum states is regulated by the Pauli exclusion principle. It is, therefore, of considerable interest that structures known as *hypernuclei* exist in which the control exercised by the exclusion principle has been removed. A hypernucleus is one in which a nuclear neutron has been replaced by a neutral hadron such as the Λ^0 or Σ^0; such systems were first seen in events induced by cosmic rays in nuclear emulsions. They have been useful in checking nuclear structure calculations and provide examples of nuclei carrying a strangeness quantum number.

Observable nuclear properties that contribute information on its structure are: ground-state mass and energy of excited states; nuclear radius or potential distribution in all states; intrinsic spin and isobaric spin of all states; parity and electromagnetic moments of all states; matrix elements for decay of all unstable states, and branching ratios for alternative modes; and spectroscopic factors, that govern the participation of the nucleus in reaction processes. These properties will be discussed in Part II of this book.

2.5 Summary

A partly historical outline has been given of the nature of the particles that may be concerned in nuclear structure and reactions and of the general types of force that operate between them.

Examples 2

2.1 Show that if two consecutive radiations from a long-lived parent have decay constants λ_1 and λ_2, the apparent mean life of the second radiation is $1/\lambda_1 + 1/\lambda_2$.

2.2 A thin ^{210}Po source of strength 50 μCi is deposited on a small sphere of radius 1 mm and suspended at the centre of an evacuated metal sphere of radius 100 mm by a thread of resistance 10^{13} ohms. Find the equilibrium potential difference between the source and the sphere and the time taken for half this value to be reached. $(1 \text{ Ci} = 3 \cdot 7 \times 10^{10}$ disintegrations per second.) [5·9 V, 0·8 s]

2.3* Examine the possibility of conservation of four-momentum: (*a*) in the materialization of a photon into an electron pair in free space and in the converse process of single-quantum annihilation; and (*b*) in the emission of a real bremsstrahlung photon by an electron in free space and in the converse process of the absorption of a photon by an electron.

3

The interaction of particles and γ-radiation with matter

The penetration of ionizing radiations into matter has been of theoretical interest and of practical importance for nuclear physics since the early days of the subject. The classification of the radiations from radioactive substances as α-, β- and γ-rays was based on the ease with which their intensity could be reduced by absorbers. Range or absorption coefficient provided, and still provides, a useful method of energy determination, and the associated processes of ionization and excitation underlie the operation of nearly all present-day particle and photon detectors. In the present chapter the basic processes of the interaction of radiation will be examined with particular reference to their role in nuclear techniques.

The interaction of nuclei with matter may also be taken to include hyperfine effects in spectroscopy and perhaps even the formation of exotic atoms in which an ordinary electron is replaced by a heavier particle. These structures give much information both on nuclei and particles and will be briefly reviewed.

3.1 Passage of charged particles and radiation through matter

3.1.1 General

For energies up to a few hundred MeV, the main process by which a charged particle such as a proton, α-particle or meson loses energy in passing through matter is the transfer of kinetic energy to atomic *electrons*, with very little deviation from the original path. The more energetic recoil electrons produced in this way are visible along the track of a charged particle in an expansion chamber as *delta rays*. In special cases, e.g. for fission fragments or other heavy ions, transfer of energy to the nuclei of the medium may also be important. In such cases of high charge, or low velocity or both, *large-angle scattering* by Coulomb interaction with the nuclei may be observed, but for fast, light ions this is a rare event. It is consequently reasonable, at least for such particles, to define a definite *range* in matter (Fig. 3.1a).

Fast electrons and positrons also lose energy by collision with other electrons and may be scattered through angles up to 90° in such collisions. They lose very little energy to nuclei, because of the disparity in mass, but can suffer large deflections, >90°. In all collisions in which accelerations are experienced, electrons and positrons also lose energy by the radiative process known as *bremsstrahlung*. Heavy particles do not radiate appreciably in this way

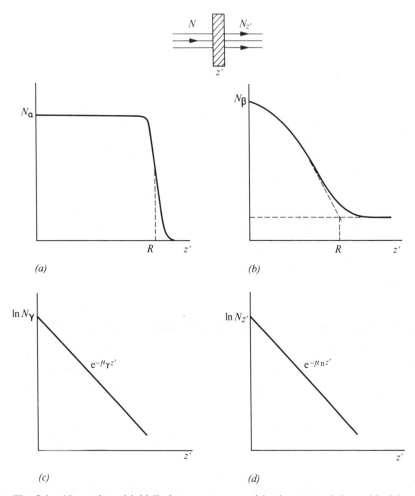

Fig. 3.1 Absorption of initially homogeneous particles by matter (schematic): (*a*) Light ions (e.g. α, p, muons) showing mean range *R*. (*b*) Electrons, showing extrapolated range *R*. (*c*) Photons, $E_\gamma < 10$ MeV. (*d*) Neutrons. Absorber thicknesses z' and ranges *R* are conveniently measured as mass per unit area of absorber (ρ_s), and μ in (*c*) and (*d*) is then a mass absorption coefficient.

because their accelerations are much smaller. As a result of scattering and radiative energy loss, a group of initially monoenergetic electrons may traverse considerably different thicknesses of matter before being brought to rest, and range can best be defined by an extrapolation process (Fig. 3.1*b*).

Electromagnetic radiation interacts with matter through the processes of elastic (Rayleigh and nuclear) scattering, photoelectric effect, Compton effect and pair production. These processes reduce the *number* of photons in a beam in proportion to the number incident on an absorber; thereby creating an exponential attenuation (Fig. 3.1*c*). No definition of range is appropriate, although the attenuation coefficient is energy-dependent. For radiation of energy above say 10 MeV, the electrons and positrons from the pair-production process themselves generate bremsstrahlung and annihilation quanta and an electromagnetic shower builds up characterized by a linear dimension called the *radiation length* (Sects. 2.1.5 and 3.1.3).

At energies above about 100 MeV the *nuclear interactions* of charged particles, other than leptons, are no longer negligible within the expected range. These interactions remove particles from a beam and above 1 GeV the concept of range must be replaced in effect by an attenuation coefficient based on total nuclear cross-sections (including scattering). An absorption plot resembling Fig. 3.1*c* then replaces the range curve Fig. 3.1*a*. Such plots are very clearly shown in the case of *neutrons* of all energies for which there is only nuclear attenuation (Fig. 3.1*d*).

The relative importance of collision and radiative energy loss for electrons and protons in lead is shown in Fig. 3.2.

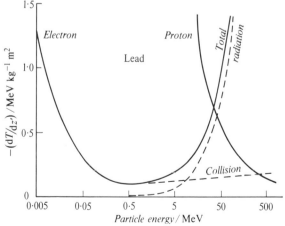

Fig. 3.2   Energy loss for electrons and protons in lead as a function of kinetic energy (Heitler, W., *The Quantum Theory of Radiation*, Clarendon Press, 1954).

3.1.2 Energy loss by collision

If a particle of charge $Z_1 e$ and velocity $v = \beta c$ passes through a substance of atomic weight A (kg) and atomic number Z the energy loss dT in a path dz' (where z' is measured by the surface density ρ_s, in kg m^{-2}) is given by the Bethe–Bloch formula

$$-\left(\frac{dT}{dz'}\right)_{\text{coll}} = \frac{1}{(4\pi\varepsilon_0)^2} \frac{2\pi N_A Z Z_1^2 e^4}{m_e v^2 A}$$

$$\times \left[\ln \frac{2m_e v^2 W_{\text{max}}}{I^2(1-\beta^2)} - 2\beta^2 - \delta - U \right] \quad (3.1)$$

where I is a mean excitation potential for the atoms of the substance. If there are no radiative losses, this is the *stopping power* of the absorber for velocity v; it has the unit energy $\times$ kg^{-1} m^2.

W_{max} is the maximum energy transfer from the incident particle to an atomic electron, given by

$$W_{\text{max}} = 2m_e v^2/(1-\beta^2) \text{ for energies} \ll M_1^2 c^2/2m_e \quad (3.2)$$

δ is a correction for polarization of the medium and U allows that inner shell electrons do not necessarily behave as free electrons.

The rigorous derivation of formula (3.1) is a quantum mechanical problem of some complexity and will not be attempted here. The general form of the main term, however, can be indicated classically as follows:

Assume (Fig. 3.3) that the incident particle moves in a straight line at a perpendicular distance b (the impact parameter of Sect. 1.2.6) from a stationary electron. The impulse conveyed to the electron by the passage of the particle is then

$$q = \int F \, dt \approx 1/4\pi\varepsilon_0 \,.\, Z_1 e^2/b^2 \,.\, 2b/v \quad (3.3)$$

where F is the perpendicular component of the force acting on the electron during passage from A to B, which may crudely be assumed to represent the distance over which a constant force

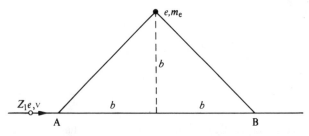

Fig. 3.3 Energy loss of a charged particle to an electron at impact parameter b.

$Z_1 e^2/b^2$ is effective. The energy transferred to the electron is

$$q^2/2m_e = (1/4\pi\varepsilon_0)^2(2Z_1^2 e^4/m_e v^2 b^2) \tag{3.4}$$

The number of electrons per unit thickness with impact parameters between b and $b + db$ is $2\pi b \, db \times N_A Z/A$ so that the differential energy loss per unit thickness for this range of b is

$$-d^2 T/dz' \, db = (1/4\pi\varepsilon_0)^2(4\pi N_A Z Z_1^2 e^4/m_e v^2 Ab) \tag{3.5}$$

Integration between limits b_{max} and b_{min}, corresponding to energy losses for which a simple theory is valid, gives

$$-dT/dz' = (1/4\pi\varepsilon_0)^2(4\pi N_A Z Z_1^2 e^4/m_e v^2 A) \ln (b_{max}/b_{min}) \tag{3.6}$$

which agrees with the first term of (3.1) using the value of W_{max} given by (3.2) if

$$b_{max}/b_{min} = 2m_e v^2/I(1 - \beta^2) \tag{3.7}$$

Reasons can be given for this ratio of extreme impact parameters (Ref. 3.1). Thus b_{max} is determined by the fact that no energy transfer takes place when the time of collision $\approx 2b/v$ becomes long compared with atomic frequencies so that the collision is adiabatic. Also, b_{min} is related to the validity of the classical treatment of the collision, which will break down when the de Broglie wavelength of the electron in the c.m. system approaches b.

The stopping-power formula, equation (3.1), predicts a variation of dT/dz' with energy similar to the experimental results for protons in aluminium which are shown in Fig. 3.4; it will be noted that m_e is the mass of the *electron* and not the mass M_1 of the incident heavy particle. The simple theory is not strictly valid at low energies, partly because of the neglect of the effect of capture and loss of electrons by the moving particle. The appearance of a maximum in the observed collision loss is at least partly due to this, although it is also predicted by the velocity dependence shown in equation (3.1). At higher energies the energy loss falls off as $1/v^2$ with increasing v to the point of *minimum ionization*, beyond which is seen the *relativistic rise* due to the $(1 - \beta^2)$ term, with $\beta \approx 1$ and γ (Fig. 3.4) = $(1 - \beta^2)^{-1/2}$.

In placing equation (3.1) on an absolute scale the principal uncertainty is in the atomic excitation potential I. It may be obtained from measurements of stopping power or from calculations based on an atomic model; both indicate that

$$I = kZ \text{ eV}$$

where $k \approx 11$, with some dependence on Z for very light and very heavy atoms.

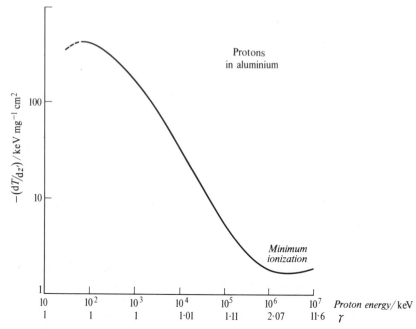

Fig. 3.4 Specific energy loss, or stopping power, of aluminium for protons. The quantity $\gamma = (1 - \beta^2)^{-1/2}$ is the Lorentz factor.

Observable phenomena basically governed by the collision-loss formula for a single particle are:

(*a*) Production of ion pairs in solids, liquids and gases and of electron-hole pairs in semiconductors.
(*b*) Production of light in scintillator materials.
(*c*) Production of bubbles in a superheated liquid.
(*d*) Production of developable grains in a photographic emulsion.

Each of these has been made the basis of an energy-sensitive particle detector.

3.1.3 Energy loss by radiative processes

Electrons of kinetic energy T and total energy E passing through a thin absorber give rise in each radiative collision to a bremsstrahlung photon of any energy between 0 and T. The process may be envisaged as in Fig. 3.5, in which the Coulomb field existing between the incident electron and a nucleus, or another electron, of the absorbing medium is represented by an emission of virtual photons. The real electron makes a Compton scattering collision

(Sect. 2.1.4) with a virtual photon, creating a virtual lepton, which then gives rise to a real photon, i.e. the bremsstrahlung quantum and a real electron of reduced energy. The bremsstrahlung process $e \rightarrow e' + \gamma$ cannot take place in free space because momentum and energy cannot then be conserved. Photons can however also be emitted by electrons moving in the magnetic field of an orbital accelerator (Sect. 4.3.3) and this emission is known as *synchrotron radiation*.

The energy distribution of the bremsstrahlung is given by the expression (Ref. 3.1)

$$\phi(E, \nu) \, d\nu \propto Z^2 f(E, \nu) \, d\nu / \nu \tag{3.8}$$

where for energies $T \approx 100$ MeV at least $f(E, \nu)$ varies only slowly with the frequency ν. The energy *loss* $h\nu\phi \, d\nu$ in each frequency interval is then constant to a first approximation. By integrating over the available frequency spectrum the energy loss per unit path in kg m^{-2} at kinetic energy T is

$$-(dT/dz')_{rad} = N_A/A \int_0^{T/h} h\nu\phi(E, \nu) \, d\nu \tag{3.9}$$

and because of the slow variation of $h\nu\phi$ with ν the integral is

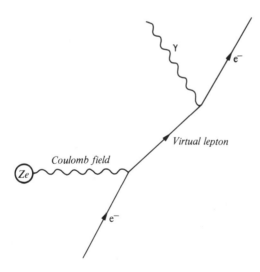

Fig. 3.5   Production of bremsstrahlung by an electron in the field of a nucleus Ze; the virtual lepton is an electron 'off the mass shell'. This and subsequent similar figures are not intended to display the precise order of events, and other diagrams showing a different order may be drawn. Time increases to the top of the page, and a fixed Coulomb field is shown as a wavy horizontal line. Rules exist for calculating cross-sections from such graphs.

approximately proportional to the incident (total) energy E, i.e.

$$-(\mathrm{d}T/\mathrm{d}z')_{\mathrm{rad}} = E \cdot N_A/A \cdot \bar{\phi} \qquad (3.10)$$

where $\bar{\phi}$ is a calculable function.

At high energies, collision loss (eqn (3.1)) varies as $\ln (1-\beta^2)^{-1}$ i.e. as $\ln E$, while radiative loss (eqn (3.10)) is proportional to E and therefore predominates. The electron energy at which the two losses are equal is the *critical energy* E_c. When radiative loss is dominant it is useful to define a *radiation length* X_0 by the equation

$$-\mathrm{d}z'/X_0 = \mathrm{d}E/E \qquad (3.11)$$

so that X_0 is the absorber thickness over which the electron energy is reduced by a factor e. Table 3.1 gives critical energies and radiation lengths for a few common materials.

TABLE 3.1 Critical energy and radiation length (from Ref. 3.1)

Material	E_c/MeV	X_0/kg m^{-2}
H	340	580
C	103	425
Fe	24	138
Pb	6·9	58
Air	83	365

Classically, bremsstrahlung originates in the acceleration of a charged particle as a result of Coulomb interaction, and its intensity is therefore inversely proportional to the square of the particle mass according to Maxwell's theory. The muon has 200 times the electron mass and its radiative energy loss is therefore negligible. This holds *a fortiori* for all other charged particles, since they are heavier still.

All charged particles independently of their mass do, however, suffer a small electromagnetic loss in matter through *Cherenkov radiation*. This depends on the gross structure of the medium through which the particle passes, and in which a macroscopic polarization is set up. If the velocity of the particle exceeds the velocity of light in the medium, the polarization is longitudinally asymmetric and secondary wavelets originate along the track AB of the particle (Fig. 3.6). These form a coherent wavefront propagating in a direction at angle θ with the path of the particle where

$$\cos \theta = c/n \div \beta c = 1/\beta n \qquad (3.12)$$

n being the refractive index. The process is analogous to the formation of the bow wave of a ship. It has the important features for nuclear detectors that coherence does not appear until $\beta = 1/n$, i.e. there is a *velocity threshold* for observable radiation, and also that the direction of emission is very well defined.

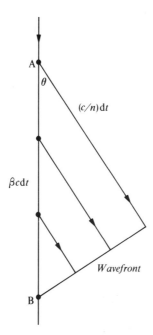

Fig. 3.6 Formation of a coherent wavefront of Cherenkov radiation.

Cherenkov loss is usually only about 0·1 per cent of collision loss. Explicitly, the loss for a singly charged particle may be written

$$-\left(\frac{\mathrm{d}T}{\mathrm{d}z}\right)_{\mathrm{Ch}} = \frac{\pi e^2}{c^2 \varepsilon_0} \int \left(1 - \frac{1}{\beta^2 n^2}\right) \nu \, \mathrm{d}\nu \qquad (3.13)$$

where the integration extends over all frequencies for which $\beta n > 1$. Within these bands the spectral distribution of loss is proportional to $\nu \, \mathrm{d}\nu$ in contrast with the $\mathrm{d}\nu$ proportionality of bremsstrahlung. Cherenkov light is thus found mostly at the blue end of the visible spectrum.

3.1.4 Absorption of electromagnetic radiation

As discussed in Section 1.2.5 the attenuation of a beam of photons by matter is described by equation (1.35a), namely

$$n = n_i \exp\left(-\mu_m \rho_s\right) \qquad (3.14)$$

where μ_m is the mass attenuation coefficient ($\mathrm{m}^2 \, \mathrm{kg}^{-1}$) and ρ_s is the mass per unit area of the absorber. The same formula describes the diminution of the *intensity* $n_i \times h\nu$, or energy flux.

To relate the measured μ_m to calculated cross-sections for interactions with absorber atoms, it may be necessary first to correct for the fact that both elastic (Rayleigh) and inelastic (Compton) scattering give scattered photons of the incident energy in the forward direction, i.e. the direction of the incident beam. The Rayleigh scattering may in many cases be neglected since it is sharply forward-peaked at high energies and does not remove photons from the beam unless collimation is very fine. Compton scattering, on the other hand, does not depend sharply on angle at small angles and may result in some solid-angle dependence of the absorption coefficient. This cannot easily be avoided by energy discrimination because the energy change is small near the forward direction. It may be corrected both experimentally, by making measurements in different geometries, and theoretically.

When such corrections have been made we may write $\mu_m = N_A \sigma / A$ with

$$\sigma = \sigma_{PE} + Z\sigma_C + \sigma_{PP} \tag{3.15}$$

The individual cross-sections relate to the photoelectric effect, the Compton effect and pair production respectively, and the factor Z embodies the assumption that all the atomic electrons contribute individually (and incoherently) to Compton scattering. This will be true if the photon energy is much greater than the K-shell ionization energy of the atom.

The relative importance of the three absorption processes as a function of energy and of atomic number is shown in Fig. 3.7 and the actual mass absorption coefficient μ_m for lead is given as a function of energy in Fig. 3.8. This curve shows: (a) sharp peaks at

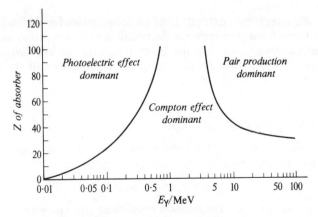

Fig. 3.7 Relative importance of the three major types of γ-ray interaction (Ref. 2.1).

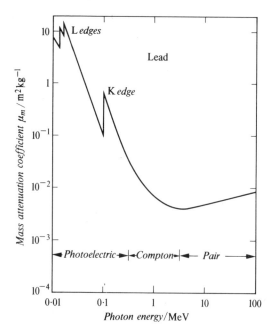

Fig. 3.8 Mass attenuation coefficient for electromagnetic radiation in lead (Ref. 2.1).

energies corresponding with the atomic absorption edges and indicating the onset of increased photoelectric absorption as a new electron shell becomes available; and (*b*) a minimum absorption coefficient due to the compensation of photoelectric and Compton cross-sections that fall with energy by the pair production cross-section that rises. Both effects mean that the energy is not a single-valued function of μ_m throughout its range for a given absorbing material.

The individual processes indicated in equation (3.15), together with elastic scattering, are all part of the total electromagnetic interaction between radiation and matter and are therefore connected. For most practical purposes, in the energy range of interest to nuclear physics, it is possible to regard them as independent, each contributing to the total cross-section to a degree determined by the quantum energy. The salient features of the processes are as follows:

(i) *Photoelectric effect* (Fig. 3.9). Because of the necessity to conserve energy and momentum, a free electron cannot wholly absorb a photon. Photoelectric absorption, therefore, tends to take place most readily in the most tightly bound electronic shell of the atom available, since then momentum is most easily conveyed to the

83

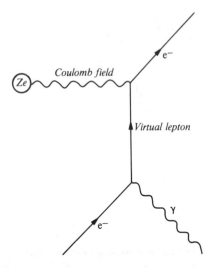

Fig. 3.9 Photoelectric effect; conservation of momentum and energy require interaction with the atom, Ze.

atom. The kinetic energy of a photoelectron from the K-shell is

$$T = h\nu - E_K \qquad (3.16)$$

where E_K is the K-ionization energy. The cross-section depends on atomic number Z and wavelength of radiation λ approximately as

$$\sigma_{PE} \approx Z^5 \lambda^{7/2} \qquad (3.17)$$

in the energy range 0·1 to 0·35 MeV. The vacancy in the atom created by the ejection of a photoelectron leads to the emission of characteristic X-rays or of electrons known as *Auger electrons* from less tightly bound shells as an alternative process favoured for light atoms.

(ii) *Compton effect* (Figs. 2.3 and 3.10). The scattering of photons by atomic electrons that may be regarded as free should be treated relativistically. In Section 2.1.4 this collision process was described as evidence for the photon hypothesis, and it was shown that conservation of 4-momentum leads to a relation between the wavelength of the incident photon and that of the scattered photon emerging at an angle θ with the original direction (eqn (2.7)):

$$\lambda' - \lambda = c/\nu' - c/\nu = (h/m_e c)(1 - \cos\theta) \qquad (3.18)$$

where $h/m_e c$ is the Compton wavelength of the electron ($2\cdot43 \times 10^{-12}$ m). The energy of the recoil electron is $h(\nu - \nu')$; it is zero for $\theta = 0$ and a maximum for $\theta = \pi$.

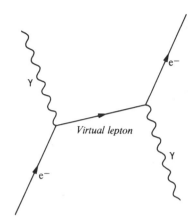

Fig. 3.10   Compton effect, with free electron.

The Compton wavelength shift is independent of wavelength and of the material of the scatterer. The differential cross-section for the process is given by a formula due to Klein and Nishina, which predicts that at a given angle the cross-section decreases with increasing photon energy. The angular variation for a range of energies is shown in Fig. 3.11; in all cases there is a finite cross-section for $\theta = 0$ and equation (3.18) shows that over a small angular range about zero the wavelength is essentially unmodified.

As $h\nu/m_e c^2$ tends to zero the angular distributions shown in Fig. 3.11 become more symmetrical about 90° and finally reach the form $(1 + \cos^2 \theta)$ characteristic of classical *Thomson scattering*. However, the assumption of scattering from free electrons then becomes less tenable for matter in its normal state, and coherent *Rayleigh scattering* from the bound electrons forming the atomic charge distribution takes over.

(iii) *Pair production* (Fig. 3.12).   The process

$$h\nu \rightarrow e^+ + e^- \qquad (3.19)$$

like the bremsstrahlung process (Sect. 3.1.3), cannot take place in free space because 4-momentum conservation cannot be satisfied. If a third electron or, better still, a nucleus is present to absorb recoil momentum, pair production occurs in accordance with the diagram of Fig. 3.12. For pair production in a nuclear field the threshold energy is $2m_e c^2$, i.e. 1·02 MeV.

Following Dirac's theory of holes (Sect. 2.1.5), the process may be pictured as the elevation of an ordinary electron from a *negative energy state* $(E \leqslant -m_e c^2)$ to a normal positive state $(E \geqslant m_e c^2)$, for which an energy consumption of at least $2m_e c^2$ is required. The hole

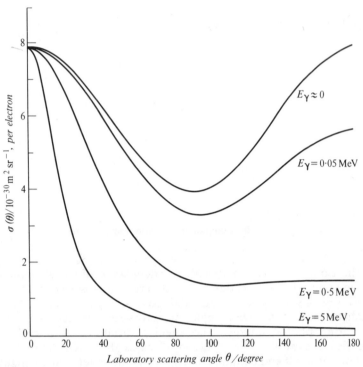

Fig. 3.11 Differential cross-section for Compton scattering, giving *number* of photons scattered per unit solid angle at angle θ, for energies 0–5 MeV.

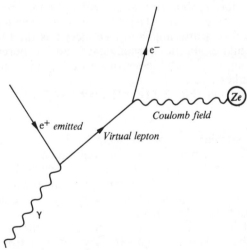

Fig. 3.12 Pair production, in nuclear field. The motion of an electron backwards in time corresponds to the *emission* of a positron.

86

left in the normally filled band of negative states then behaves as a positive electron. Alternatively, in terms of Fig. 3.12, the antiparticle or positron may simply be regarded as a normal particle moving backwards in time.

The exact converse of process (3.19) is the transition of an ordinary electron to a state of negative energy with the emission of a single quantum of annihilation radiation, rather than the two quanta normally produced. This is just a bremsstrahlung process, as may be seen by comparing Figs. 3.5 and 3.12, and indeed the two processes are both described by the theory due to Bethe and Heitler. In each case the cross-section varies as Z^2, and increases at first as the primary photon (or electron) energy increases. This corresponds to the production of pairs at larger and larger distances from the nucleus, but a limit is set by the screening of the nuclear charge by the atomic electrons. The variation of σ_{PP} with energy is shown for lead, with and without screening, in Fig. 3.13.

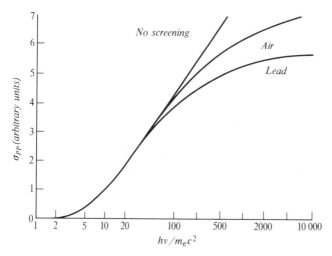

Fig. 3.13   Effect of screening on pair production cross-section for air and lead (Ref. 3.1).

3.2 Experimental studies

The theoretical results outlined in Section 3.1 are applied mainly in the determination of the energy of known particles, including photons, and in identifying unknown particles and finding their mass. In such work several more detailed features of the general energy-loss process and deductions from it must be considered, and these are treated in the following sections.

3.2.1 Multiple Coulomb scattering

In the passage of a charged particle through matter, small-angle deflections are continually arising because of distant Coulomb interactions with nuclei and electrons. These accumulate in a random manner to produce a resultant deviation which has a Gaussian distribution for a group of particles. This distribution of angles may be written

$$P(\theta)\, d\theta = (2\theta/\langle\theta^2\rangle) \exp\left(-\theta^2/\langle\theta^2\rangle\right) d\theta \qquad (3.20)$$

and according to Rossi and Greisen

$$\langle\theta^2\rangle = (Z_1^2 E_s^2/p^2\beta^2 c^2)z'/X_0 \qquad (3.21)$$

where E_s is a constant (21 MeV) independent of the mass of the particle and the nature of the medium, p is the particle momentum, z' is the path in the medium, assumed large, and X_0 is the radiation length (Sect. 3.1.3).

Multiple scattering limits the precision of momentum and angle measurements on particle tracks. For a proton of momentum 10 GeV/c, observed in a hydrogen bubble chamber in a magnetic field, multiple scattering introduces an uncertainty comparable with errors of measurement. For heavy ions there are noticeable deviations from the simpler theories.

3.2.2 Range–energy curves; straggling

The theoretical range of a heavy charged particle is equal to its path length in matter because over most of the path scattering is negligible. The range may formally be obtained by integration of the expression for energy loss, giving

$$R = \int_0^T dT(dT/dz')^{-1} \qquad (3.22)$$

but in practice this cannot be carried out for the full range of the particle because of the corrections necessary to the explicit formula for dT/dz' at low energies. Range–energy curves are therefore constructed semi-empirically by combining observations of the range of particles of known energy with integrations of the energy-loss formula over some particular region.

The first range–energy curve to be established was for α-particles in air. Energies of α-particles from radioactive elements were obtained from the classical magnetic deflection experiments of Briggs, Rosenblum and Rutherford with a precision of 1 part in 10^5 in favourable cases. Ranges were measured by observing tracks in an expansion chamber or by allowing the α-particles to pass through a shallow ionization chamber which was then moved along until the particle terminated its range in the chamber. The ranges displayed in range–energy curves are by convention mean ranges, i.e. the average range of a group of particles initially of homogeneous velocity. The range–energy relation for α-particles of energy up to 14 MeV is given in Fig. 3.14a. For other materials, these α-particle ranges are found empirically to vary as $A^{1/2}/\rho$, where A is the atomic weight and ρ the density of the absorber.

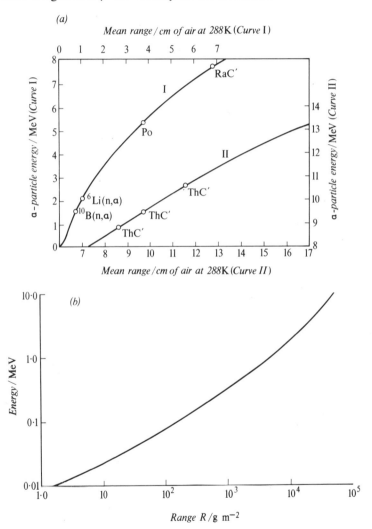

Fig. 3.14   Range–energy curves: (a) α-particles 0–14 MeV, showing reactions or radioactive bodies used to provide calibration points (Ref. 2.1). (b) Electrons 0·01–10 MeV (Katz, L. and Penfold, A. S., *Rev. Mod. Phys.*, **24**, 28, 1952).

The range–energy curves for different particles may be connected by use of the stopping-power formula, equation (3.1). Since dT/dz' depends only on the charge (Z_1) and the particle velocity (v) for a given stopping material, it follows that the range for non-relativistic particles of given velocity is proportional to M/Z_1^2; e.g. protons, deuterons and tritons of energies 10, 20 and 30 MeV have ranges in the ratio $1:2:3$.

The range of a slow electron in matter, as noted in Section 3.1.1, will be much less than its path length because of large-angle scattering. Ranges are therefore normally defined by the extrapolation procedure illustrated in Fig. 3.1b; this procedure is sufficiently reproducible to permit such ranges, both for homogeneous electrons and β-ray distributions, to be used in establishing a range–energy curve (Fig. 3.14b). To a good approximation, ranges expressed in $g\,m^{-2}$ (i.e. as mass per unit area, which is nearly proportional to the total number of electrons per unit area) are independent of the material of the absorber. The relation $R/g\,m^{-2}\,Al = 5430(E_0/MeV) - 1600$, first given by Feather, has been much used for finding the maximum energy in a β-ray spectrum.

An average α-particle from a radioactive source makes about 10^6 collisions resulting in small energy transfers before coming to rest in an absorber. A group of such particles initially of uniform velocity will show a distribution of velocities about a mean value after passage through a certain thickness of matter owing to the statistical nature of the energy loss. The ranges of the particles will, therefore, be grouped about a mean value, and since the number of collisions is large, the distribution of ranges may be well represented by a Gaussian curve. Figure 3.15a shows the results of an experiment to investigate the energy straggling of [241]Am α-particles; the standard deviation of the range distribution is about 1 per cent.

Most measurements of the energy of particles from nuclear reactions (say up to 50 MeV) are now based on complete absorption of the particle in a solid-state detector, rather than on measurements of range. In such detectors the number of electron–hole pairs, from which the output signal is derived, follows a Poisson distribution with standard deviation $\sigma = \sqrt{N}$ where N is the average number. In fact, as pointed out by Fano, the individual events are not independent because of the fixed total energy that may be lost and the resulting correlation narrows the distribution. For α-particles of energy 6 MeV a line width of $0\cdot2$ per cent is obtainable.

The slowing down of electrons in matter inevitably introduces a large straggling which is not only considerably greater in proportion than that for heavy particles but is also asymmetric, because of the high probability of loss of a large amount of energy in a single collision. The distribution of energy loss has been calculated by Landau and by others and is illustrated by the results of Goldwasser

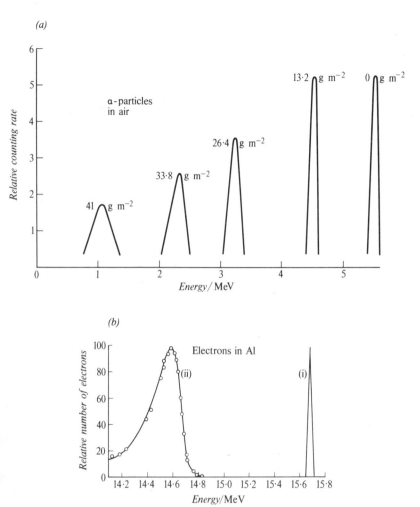

Fig. 3.15   (*a*) Energy distribution of 5·5-MeV α-particles from ²⁴¹Am after passing through indicated thicknesses of air. (*b*) Energy distribution of electron beam after passing through 8·6 kg m⁻² of aluminium: (i) incident beam, (ii) transmitted beam (Goldwasser, E. L. *et al.*, *Phys. Rev.*, **88**, 1137, 1952).

et al. on the energy loss of 15·7-MeV electrons in a thin metal foil (Fig. 3.15*b*). The width of the Landau distribution has been measured and compared with theory for protons, electrons and muons and must always be taken into account in computing the most probable energy loss of a fast particle. As the velocity of the particle decreases, the relative importance of small energy transfers increases and the straggling tends to the symmetrical Gaussian form.

3.2.3 Capture and loss: stopping power for heavy ions

Near the end of the range of an α-particle in matter, the capture and loss of electrons by the ion, or charge exchange, i.e.

$$\left. \begin{array}{l} e^- + He^{++} \leftrightarrows He^+ \\ e^- + He^+ \rightleftarrows He^0 \end{array} \right\} \tag{3.23}$$

occurs very frequently. Experimentally observed cross-sections for helium ions of energy less than 1 MeV suggest mean free paths for processes (3.23) of a few thousandths of a millimetre for air at standard temperature and pressure. Charge equilibrium between the He^{++}, He^+ and He^0 components of a low-energy α-particle beam is thus reached for extremely thin layers of matter. Similar considerations apply to protons of low velocity, which may capture electrons to form neutral atoms.

Charge exchange is important for heavy ions, including fission fragments, and has been extensively studied because of the interest in establishing range–energy relations for such particles. These are required, for instance, in recoil methods of finding nuclear lifetimes (Sect. 9.4.1). For heavy ions, the velocity range over which electron exchange processes take place is much greater than for protons or α-particles and the energy loss per unit path is determined not by the atomic number Z_1 of the ion but by something approaching a root-mean-square (r.m.s.) value of the actual instantaneous charge as reduced by the number of retained electrons.

The r.m.s. charge of a heavy ion is highly velocity dependent and is governed by the principle that on average, and allowing for the essentially fluctuating nature of charge exchange processes, electrons whose orbital velocity in the ion is less than the ionic velocity will normally be lost. As an ion slows down from high velocities, at which the normal fully stripped rate of energy loss given by equation (3.1) is observed, more and more electrons are retained after capture from atoms of the stopping medium until finally a neutral atom state is reached. Ionization and excitation processes along the particle path diminish in accordance with the r.m.s. charge. This is clearly shown by the tapered appearance of heavy-ion tracks in nuclear emulsion, which is due to diminution in the number of δ-ray electrons produced by collision. Because of severe charge reduction the stopping power for heavy ions passes through a maximum as the ion velocity decreases, which accentuates the similar effect observed for protons and α-particles.

At very low velocities, it is predicted from calculations based on atomic models that the stopping power of matter for heavy ions should vary directly as the velocity $v = \beta c$. Especially at velocities for which this is true, but to some degree at all energies, there will be a loss of energy directly to the nuclei of the absorbing medium, since such energy transfer varies as β^{-2}, from equation (3.1). Eventually, the nuclear stopping also passes through a maximum and decreases to zero as the heavy ion gradually becomes a neutral atom. The transfer of energy to neutral atoms leads to little additional ionization, and in contrast with transfer to electrons, is subject to very large statistical fluctuations, because the energy transfers are relatively larger and fewer in number. This seriously impairs the resolution of solid-state detectors (Sect. 4.5.2) for heavy ions.

Figure 3.16 shows the variation of stopping power with energy for ^{16}O ions in aluminium.

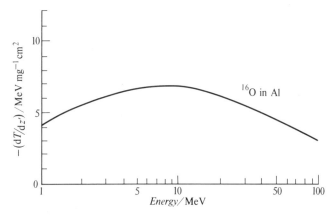

Fig. 3.16 Stopping power of aluminium for oxygen ions (Booth, W. and Grant, I. S., *Nuclear Physics*, **63**, 481, 1965).

3.2.4 *Passage of charged particles through crystals; channelling*

In Section 3.1.2 the theory of energy loss by collision was developed on the assumption of an isotropic medium in which the distribution of impact parameters b for collision with electrons could be written in the form $2\pi b \, \mathrm{d}b$. If a charged particle moves through a single crystal, however, the crystal structure defines certain directions or planes for which the electron density may be especially low, so that within a small range of angles ($\approx \frac{1}{10}°$ about these directions in typical cases) the impact parameter distribution may differ considerably from the isotropic form. In these directions, e.g. directions of high symmetry, there is increased penetration and finely collimated incident ions emerge from a thin crystal with an energy loss smaller than that which they experience in other directions. This phenomenon is known as *channelling*; for random angles of incidence and in amorphous solids the energy loss corresponds with the average electron density normally used in stopping-power calculations.

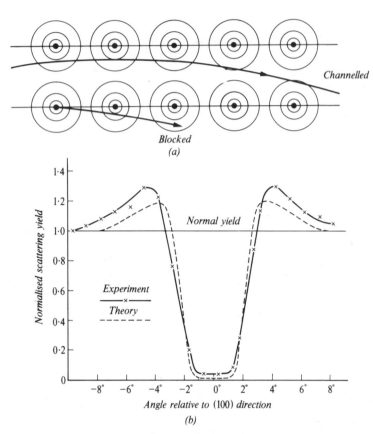

Fig. 3.17   Channelling and blocking: (*a*) Trajectories of ions between rows of atoms. A particle incident nearly parallel to the row is channelled; a particle emitted from a lattice site is scattered away from the channelling direction. The circles represent the atomic electron density. (*b*) Blocking dip in 135° Rutherford scattering of 480-keV protons incident, with an angular spread of 0·1°, along the ⟨100⟩ direction of a tungsten crystal (Andersen, J. U. and Uggerhoj, E., *Can. Jnl. Phys.*, **46**, 517, 1968).

The occurrence of channelling in a simple two-dimensional case is illustrated in Fig. 3.17*a*. An ion travelling in approximately the direction of a crystal axis approaches an atom and finds itself in a net repulsive field deriving from the resultant of the ion–nucleus and ion–electron interactions. This deflects the incident particle slightly towards the region of low electron density between the atomic rows. At the next atom of the row a similar effect occurs and as a result of the correlated series of Coulomb scatterings the ion moves from side to side between the rows of atoms. It is therefore steered through the crystal mainly in a region of low electron density and its energy loss per unit path is correspondingly reduced. It also tends to avoid close nuclear collisions. Such effects will be enhanced for heavy ions, for which it has already been noted that nuclear stopping may be important in comparison with electronic loss. Channelling is also observed, however, for light particles of an energy of several MeV, for electrons and for mesons.

An effect closely allied to channelling is the *blocking* of the emission of charged particles from lattice sites. Because of the strong Coulomb deflections, particles will not be able to enter a direction of crystal symmetry over a small range of angles. Nuclear reactions and large-angle Coulomb scattering provide convenient sources of particles leaving lattice points and yields show a marked dip as a crystal target is rotated round a beam direction because close collisions are inhibited over a small angular range. Figure 3.17b shows results obtained for back-scattered protons.

3.2.5 Examples of the use of the stopping-power formula

In the series of experiments leading to our first knowledge of the mass of the pion m_π use was made of the following quantities related to energy loss, mainly in nuclear emulsion, and for non-relativistic particles:

(a) range R;
(b) grain density g;
(c) mean multiple-scattering angle.

In addition, particle momenta p were known from magnetic deflection experiments. It was assumed from the visual evidence of pion decay to a muon and then to an electron that the pionic charge was equal to that of the electron.

To obtain a mass value, two observed quantities must be used, since each quantity is velocity dependent. From the stopping-power formula (eqn (3.1) and Sect. 3.2.2), the range of a pion of velocity β is

$$R_\pi = m_\pi/Z_\pi^2 \cdot f_1(\beta) \qquad (3.24)$$

and the corresponding grain density, proportional to energy loss, is

$$g_\pi = Z_\pi^2 f_2(\beta) \qquad (3.25)$$

Similar equations can be written for protons, and if the pion is compared with a proton of the same β, as defined by grain density, then $R_\pi/R_p = m_\pi/m_p$.

Alternatively, the mean multiple-scattering angle per unit path length, which is proportional to $(p\beta)^{-1}$ by equation (3.21), may be found for a pion of known range. If a range–energy relation for protons in emulsion is known then that for pions may be deduced in

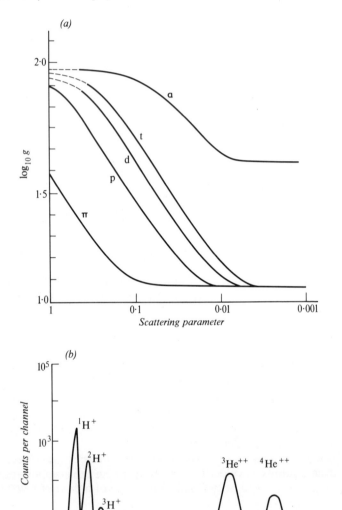

Fig. 3.18 Identification of charged particles: (a) Relation between number of grains g per 50 μm of a particle track in nuclear emulsion and the average angular displacement per 100 μm of track (Powell, C. F., *Reports on Progress in Physics*, **13**, 350, 1950). (b) Mass spectrum of particles from the reaction ^{27}Al + ^{3}He at 30 MeV using the $T - \Delta T$ identification system (England, J. B. A., *Nucl. Instrum. Meth.*, **106**, 45, 1973).

terms of m_π. The observations of $p\beta$ and R_π then permit elimination of the velocity β and deduction of the mass. Instead of range measurements, it is also possible to use grain density observations since this quantity is also velocity dependent. Figure 3.18a illustrates the discrimination between particles obtained in this way.

Similar use may be made of direct measurements of the momentum p and residual range R_π, but this, like the former method, is dependent on knowledge of the proton range–energy relationship. This may be avoided if magnetic deflection experiments are used to compare the momenta p_π and p_p of pions and protons of known range. If the initial velocities β_π and β_p are the same we have from equation (3.24)

$$R_\pi/R_p = m_\pi/m_p = p_\pi/p_p \qquad (3.26)$$

so that the mass ratio is known when the ratio of ranges is equal to the ratio of momenta. Such an experiment was conducted with a cyclotron producing both protons and pions with a range of energies, and momenta were obtained by noting the deflection of these particles in the magnetic field of the cyclotron itself.

A more modern technique widely used in particle detection systems for mass discrimination is to combine an energy-loss signal from a thin 'passing' counter with a total energy signal from a thick counter in which the particles stop. The product $T\,dT/dz'$ is proportional, for non-relativistic particles, to $(\frac{1}{2}M_1v^2)Z_1^2/v^2$, i.e. to $M_1Z_1^2$, independently of energy; Fig. 3.18b gives a mass spectrum obtained in this way in a nuclear reaction experiment.

3.3 Exotic atoms

The final state of an ion slowed down in matter is an atom formed by complete neutralization of the ionic charge by electron capture. The behaviour of elementary particles coming to rest in matter reflects their intrinsic nature. Thus positrons, which are stable, disappear by annihilation with electrons, but before they do so, may form a hydrogen-like atom of positronium (e^+e^-) some of whose properties can be studied during times of the order of the triplet ^{3}S-state lifetime of $\approx 10^{-7}$ s. Antiprotons are hadronic and only reach low velocities if they escape nuclear interactions in matter. In hydrogen they may form *protonium* ($p\bar{p}$) as a preliminary to annihilation, mainly into pions.

Muons and charged pions, kaons and hyperons have lifetimes between 10^{-6} and 10^{-10} s and some therefore survive a slowing-down process which takes 10^{-10} to 10^{-9} s. The positively charged particles are kept away from the nuclei of the stopping material by Coulomb repulsion and disappear by normal decay processes. The negative particles, however, may enter a hydrogen-like bound atomic state in the Coulomb field of a nucleus, forming an *exotic atom.* The particle may exist in this atomic state until decay takes place or it may promote a nuclear interaction. The structures discussed here are, of course, essentially different from hypernuclei (Sect. 2.4) in which the nucleus itself is unusual.

In exotic atom formation the slowed-down particle is normally captured into a state of high principal quantum number and reaches lower states in perhaps 10^{-13}–10^{-14} s by a series of transitions in which radiation (or Auger electrons) is emitted. The Bohr radius $a_0 = 4\pi\varepsilon_0^2\hbar^2/m_x e^2$ for an exotic atom is smaller by a factor m_x/m_e than that for electronic hydrogen and this factor is at least 207, its value for the muon. Similarly, the transition energies for the exotic atom are *greater* by the same factor and may be in the X-ray (keV) or nuclear (MeV) range of energies. The importance of exotic atoms lies in the fact that the wavefunction of the bound particle in all states, and especially in the lowest s-state if this is reached, overlaps the nucleus much more strongly than do the normal electronic wavefunctions. The finite size of the nucleus results in a diminution in transition energies compared with what would be expected for a point charge, and the diminution for the lower states is a major effect that is easily observed, in contrast with the same effect for electrons.

Muons interact only weakly with nuclei and normally reach the ground state of the muonic atoms that they form. The highest energy radiations observed are the muonic K X-rays due to the transition $2p \to 1s$; for the element titanium, for instance, these have an energy of 1 MeV. Extensive and accurate measurements of μK X-ray energies have contributed much information on the radius of the nuclear charge distribution. Precision measurements of transitions between the outer muonic levels, e.g. $3d \to 2p$ for which the nuclear size effect is smaller, give an accurate value for the muon mass.

Pionic atoms have been used in the same way to determine the mass of the pion, e.g. from the energy of the transition $4f \to 3d$, which has a value of $87 \cdot 622 \pm 0 \cdot 001$ keV in pionic titanium. The pion interacts strongly with nuclei and for atoms with Z greater than about 12, nuclear absorption from the 2p state takes place with greater probability than radiation, so that the πK X-rays are not observed. For a similar reason, pionic atoms are less useful than muonic atoms for nuclear size determinations.

In kaonic atoms, nuclear absorption is an even stronger effect and takes place from states with high quantum numbers. The cut-off of kaonic X-rays of a particular atomic sequence, e.g. $5g \to 4f$, as Z increases, shows the onset of absorption from a particular state, e.g. 5g, and provides a sensitive probe of the nuclear density distribution at large radii. Identification of the reaction products shows whether the kaon has interacted with a neutron or a proton.

3.4 Hyperfine effects in atoms

The interaction of nuclei with the extra-nuclear structure of electrons (and other particles in the case of exotic atoms) that surrounds them is an important source of information on nuclear angular momentum, nuclear moments and nuclear size. The effects of these nuclear properties are manifest in optical, microwave and radio-frequency spectroscopy. Because of the small scale of the energy splittings compared with the fine structure of optical lines, they are generally described as *hyperfine* effects. They are additional, of course, to the effect of the basic Coulomb attraction that binds the atom and defines the gross structure of the atomic energy spectrum, and to the fine structure of these levels arising from the electronic spin–orbit interaction.

3.4.1 Definitions

The following quantities, some of which have been introduced in Section 1.2.6, are required in a discussion of hyperfine effects:

(*a*) *The nuclear angular momentum vector* I, with absolute magnitude

$$|I| = \sqrt{I(I+1)}\, \hbar \tag{3.27}$$

where I, the nuclear spin quantum number, is integral or half-integral. In accordance with the uncertainty principle, the direction in space of the vector I cannot be determined, but if an axis Oz of quantization is defined, e.g. by an external magnetic field, the component of I along Oz will be observable and will have the value $m_I \hbar$ where m_I has one of the $2I+1$ values I, $(I-1), \ldots, -(I-1), -I$. In a vector diagram I is considered to lie on a cone described round the axis Oz with semi-angle β given by

$$\cos \beta = m_I \hbar / |I| = m_I / \sqrt{I(I+1)} \tag{3.28}$$

The maximum value of the component of I is $I\hbar$ and this is usually known as the nuclear spin.

(*b*) *The nuclear magnetic moment vector* μ_I, which is a vector parallel or anti-parallel to I. It may be expressed in absolute units or in nuclear magnetons, defined as $\mu_N = eh/4\pi m_p$ in analogy with the Bohr magneton, but smaller by a factor of $m_e/m_p = \frac{1}{1836}$. The observable nuclear magnetic moment μ_I is the expectation value of the z-component of the vector operator μ in the state $(I, m_I = I)$ in which m_I has its maximum value.

99

(c) *The nuclear gyromagnetic ratio* γ_I, equal to the nuclear moment in absolute units divided by the nuclear spin also expressed in these units, i.e.

$$\gamma_I = \boldsymbol{\mu}_I/\boldsymbol{I} = (\boldsymbol{\mu}_I/\boldsymbol{I})_z \quad \text{or simply} \quad \mu_I/I\hbar \qquad (3.29)$$

The ratio of the nuclear moment expressed in nuclear magnetons to the nuclear spin expressed in units of $\hbar$ is defined to be the dimensionless nuclear *g-factor* g_I and is related to γ_I by the equation

$$g_I = \gamma_I 2m_p/e \qquad (3.30)$$

If the nuclear moment and spin are oppositely directed, as in the case of the neutron, μ_I, γ_I and g_I are negative. The g-factor defined in (3.30) has a value of the order of unity. For some purposes it has become customary to use nuclear g-factors defined in terms of the Bohr magneton and these g-factors are then of the order of $\frac{1}{2000}$. From (3.29)

$$\mu_I = \gamma_I I\hbar = g_I I\mu_N \quad \text{or} \quad g_I I\mu_B \qquad (3.31)$$

(d) *The nuclear electric quadrupole moment*, Q_0, which measures the deviation of the nuclear charge distribution from a spherical shape. If the nucleus has symmetry about an axis Oz' fixed in itself (Fig. 3.19) and if the charge density is ρ then the *intrinsic* quadrupole moment Q_0 is defined by the equation

$$eQ_0 = \int \rho(3z'^2 - r'^2)\,\mathrm{d}^3 r' = \int \rho r'^2 (3\cos^2\theta' - 1)\,\mathrm{d}^3 r' \qquad (3.32)$$

where $\mathrm{d}^3 r'$ is the volume element and $\int \rho\,\mathrm{d}^3 r'$ for the whole nucleus is equal to Ze.

This is a purely classical definition and the quantity multiplying the element of charge is directly derivable from an expansion of the electrostatic potential at a distant point due to a charge distribution. This expansion also contains an electric dipole term, but if the nucleus (or atom) is in a state of well-defined parity the electric dipole moment vanishes. This is because in the quantum mechanical expression for the dipole moment the charge density ρ is replaced by the even function $e|\psi|^2$ and this is multiplied, not by a quantity containing squared terms such as $\cos^2\theta'$ as in (3.32), but by $\cos\theta'$. Integration over the nuclear volume then gives zero.

The observable electric quadrupole moment Q_I is the expectation value of a quadrupole operator (similar to 3.32) in the state I, $m_I = I$ in which m_I has its maximum value. It is less than the intrinsic moment Q_0 because the axis of nuclear spin cannot be completely aligned with the space-fixed axis Oz. In terms of the angle β given in equation (3.28) with $m_I = I$, it may be

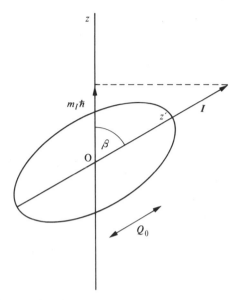

Fig. 3.19 Nuclear quadrupole moment. In this diagram both the intrinsic angular momentum I and the intrinsic quadrupole moment Q_0 are related to the body-fixed axis Oz'. Observable values I $(=m_I\hbar)$ and Q_I are defined with respect to the space-fixed axis Oz.

shown that

$$Q_I = \tfrac{1}{2}(3\cos^2\beta - 1)Q_0 = (2I - 1)/2(I + 1)Q_0 \qquad (3.33)$$

Arguments similar to that given for the electric dipole moment above show that Q_I exists only for states with $I \geqslant 1$, even when Q_0 is finite.

The sign of Q_I is that of Q_0, i.e. negative for an oblate and positive for a prolate (cigar-shaped) nucleus. It is measured in m^2 or *barns* $(10^{-28}\,m^2)$. Usually, tables give the values of Q_I rather than Q_0.

3.4.2 *The magnetic interaction*

In the absence of external fields, magnetic interaction arises from the fact that the nuclear magnetic moment finds itself in a field of magnetic induction B_i arising from the electronic motion in the rest of the atom (or crystal), with associated angular momentum J. The intrinsic magnetic moment of the electron contributes to B_i and, in fact, wholly determines it in the case of a hydrogen-like atom with a single s-electron. We shall consider only atoms with a single valence electron.

The total mechanical angular momentum of the atom is F, where

$$F = I + J \qquad (3.34)$$

This vector sum can be performed in either $2I+1$ or $2J+1$ different ways according to the relative magnitudes of I and J, and each arrangement corresponds to a different relative orientation of the vectors I, J. In such a vector model, as in the case of the coupling of L and S vectors in the atom, the individual vectors I, J are taken to precess round the direction of their resultant F and this may be understood as a result of the couple $\boldsymbol{\mu}_I \times \boldsymbol{B}_i$ arising between the magnetic moment $\boldsymbol{\mu}_I$, parallel to I and the internal field $\boldsymbol{B}_i$, assumed parallel to J.

In a state of total angular momentum F the interaction energy, which is a component of the total energy of the atom, is

$$W_F = -\boldsymbol{\mu}_I \cdot \boldsymbol{B}_i = -a\boldsymbol{I} \cdot \boldsymbol{J} \qquad (3.35)$$

where a is a magnetic interaction constant. The value of the scalar product $I \cdot J$ can be evaluated for an eigenstate of the system by noting that

$$F^2 = I^2 + J^2 + 2I \cdot J$$

and inserting the eigenvalues of the squared vector operators

$$I \cdot J = \tfrac{1}{2}[F(F+1) - I(I+1) - J(J+1)]\hbar^2$$
$$= \tfrac{1}{2}C, \quad \text{say}, \qquad (3.36)$$

so that

$$W_F = -aC/2 \qquad (3.37)$$

and the energy levels of the atom are split into a group of states whose number may be determined by the nuclear spin (when $|I| < |J|$) and whose separation is dependent on the nuclear moment (through a). This group of states is known as the *hyperfine structure* and is shown for the $^2S_{1/2}$ ground state of hydrogen and for the $^2S_{1/2}$ ground state and $^2P_{1/2}$ excited state of sodium in Fig. 3.20a. The order of the levels follows from the fact that because of the negative electronic charge $\boldsymbol{B}_i$ is generally of opposite sign to J; the nuclear moment is positive in both cases, so that a in (3.35) and (3.37) is negative.

The *hyperfine splitting* ΔW between the states F and $F-1$ is equal to aF from (3.37). The interaction constant a can be calculated for a simple atomic structure; for the ground state of hydrogen it corresponds to a frequency difference

$$\Delta \nu = h^{-1} \Delta W = 1420 \text{ MHz}$$

and this transition gives rise to the celebrated 21-cm line of astrophysics, which is an indicator of the presence of atomic hydrogen.

For the ground state of sodium the corresponding frequency is about 1800 MHz; the hyperfine structure of the ground and first excited state is elegantly shown (Fig. 3.20*b*) by observing the intensity of tunable dye-laser light scattered from a sodium atomic beam as the laser frequency is varied near the frequency of one of the D-lines.

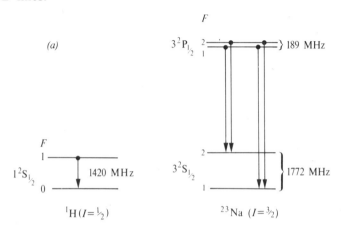

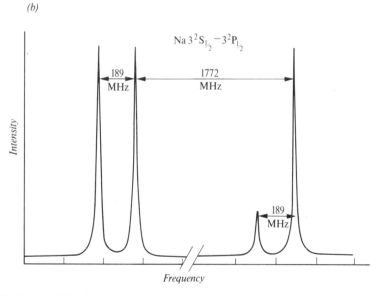

Fig. 3.20   (*a*) Hyperfine structure of ground state of hydrogen and of $3^2S_{1/2}$, $3^2P_{1/2}$ states of sodium. (*b*) Study of sodium hyperfine structure by scattering of laser light from a sodium atomic beam.

Magnetic hyperfine effects are observable in the spectra of muonic atoms and also, on a magnified scale, in positronium for which the 'nucleus' has a moment of μ_B.

3.4.3 *The electrostatic interaction*

Hyperfine effects arising from the interaction of the nuclear electric quadrupole moment with an electrostatic field *gradient* at the nucleus due to the electronic structure were first observed as deviations of hyperfine structure separations from the prediction $\Delta\nu = aF/h$ based on an exclusively magnetic interaction.

The theory of the electrostatic interaction is more complicated than that for the magnetic interaction (Sect. 3.4.2) and will not be given here. Deduction of the moment Q_I from the observed quadrupole displacements in atomic spectra requires knowledge of the internal electric field gradient, which must be predicted from an atomic model.

Electrostatic hyperfine effects in ordinary atoms tend to be smaller than the magnetic hyperfine structure, but in muonic atoms the muonic magnetic moment is scaled down by m_e/m_μ with respect to the Bohr magneton and the quadrupole effect is more important.

3.5 Summary

This chapter surveys the interactions between particles and radiation and their environment, as far as phenomena of relevance to nuclear physics are concerned. Charged particles lose energy by collisions with electrons; in addition, fast electrons may radiate bremsstrahlung. Electromagnetic radiation itself for energies of the order of 1 MeV and above undergoes the characteristic photoelectric, Compton and pair production interactions. In experimental measurements, account must be taken of the fluctuation phenomena leading to multiple scattering, straggling and charge exchange.

The electromagnetic interaction dominates energy-loss processes. Since it also binds ordinary electrons to nuclei in atoms, the corresponding bound states of some fundamental particles are examined. In all atoms hyperfine effects due to nuclear moments are also mediated by the electromagnetic field.

Examples 3

3.1* Verify the statement (eqn (3.2)) that under a certain condition the maximum transfer of energy from a charged particle of velocity v to an electron is $2m_e v^2/(1-\beta^2)$ (use the four-vector expression for energy-momentum).

3.2 Assuming that equation (3.1) were rigorously true at low energies, show that the stopping power reaches a maximum value as the velocity decreases. Find the energy of protons in aluminium for which this maximum would be expected, assuming that $I = 150$ eV. [187 keV]

3.3 From the simple theory outlined in Section 3.1.2 show that the cross-section for transferring an energy between Q and $Q + dQ$ to a free electron in the collision is given by

$$d\sigma = (Z_1^2 e^4 / 8\pi\varepsilon_0^2 m_e v^2) \, dQ/Q^2$$

3.4 An α-particle is found to have a range of 300 μm in a nuclear emulsion. What range would you expect for (a) a ^{3}He nucleus, (b) a ^{3}H nucleus, each of the same initial velocity as the α-particle? [225 μm, 900 μm]

3.5 Figure 3.1a suggests that a beam of α-particles suffers no loss by scattering at least until the particles are near the end of their range. Using an averaged scattering cross-section of 3×10^{-26} m^2, investigate this assumption for a beam of 5-MeV α-particles in nitrogen.

3.6 A thin ionization chamber is traversed successively by a 10-MeV proton, a 20-MeV deuteron, a 30-MeV ^{3}He particle and a 40-MeV α-particle. Calculate the relative size of the ionization pulses produced (omit the logarithmic term in eqn (3.1)). [1, 1, 4, 4]

3.7 At what energies have the following particles a collision loss in Al of 175 keV mg^{-1} cm^2: electron, muon, proton, deuteron, α-particle? (Use Fig. 3.4 and eqn (3.1) and omit the variation due to the logarithmic term.) [0·5 keV, 0·11, 1, 2, 16 MeV]

3.8 A kaon of kinetic energy 0·1 GeV slows down in an aluminium absorber. Using the collision-loss formula, and assuming an average value of 5 for the logarithmic term, find how long it takes to reach an energy of 1000 eV. Make any reasonable assumptions necessary. [≈10^{-10} s]

3.9* Calculate the momentum transfer required in the absorption of a pion by a nucleon in a complex nucleus. What conclusion may be drawn from the magnitude of this transfer, if the Fermi momentum in the nucleus is given by $k_F = 1·36$ fm^{-1}?

3.10 In an experiment with a source of X-radiation the mass attenuation coefficients μ/ρ for the elements Al, Ti, Cu are found to be 9·97, 37·7 and 9·97 m^2 kg^{-1} respectively (the atomic numbers are 13, 22 and 29). Explain qualitatively the result for Ti.

3.11 A beam of X-radiation is attenuated by a factor of 0·64 in passing through a block of graphite of thickness 0·01 m. Assuming that the attenuation is due only to classical Thomson scattering, with a total cross-section of $6·7 \times 10^{-29}$ m^2 per electron, estimate the atomic number of carbon.

3.12 Calculate for the Compton scattering of X-rays of wavelength 0·01 nm: (a) the wavelength of the scattered radiation at 45°; (b) the velocity of the corresponding recoil electron; (c) the momentum transfer to the electron. [0·0107 nm, 5·3 × 10^7 m s^{-1}, 93 keV/c]

3.13 Calculate the energy of a 100-keV photon scattered backwards ($\theta = 180°$) from an electron of energy 1000 keV moving towards the photon. [1015 keV]

3.14 What is the energy of Compton-scattered photons at 90° when $h\nu \gg m_e c^2$? [511 keV]

3.15* Show that the threshold quantum energy for the production of an electron-positron pair in the field of a free electron is $4m_e c^2$.

3.16 A triton (^{3}H) of energy 5000 MeV passes through a transparent medium of refractive index $n = 1·5$. Calculate the angle of emission of Cherenkov light. [44°]

3.17 Calculate the number of photons produced per metre of air ($n = 1·000\,293$) by the Cherenkov effect from the path of a relativistic electron in the wavelength range $\lambda = 350$ to 550 nm. [27]

3.18 Show that in the annihilation of positrons at rest by electrons of momentum p, the angle between the two annihilation quanta is $\pi - p_\perp/m_e c$ where $p_{\perp r}$ is the momentum of the electron at right angles to the line of flight of the quanta.

3.19* For the two-quantum annihilation of a fast positron of kinetic energy T by a free electron at rest, show that the energy of the forward going annihilation radiation is approximately $T + \frac{3}{2}m_e c^2$.

3.20* Calculate the angle between coupled I and J vectors for the case $I = \frac{3}{2}$, $J = \frac{1}{2}$, $F = 2$ or 1 (e.g. ^{23}Na).

4 Some apparatus of nuclear physics

In 1932 Cockcroft and Walton succeeded in producing artificial transmutation by accelerated ions. They observed the (p, α) process

$$^7\text{Li} + \text{p} \rightarrow {}^4\text{He} + {}^4\text{He} \quad (Q = 17\cdot35 \text{ MeV}) \tag{4.1}$$

using protons of energy 100–500 keV. Since their work, the technology of accelerators has developed extensively and much of our knowledge of nuclear structure and reactions derives from experiments using accelerated ions and electrons. Only in neutron physics does another type of radiation source, the nuclear reactor, provide comparable amounts of information.

Despite the fact that several fundamental particles were discovered in cosmic radiation (Ch. 2), the properties of particles and resonances have been established in the main by accelerator experiments. Beyond the limit of accelerator energies, cosmic rays have regained their original role as a source of new information.

The requirements of an accelerator for nuclear structure studies are, broadly speaking, an energy sufficient to overcome the Coulomb repulsion between the accelerated ion and its target nucleus and a resolution and an energy variability capable of responding to details of the nuclear-level spectrum. The Coulomb barrier height for the uranium nucleus is approximately 15 MeV for an incident proton and the precision of nuclear reaction data is now often better than ±1 keV. The electrostatic generator, the cyclotron and the linear accelerator have all been developed to meet the specification in whole or in part. In particle physics, centre-of-mass energy, coupled with reasonable intensity, is still a paramount obligation. The requirement is best met by the proton or electron synchrotron or the electron linear accelerator, although if intensity can be sacrificed a related storage-ring system provides a very large increase in available energy.

Detecting systems have similarly evolved to meet specific requirements and to match accelerator performance. In nuclear structure studies, energy-loss and full energy signals from solid-state detectors provide particle identification and energy measurement for light charged particles with a line width of <10 keV at 30 MeV, for instance. The fabrication of large semiconductor detectors has permitted line widths of about 1 keV to be reached in the determination of photon energies. Visual methods of detection, having almost vanished from low-energy experiments, have assumed a vital role in particle physics, above all in cases in which the maximum information associated with an interaction vertex is required. Most neutrino experiments could hardly be conceived without the bubble or spark chamber. Hybrid systems combining electronic triggers with large-aperture detectors are providing increased data rates. Position-sensitive detectors are used in both high- and low-energy physics to read events in the focal plane of spectrometers or to indicate particle trajectories.

Vital ancillaries to the accelerator-detector systems are the beam transport and data acquisition systems, which are themselves large and complex installations, designed always to increase the effectiveness of overall performance.

Finally, mention should be made of the energies lying between those of particle physics and of nuclear structure, the intermediate energy range from, say, 100 MeV to 2000 MeV. This is the domain of the high-energy probes of nuclear structure throwing light on short-range correlations between nucleons and has produced its own special group of accelerators, within which are found the so-called 'meson-factories'.

References 4.1–4.4 cover the material of this chapter, and much more, in depth.

4.1 The tandem electrostatic generator (Ref. 4.1)

Cockcroft and Walton employed a transformer-rectifier-condenser voltage multiplying unit to accelerate protons to a velocity v given by

$$\tfrac{1}{2}m_\mathrm{p}v^2 = eV \qquad (4.2)$$

where V is the output voltage of the generator on load. Generation of a voltage up to about 5 MV by cascade multiplying circuits operating at high frequencies and enclosed in pressure vessels is still an important technique when high-intensity beams of both protons and electrons are required. For general nuclear structure research, however, by far the most versatile and powerful accelerator is the electrostatic generator developed by Van de Graaff in 1931.

This generator is a direct illustration of the definition of the potential of a conductor as the work done in bringing unit charge from a standard reference point to the conductor. In the electrostatic generator the conductor forms the high-potential terminal of an evacuated accelerating tube (Fig. 4.1, lower half) in which ions of the required type move, emerging with a high velocity at the earth potential end of the tube. Charge is conveyed to the terminal by an endless insulating belt, or in recent systems by a series of conductors insulated from each other but forming a flexible chain. Transfer of charge is made to the belt or chain at the low-potential end by a corona discharge from spray points, as indicated in Fig. 4.1 (or by electrostatic induction in the case of a chain). At the terminal end the transfer process is reversed, and by allowing the terminal pulley to reach a higher potential than the terminal itself, negative charge may be conveyed to the downgoing belt. In one large modern installation it is planned to deliver a charging current of up to 550 μA to the terminal; this has to supply not only the accelerated beam current but also all losses due to corona discharges and leakage through the insulators forming the 'stack' structure necessary to support the accelerating tube and terminal. The stack structure itself is essentially a series of equipotential surfaces separated by rigid insulators and connected by resistors. Its electrical function is to provide a uniform axial electric field, within which the accelerator tube is contained, and which avoids regions of high field strength that may lead to breakdown.

Electrostatic generators are normally enclosed in a pressure vessel filled with an insulating gas, to permit the terminal voltage to be raised to a level corresponding to an overall axial voltage gradient of about $2 \, MV \, m^{-1}$, which can be withheld by a long accelerating tube. This is generally the weakest part of the electrical structure, and to attain the necessary gradient magnetic suppression of secondary electrons produced within the tube may be necessary.

The radial voltage gradient at the terminal for cylindrical geometry is given by the formula

$$E_r = V/(r \ln (R/r)) \qquad (4.3)$$

where V is the potential of the central terminal, r its radius and R the inner radius of the pressure vessel. For a given E_r (of perhaps $20 \, MV \, m^{-1}$, determined by breakdown in the insulating gas), V is a maximum when $R/r = e$. The most effective insulating gas is sulphur hexafluoride (SF_6) and pressures of up to 100 atmospheres are used in large machines.

In the early electrostatic accelerators ions were injected into the accelerating tube from a gaseous discharge tube housed in the terminal. However, in the *tandem* accelerator, shown in Fig. 4.1, the ion source is at earth potential and produces negative ions, e.g. H^-, He^-, $^{16}O^-$. These are accelerated to the terminal through an extension of the accelerating tube already mentioned. In the terminal, H^- ions, for instance, are moving with a velocity v given by (4.2) and passage through a thin *stripper*, e.g. a carbon foil of thickness about $50 \, \mu g \, cm^{-2}$ or a tube containing gas at low pressure, removes the

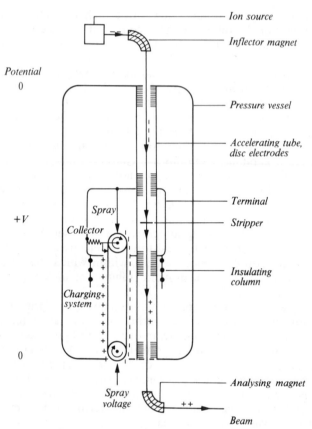

Fig. 4.1 The electrostatic generator (schematic). The lower half is effectively a single-ended machine and the addition of the upper half with an extra tube and a stripper converts it to a tandem accelerator.

extra electron and a further electron as well so that a positive ion H^+ is available for acceleration to earth potential from the terminal, yielding an energy of $2\,eV$. Negative ions of heavier elements may be stripped to a positive charge state qe in the terminal and the final velocity in this case corresponds to acceleration through a potential $(1+q)\,V$; additional strippers in the positive-ion tube can be used to increase q.

The energy of an electrostatic generator may be stabilized by controlling the charging current from an error signal derived from the accelerated beam itself. This passes through an analysing magnet (Fig. 4.1) and an exit slit with insulated jaws, from which a difference signal is derived if the beam deviates from the axis.

Energy definition to about $0\cdot5$ keV is possible in this way with a spread about the mean energy of not more than $1–1\cdot5$ keV even in

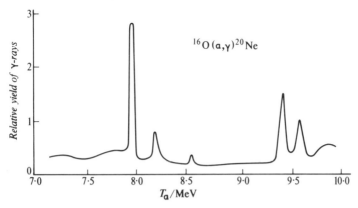

Fig. 4.2 Yield of capture gamma rays in the reaction $^{16}O(\alpha, \gamma)^{20}Ne$ as a function of incident α-particle energy (Pearson, J. D. and Spear, R. H., *Nuclear Physics*, **54**, 434, 1964).

the largest accelerators of, say, 20 MV terminal voltage supplying proton currents of say, 10 μA. This type of performance is ideal for nuclear reaction experiments. Figure 4.2 is a yield curve displaying nuclear resonances obtained in the $^{16}O(\alpha, \gamma)$ reaction.

The tandem generator has the obvious advantage of easy access to its ion source, which can be a complex, multipurpose installation at ground potential. A further advantage, shared with other direct-current machines, is its continuous output, which is convenient for coincidence counting. For some experiments, however, timing is important and in the determination of reaction neutron energies by time-of-flight methods a pulsed beam may be obtained by deflecting the continuous beam at a frequency of about 5 MHz backwards and forwards across a narrow slit near the ion source. Further compression, to pulse widths of about 2 ns, may be obtained by velocity modulation (bunching) techniques.

The energy of the proton beam from a tandem accelerator operating in the 10–25 MeV range has been determined with a precision of 1 part in 10^5 by a time-of-flight method. The beam is passed through two radiofrequency deflectors separated by an accurately measured distance of about 150 m and it is arranged that a detector signal is obtained only when the time of flight corresponds with a precisely determined number of radiofrequency periods plus a known phase displacement.

4.2 Linear accelerators (Refs. 4.2 and 4.3)

It was early recognized that the extension of direct-voltage methods to the production of very high particle energies would ultimately encounter insulation problems. The possibility of successive re-application of the same moderate electric field to a particle beam was, therefore, studied, and led to the development of the linear accelerator.

4.2.1 Drift-tube accelerators

The earliest heavy-particle accelerators (Wideroe 1928, Sloan and Lawrence 1931, Beams and Snoddy 1934) were of this type. The principle is illustrated in Fig. 4.3. In the Sloan–Lawrence accelerator, operated at approximately 30 MHz, a number of field-free drift tubes of length $L_1, L_2, \ldots, L_n$, separated by small accelerating gaps, were connected alternately to the output terminals of an oscillator of free-space wavelength λ. The length of the drift tubes is such that the field in a gap just reverses in the time that a particle takes to pass from one gap to the next. If the voltage across each gap at the time of passage of the particles is V then the particle energy at entry to the drift tube numbered n (Fig. 4.3) is neV (for

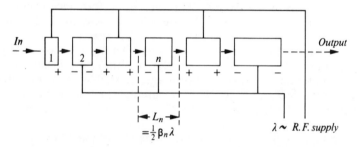

Fig. 4.3   Principle of the Sloan–Lawrence linear accelerator.

an assumed initial injection energy of eV) and the particle velocity is

$$v_n = (2neV/M)^{1/2} \qquad (4.4)$$

where M is the mass of the particles being accelerated. The frequency of the oscillator is c/λ and for a time of flight of half a cycle, the length of drift tube n must therefore be

$$L_n = \tfrac{1}{2}v_n\lambda/c = \tfrac{1}{2}\beta_n\lambda \qquad (4.5)$$

so that for non-relativistic energies, from (4.4) and (4.5)

$$L_n \propto n^{1/2}$$

It also follows that if the energy gain per gap is held constant the accelerator length is directly proportional to wavelength. The particles emerge in bunches corresponding closely with the times of appearance of the gap voltage V, at which resonance is possible.

The apparent requirement that the drift-tube structure should be designed for exact resonance with the accelerating beam was realized to be unnecessary following the enunciation of the principle of *phase stability* by McMillan and by Veksler in 1945. In its application to the linear accelerator (Fig. 4.4) this principle considers a particle that crosses a gap with a phase angle ϕ_s with respect to the accelerating voltage waveform (point A). If ϕ_s corresponds to the voltage V (eqn (4.4)) for which the drift-tube structure is designed, then the particle arrives at the next gap with the same phase angle. Late particles, however, ($\phi > \phi_s$, points B) receive a larger acceleration in the gap, traverse the drift tube more quickly and move towards point A in phase at the next gap. Similarly, early particles

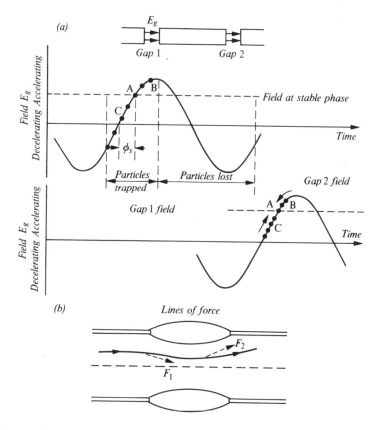

Fig. 4.4 (*a*) Phase stability in a drift-tube ion accelerator. The dots show the phase angles with respect to the gap field of a bunch of particles of uniform velocity arriving at gap 1. At gap 2 there is increased bunching about the stable phase ϕ_s. (*b*) Radial defocusing of particles passing through a cylindrical gap in a field increasing with time.

($\phi < \phi_s$, points C) will be less accelerated and will also move towards A in phase. Particles corresponding to point A thus have stable phase, and all particles with phase angles within a certain range of ϕ_s will be trapped and will oscillate about the point of stable phase. Latitude is therefore possible in the mechanical tolerances that must be applied to the drift-tube structure.

The desirable feature of axial (phase) stability leads to *radial instability* because the stable phase point is on the rising part of the voltage wave. Figure 4.4*b* illustrates this; because the field increases as the particle traverses the gap, the defocusing force predominates. Radial stability is now generally restored by means of quadrupole magnets within the drift tubes themselves.

The original mode of excitation indicated in Fig. 4.3 is now used only for low-energy accelerators. For energies above, say, 5 MeV for protons, the availability of high powers at microwave frequencies makes it highly desirable to use the resonant-cavity type of excitation introduced by Alvarez (Fig. 4.5). The cavity is tuned by radius, rather than length, and is excited in the lowest longitudinal mode, with a uniform axial electric field throughout its length. The drift tubes divide the cavity into n sections and the particles traverse the successive drift tubes while the field is in the decelerating phase,

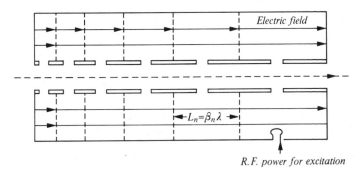

Fig. 4.5   Alvarez resonant accelerator.

receiving acceleration at the gaps. The section length L_n increases with particle velocity according to the equation

$$L_n = v_n \lambda / c = \beta_n \lambda \qquad (4.6)$$

since the particles traverse one complete section in each cycle. The high radiofrequency power required to excite the cavity is not usually supplied continuously and the accelerator is operated from a pulsed transmitter with a duty cycle (on–off ratio) of about 1 per cent. A d.c. injector supplies ions of energy 500–4000 keV to the main accelerator, and an improvement in intensity is sometimes obtained by incorporating a special cavity to 'bunch' the injected beam at approximately the selected stable phase angle of the main radiofrequency field.

Linear accelerators have been built both for protons and for heavier ions; the principle is similar in each case. For a given structure and wavelength, equation (4.6) requires that the velocity increments at each accelerating gap should be the same for each particle. A range of values of Ze/M, corresponding to different heavy ions, in different charge states, can therefore be accelerated by adjusting the radiofrequency voltage so that the gap field E is proportional to M/Z. The performance of the Alvarez proton accelerator at Berkeley is shown in Table 4.1.

The main advantages of the linear accelerator as a source of nuclear projectiles are the good collimation, the high homogeneity, the relatively high intensity of the beam and the possibility of extension of the machine to extremely high energies. A disadvantage for experiments requiring coincidence counting is the sharply bunched nature of the output, which increases the ratio of random to real coincidences in the counter systems. It is also difficult, though not impossible, to vary the output energy and in this and the preceding respect the linear accelerator is much inferior to the electrostatic generator. The major technical limitation in the extension of linear accelerators towards higher energies and higher intensities is in the development (and maintenance) of the necessary high-power oscillators. It is possible, however, that the situation will be completely transformed by the development of *superconducting* (cryogenic) accelerators in which the power dissipation will be very small.

TABLE 4.1 Performance of linear accelerators

Machine	*Berkeley proton accelerator*[1]	*Stanford electron accelerator (SLA)*[2]
Energy	31·5 MeV	20 GeV
Sectionalization	47 drift tubes	960 × 3·05 m coupled sections
Length	12 m	3 km
Frequency	202·5 MHz	2850 MHz
Pulse length	400 μs	2·5 μs
Pulse repetition rate	15 Hz	1–360 Hz
Peak power input	2·3 MW	245 × 16 MW
Shunt impedance	280 MΩ	53 MΩ m^{-1}
Mean current	1 μA	30 μA
Energy spread	0·5%	1.3%

[1] Alvarez, L. W. *et al.*, *Rev. Sci. Instrum.*, **26**, 111, 1955.
[2] Neal, R. B., *Physics Today*, **20**, April 1967, p. 27.

4.2.2 Waveguide accelerators

A standing-wave pattern in a cavity, such as that developed in the ion accelerator (Sect. 4.2.1), may also be regarded as a superposition of two progressive waves moving in opposite directions. One of these waves travels with the particles and accelerates them. This suggests the feasibility of an equivalent form of accelerator in which particles are continuously accelerated by a progressive wave in a metal guide. If the particles are moving with a relativistic velocity, the wavelength of the radiofrequency field is constant; the energy conveyed to the particles increases their total mass rather than accelerating them. Waveguide accelerators are especially suitable for electrons since these particles have a velocity of $0 \cdot 98c$ for an energy of only 2 MeV, which may easily be provided at injection by an electrostatic accelerator.

In familiar types of waveguide the phase velocity of the travelling wave is always greater than the velocity of light but may be reduced by loading the guide with a series of diaphragms (Fig. 4.6). Electrons move in bunches near the peak field of the travelling wave and

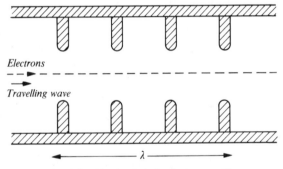

Fig. 4.6 Disc-loaded circular waveguide.

gain energy continuously from the wave. At velocities approaching the velocity of light radial defocusing forces vanish (since there is compensation between effects due to the electric and magnetic fields in the guide) and there is no definite axial stability, so that good quality of mechanical construction is necessary.

Travelling-wave accelerators usually operate from pulsed magnetrons or klystron amplifiers with a wavelength of about $0 \cdot 1$ m. In the remarkable 3-km Stanford linear accelerator (SLA), whose performance is shown in Table 4.1, power is fed into a disc-loaded waveguide of high mechanical precision from 245 klystron amplifiers driven in synchronism from a master oscillator. Electrons are injected into a short tapered section in which the phase velocity and longitudinal accelerating field both increase, and are bunched at a phase near the peak field of the travelling wave. Phase oscillations are damped by the effect of the rapidly increasing mass and the bunch moves through the main length of uniformly loaded guide with the velocity of light. No auxiliary focusing is used but despite this a beam is obtained, once perturbing magnetic fields have been neutralized, at the end of the guide. This is possible because of the relativistic shortening of the guide in the electron frame of reference; from the point of view of an observer at rest with respect to the electron the guide appears to be less than 1 m long and there is not much time for lateral divergence. The machine is supplied with microwave power from the klystrons in pulses of 2·5 μs length. About half this time is required to build up the fields in the guide and electrons are injected and accelerated during the remaining part of the pulse.

Electron linear accelerators may also be used to produce and to accelerate positrons. A heavy-metal radiator is placed in the electron beam at an intermediate point along the accelerator and the positrons emerging (as a result of direct production or photoproduction via bremsstrahlung) can be accelerated if the phase of the travelling wave is adjusted. The positrons and electrons from the same accelerator, can be used to provide bunches of particles in storage rings for the study of (e^+e^-) collisions. Two-quantum annihilation of the positrons in flight can be used to furnish a beam of homogeneous photons.

The electron linear accelerator, as shown by the figures in Table 4.1, is a machine of poor duty cycle compared with that of orbital accelerators (Sect. 4.3) and it is expensive in radiofrequency power although superconducting techniques will improve performance substantially. Upward extension of energy is not limited by the radiative losses inherent in orbital-electron accelerators, but physical length obviously becomes excessive for energies much above that of the SLA (Table 4.1).

4.3 Orbital accelerators (Ref. 4.3)

4.3.1 General

As noted in Sections 1.2.6 and 2.1.1, a particle of mass M and charge e moving with constant speed v in a plane perpendicular to the lines of force of a uniform magnetic field of flux density B, is acted on by a constant force at right angles to its velocity vector and describes a circle of radius r given by

$$evB = \gamma M v^2 / r \qquad (4.7)$$

where $\gamma = (1 - v^2/c^2)^{-1/2} = (1 - \beta^2)^{-1/2}$. The momentum of the particle is

$$p = \gamma M v = eBr \qquad (4.8)$$

117

and its angular velocity is

$$\omega = v/r = eB/\gamma M \text{ rad s}^{-1} \qquad (4.9)$$

The frequency of rotation in the circular orbit, known as the *cyclotron frequency*, is

$$f = \omega/2\pi = eB/2\pi\gamma M$$
$$= 15{\cdot}25 \text{ MHz per tesla for protons for } \gamma = 1 \ (f = f_0) \qquad (4.10)$$

If the field is uniform azimuthally, but non-uniform radially, the axial component at radius r may be written

$$B_z = B_0(r_0/r)^n \qquad (4.11)$$

where B_0 is the flux density at a reference radius r_0. The *field index* is

$$n = -(r/B_z) \, \partial B_z/\partial r \qquad (4.12)$$

and the frequencies of axial and radial oscillation about an equilibrium orbit of radius r_0 are

$$\begin{aligned} f_z &= f_0(n)^{1/2} \\ f_r &= f_0(1-n)^{1/2} \end{aligned} \qquad (4.13)$$

For the orbit to be stable it is necessary for n to be between 0 (uniform field) and 1.

The total energy of the particle, omitting any potential energy term, and setting $c = 1$, is given by

$$E = \gamma M = (p^2 + M^2)^{1/2}$$
$$= [(eBr)^2 + M^2]^{1/2} \qquad (4.14)$$

and the kinetic energy T is given by

$$T + M = E,$$

or

$$T(T + 2M) = p^2 = (eBr)^2$$

In the non-relativistic approximation $T \ll M$ and

$$T \rightarrow p^2/2M \qquad (4.15a)$$

while for extreme relativistic velocities $T \gg M$ and

$$T \rightarrow E \rightarrow eBr \qquad (4.15b)$$

Equation (4.14) gives the total energy of a particle moving at radius r in an orbital accelerator in which the magnetic flux density at the orbit is B. The way in which the final energy is attained in two different types of accelerator will now be described.

4.3.2 *The isochronous cyclotron (fixed field, fixed frequency)*

The cyclotron is a magnetic resonance accelerator developed by Lawrence and his collaborators in Berkeley in the 1930s.

From equation (4.10) it is clear that for energies $\ll Mc^2$, i.e. when γ is very close to unity, the cyclotron frequency f for a particle of charge e and mass M moving in a field B_0 is constant and equal to f_0 (isochronous motion). An oscillating electric field with a frequency f_0 may then be applied to increase the energy of the circulating particles. This is achieved in practice by the radiofrequency excitation of a D-shaped electrode (two dees in most early cyclotrons) in the vacuum system in which the particles move (Fig. 4.7a). The particles, under present assumptions, are synchronous with the radiofrequency electric field and receive an acceleration causing an increase in orbit radius twice per revolution, at entry to and exit from the D-electrode. At the maximum radius R the kinetic energy is

$$T = p^2/2M = e^2 B_0^2 R^2/2M \qquad (4.16)$$

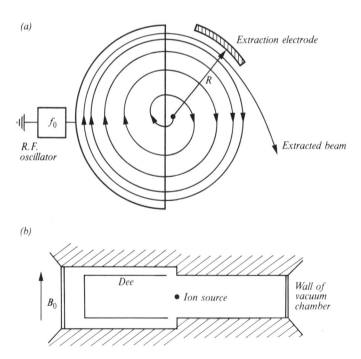

Fig. 4.7 The isochronous cyclotron. (*a*) Path of ions in a cyclotron from central ion source to extracted beam. (*b*) Vertical section showing dee and wall of vacuum chamber. The section indicates an azimuthal variation of the main field B_0 (see text).

and the particles may be extracted by the application of suitable deflecting fields. If *negative* ions are accelerated, extraction is simplified by the use of a stripping foil to reverse their charge and curvature.

There is no phase stability in a uniform-field cyclotron although bunching occurs early in the acceleration process and is maintained, at about a 10 per cent duty cycle, during the passage to maximum radius. Axial and radial stability require $0 < n < 1$ (eqn (4.13)) during the main acceleration, i.e. a radially *decreasing* magnetic field as shown in Fig. 4.8a. Unfortunately, this destroys isochronism (eqn (4.10)) and the effect is accentuated by the fact that γ is gradually increasing above 1 as the particle velocity increases. The conventional cyclotron is, therefore, limited in output energy; the performance of a typical machine is shown in Table 4.2.

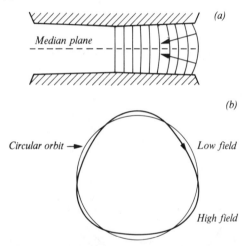

Fig. 4.8 The isochronous cyclotron. (*a*) Axial forces in radially decreasing cyclotron field. (*b*) Closed orbit in azimuthally varying field.

Although it is possible because of the principle of phase stability to design a cyclotron (*synchrocyclotron*) in which synchronism can be preserved by a programmed decrease of the frequency f of the electric field, it is preferable to use a magnetic field shape that permits the radial *increase* required for isochronism, but provides some extra focusing forces. The radial variation of mean field required is found from equation (4.10) to be

$$B = \gamma B_0 = B_0/(1 - \beta^2)^{1/2} \tag{4.17}$$

and approximately this gives

$$B = B_0(1 + \tfrac{1}{2}\beta^2) = B_0(1 + r^2\omega_0^2/2c^2) = B_0(1 + 2\pi^2 f_0^2 r^2/c^2) \tag{4.18}$$

The extra focusing is obtained, following a suggestion of L. H.

Thomas, by introducing *azimuthal variation* into the field, as already indicated in Fig. 4.7*b*. In such a field, also known as *sector-focusing*, there are alternate high and low field regions. A closed orbit in this field is non-circular, as shown in Fig. 4.8*b*, and as a particle in such an orbit crosses a sector boundary (either high→low or low→high) *radial* components of velocity arise. The field variation at the boundaries gives rise to *azimuthal* components and the new $v \times B$ force is axially focusing in both types of transition region. The sector focusing may be improved by using spiral instead of radial sector boundaries. Several such cyclotrons with design energies of the order of 100 MeV for protons are now operating.

TABLE 4.2 Performance of orbital accelerators

Machine	Particle energy /MeV	Orbit diameter /m	Orbital frequency /MHz	Pulse repetition rate/Hz	Output current or particles per pulse	Magnet weight /10^3 kg
Fixed frequency cyclotron (Birmingham)[1]	20 (d)	1·5	10·3	—	500 μA	254
AG proton synchrotron (CERN)[2]	27 000 (p)	200	—	0·3	$>2 \times 10^{12}$	3450

All accelerators operating with radiofrequency fields exhibit a fine time-structure of the beam on a scale determined by the radiofrequency.
Note: p = protons, d = deuterons
[1] *Nature* **169,** 476, 1952.
[2] Adams, J. B., *Nature* **185,** 568, 1960.

4.3.3 The synchrotron (variable field)

The construction of an isochronous cyclotron for the multi-GeV range of energies would be prohibitively expensive. It is therefore necessary to design a phase-stable accelerator in which the particles are held in a mean orbit of constant radius from initial to final energy. Only an annular magnet is then required, but the guiding magnetic field must increase to match the increasing momentum of the circulating particles; these gain energy from a radiofrequency cavity through which they pass.

The main components of a synchrotron are sketched in Fig. 4.9, together with a typical magnetic field cycle. Particles injected into the magnet ring from a linear accelerator pass through the radiofrequency accelerating cavity and are ejected by a pulsed extraction magnet if required as an external beam.

Alternatively, the circulating beam may be directed at the end of the magnet cycle to an internal target from which secondary beams of particles such as photons, pions, kaons or hyperons may be

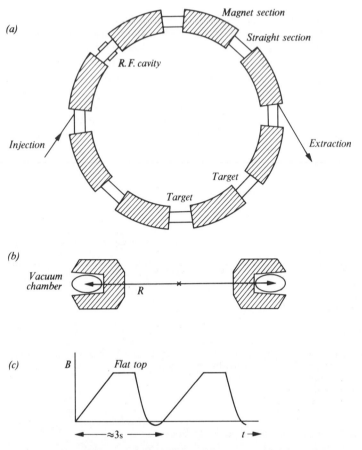

Fig. 4.9   The weak-focusing proton synchrotron. (*a*) Magnet ring. (*b*) Cross-section through magnetic sector. (*c*) Magnetic field cycle, showing flat top to give a long 'spill-time' of extracted beam.

obtained. Injection, acceleration, extraction and target equipment is conveniently located in the straight sections forming part of the magnet ring.

The momentum corresponding to a magnetic induction B and a radius R is given by equation (4.8) and inserting numerical values

$$p/(\text{Gev/c}) = (0 \cdot 3 B/\text{tesla})(R/\text{metre}) \qquad (4.19)$$

For relativistic particles this is also the total energy E in GeV, so that for 1 km radius, which applies to the 400-GeV proton synchrotron (SPS) at CERN, Geneva, a magnetic induction of 1 T corresponds to an energy of 300 GeV. The acceleration frequency is constant in the relativistic region.

In *weak focusing* (*constant gradient* or CG) synchrotrons, radial and axial stability of the beam is achieved by use of a field index between 0 and 1 uniformly in each magnetic sector of the ring except for edge regions. A major advance in accelerator design was made in 1951–53 when Christofilos, and also Courant, Livingston and Snyder realized that it is not necessary to render the axial and radial oscillations stable simultaneously. In the *strong focusing* (*alternating gradient* or AG) synchrotron alternate magnetic field sectors have reversed field gradients, i.e. $n \gg 1$ and $n \ll 1$ (negative) so that there is either axial or radial focusing but not both simultaneously in a given sector. There is, however, *net* focusing in both directions after passage through *two* sectors (cf. the achromatic lens combination in optics). The advantage of AG focusing is that because the free oscillation periods shown in equation (4.13) become much shorter the amplitudes are reduced and a considerable spread in momentum can be accommodated in a small radial space, with consequent saving of magnet and vacuum chamber costs. The engineering design of an AG synchrotron may use either *combined* function magnetic sectors in which bending and focusing both take place, or *separated* function component fields.

The particle motion is phase-stable in synchrotrons. In weak focusing machines a particle arriving early (B) at the accelerating gap (Fig. 4.10) receives a greater energy than it should and moves to a slightly greater radius for which an orbit takes *longer* to describe. At the next transit of the accelerating cavity it has, therefore, moved towards the phase-stable point A. Similarly, late particles (C) also

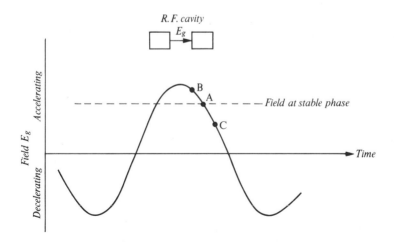

Fig. 4.10 Phase stability in a weak focusing (CG) synchrotron; particles arriving at times B and C move towards time A, relative to the voltage wave, as acceleration proceeds.

move towards A, which is on the *falling* part of the cavity-voltage waveform in contrast with the situation for linear accelerators (Fig. 4.4). While these phase changes are taking place the magnetic field of the accelerator is rising, and the radiofrequency is rising if the particles are non-relativistic, so that the phase-stable point A always corresponds with an increasing energy.

In strong focusing machines, because of the large value of n, the change of radius with momentum is rather small and the orbit for late particles at low energies may actually take a shorter time to describe. The phase-stable point is then on the *rising* part of the voltage wave, as with linear accelerators. Above a certain energy known as the transition energy, however, the orbit for late particles begins to take a longer time to describe than that for the synchronous beam and the phase-stable point then moves to the falling part of the voltage wave as for CG machines. The performance of the CERN PS (proton synchrotron) is shown in Table 4.2.

The highest-energy proton accelerators now existing (FNAL 500 GeV, CERN SPS 400 GeV) are AG machines. So is the highest-energy orbital electron accelerator (Cornell 10 GeV), but this is a lower energy than that of the Stanford linear accelerator (20 GeV) because of the severe and fundamentally unavoidable radiation loss in electron synchrotrons. Electrons moving at radius R with energy E suffer an energy loss per turn proportional to E^4 and

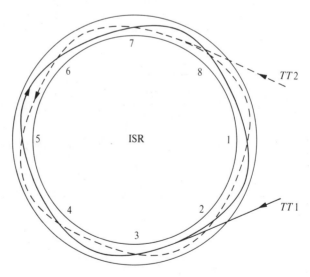

Fig. 4.11 Intersecting proton beams in the CERN storage rings (ISR). The oppositely circulating beams are sent from a proton synchrotron through the transfer tubes *TT*1 and *TT*2 into high-vacuum systems with a separate ring of magnets for each beam. Experiments are conducted at the intersections 1–8.

ultimately the radiofrequency circuit is unable to supply the losses, so that no further energy increment is possible. The *synchrotron radiation* produced in this way now forms an intense spectroscopic source from the X-ray region to the infra-red.

Because of the limitation due to radiative losses, electron synchrotrons are not designed for exceptionally high magnetic fields and the magnet cycle can be much more rapid, and the mean particle intensity greater, than in proton synchrotrons. Both machines produce a beam in which there is radiofrequency bunching within pulses of a repetition rate determined by the modulation cycle. By programmed control of the magnetic field and extraction fields, a long 'spill time' for the extracted beam, suitable for counter experiments, can be obtained (Fig. 4.9c).

The beams from high-energy proton and electron synchrotrons (and from the Stanford electron linear accelerator) have been used to fill *storage rings* in which they can circulate for long periods without loss if the pressure in the ring is low enough. If oppositely moving beams of energy E and mass M are made to intersect in storage rings (Fig. 4.11) the energy available in the centre-of-mass system is just $2E$ compared with $\sqrt{2M(E+M)}$ for a collision with a target particle at rest in the laboratory. Electrons in storage rings also radiate and losses must be made up from a radiofrequency cavity. Such electron rings are very attractive as sources of synchrotron radiation.

It is possible to inject positrons, generated electromagnetically in a target bombarded by electrons, into an electron ring in which they circulate in a direction opposite to the electrons. Study of the (e^+e^-) collision in this way has revealed much information on the new J/Ψ particles (Sect. 2.2.3).

4.4 Bending and focusing magnets

Magnets are used extensively in beam handling at all energies and also as spectrometers when the highest possible resolution is required. The design of such elements is related to that of accelerator magnetic sectors.

4.4.1 The magnetic spectrometer

Suppose that charged particles of uniform momentum p from a point source S move in a plane perpendicular to the lines of force of a uniform magnetic field in which the radius of curvature of their path is ρ so that, by equation (4.8)

$$\rho = p/eB \qquad (4.20)$$

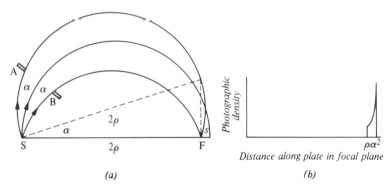

(a)   *(b)*

Fig. 4.12   First-order focusing after 180° deflection of charged particles in a uniform magnetic field.

if the particles are singly charged. Geometrically, as shown in Fig. 4.12, the paths of such particles converge to a focus F of finite width s determined by the angular width 2α of the cone of particles selected, for instance, by an aperture AB. From the figure

$$s = 2\rho(1 - \cos \alpha) \approx \rho\alpha^2 \qquad (4.21)$$

and the focus, in the plane, is said to be of *first order.*

The *resolution in momentum* is the relative change in momentum necessary to shift the line from a point source by its own width, i.e.

$$R = \Delta p/p = \Delta\rho/\rho \quad \text{for} \quad \Delta\rho = s/2$$

so that $R = \alpha^2/2$ from (4.21), while the *dispersion*, defined as the change in line position per unit relative momentum change, is

$$D = \mathrm{d}(2\rho)/\mathrm{d}p/p) = 2p\,\mathrm{d}\rho/\mathrm{d}p$$
$$= 2\rho \text{ from (4.20)}$$

In practice the line width observed includes a contribution from finite source width which generally exceeds the geometrical aberration $\rho\alpha^2$.

If both source and detector are contained within the magnetic field and if the field itself is known to high accuracy, the semicircular spectrometer can be operated as an instrument of high precision. For nuclear reaction experiments it is often desirable to sacrifice precision to the requirements of intensity, and then an inhomogeneous field (eqn (4.11)) which provides focusing both radially and axially may be used. This *double focusing* condition is achieved when the wavelengths of the radial and axial oscillation are equal, and in terms of frequencies equation (4.13) then gives $n = \frac{1}{2}$, $f_z = f_r = f_0/\sqrt{2}$. The paths converge to a focus after a deflection angle of $\sqrt{2}\pi = 254\cdot6°$.

Focusing is also obtainable, with a smaller angle of deflection, when the source and detector are outside the boundaries of the magnetic field. Such magnets are convenient for use with accelerators but require careful calibration when used in precision experiments because of the effects of fringing fields. These fields can, however, be turned to advantage in reducing aberrations if the pole of the spectrometer is split into two sections as suggested by Enge. In high-resolution studies of nuclear reaction products, the spectrometer design must also allow for the kinematic shift due to the variation of particle energy over the angular acceptance of the instrument.

All types of magnet with simple dipole fields and beam width control by slits are used extensively for beam deflection, and in high-energy physics for defining a 'momentum bite' $\Delta p/p$ when an accelerator produces a secondary beam of product particles with a range of momenta.

4.4.2 The magnetic quadrupole lens

A magnet with four poles of hyperbolic contour, arranged as shown in Fig. 4.13a, transmits axial particles without deflection. An off-axis charged particle, however, finds itself in a magnetic field whose potential is

$$V \propto r^2 \sin 2\theta \qquad (4.22)$$

where r and θ are shown in the figure. The deflecting force for paraxial particles is always proportional to the distance of the particle trajectory from the axis and is directed towards the axis for particle trajectories with a small angle θ to one of the planes separating the poles and away from the axis for trajectories with a small angle to the perpendicular plane.

A pair of such quadrupoles set with an angular displacement of $\pi/2$ may be adjusted to provide net focusing in both planes (though with different focal lengths in general) because the diverging effect produced in one plane by the first element leads to trajectories further from the axis in the second element. For these the deflecting force is stronger and of opposite sign, so that there is overall focusing.

Magnetic quadrupoles are widely used for concentrating charged particle beams. In high-resolution spectroscopy, a quadrupole element may be combined with dipole deflections to increase the solid angle of acceptance; such spectrometers are known as QD instruments. Since an inhomogeneous magnetic field deflects a magnet, quadrupoles can also be used for neutral atomic beams providing that there is a resultant electronic magnetic moment.

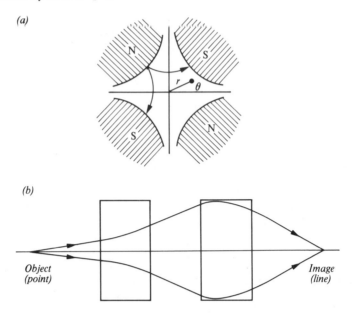

Fig. 4.13 The magnetic quadrupole. (*a*) Cross-section through poles, showing the (r, θ) coordinates of a paraxial trajectory. (*b*) Quadrupole doublet, showing net focusing in one plane by the diverging–converging fields. In the perpendicular plane the fields are converging–diverging and the focal length is different so that the lens is in general astigmatic, with two focal lines.

Electrostatic quadrupole lenses operate in a similar way to magnetic elements and are used, for instance, in the drift tubes of linear accelerators.

4.5 Detectors (Ref. 4.4)

Nuclear detectors of single particles are required to determine one or all of the quantities position, angle of trajectory, momentum, velocity, energy and charge. Some of the general techniques used, often involving combinations of detectors, have already been described in Section 3.2.5 as examples of the application of the stopping-power formula. In this section a brief account is given of some of the main types of individual instrument.

4.5.1 The scintillation counter

The development of the modern scintillation particle-counting system from the original zinc sulphide screen viewed by a microscope, as used by Rutherford, is due firstly to the application of the photomultiplier as a light detector (Curran and Baker) and secondly to the discovery of scintillating materials that are transparent to their own fluorescent radiations (e.g. anthracene by Kallman and sodium iodide by Hofstadter).

The processes resulting in the emission of light when a charged particle passes through matter depend on the nature of the scintillating material. Inorganic crystals such as NaI are specially activated with impurities (e.g. thallium) to provide luminescence centres in the band gap. Electrons raised from the valence band of the crystal by Coulomb interaction may themselves excite an electron in a luminescence centre. This excitation will take place, preferentially, in accordance with the Franck–Condon principle, so that the relevant interatomic spacing in the centre is left unchanged (Fig. 4.14,

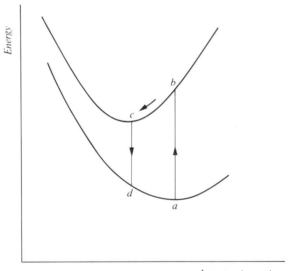

Fig. 4.14 Franck–Condon principle. The two curves are the energies as a function of an interatomic spacing of different electronic levels of a luminescence centre. The principle asserts that radiative transitions tend to leave the interatomic distance unchanged. The transition (*bc*) is non-radiative.

transition *ab*). It is followed by a radiationless transition *bc* and then by photon emission *cd*, also preserving interatomic spacing. The fluorescent radiation is of a longer wavelength than that corresponding to the absorption transition *ab* and is thus not usually reabsorbed by the scintillator. In the case of organic scintillators, molecular electrons are excited directly to one of several singlet levels and make radiationless transitions to the lowest excited singlet level. From this, fluorescent radiation to the vibrational substates of the ground state takes place, independently of the initial excitation process, and without the intervention of any activator, so that the light pulse is generally faster than with inorganic materials.

The scintillators chiefly used for particle or photon detection are:
(i) *Sodium iodide* (thallium activated).   Because of high density this is an efficient scintillator for gamma-ray detection. The decay time of the light pulse from the luminescence centres is about 0·25 μs, with a longer component of about 1·5 μs.
(ii) *Plastic and liquid scintillators.*   These are readily obtainable in very large volumes and can be adapted to many different geometrical arrangements, including counting of a source over a solid angle of 4π. The scintillator often contains a wavelength shifter to degrade the emitted spectrum to a wavelength that is transmitted efficiently through the crystal and that matches a photomultiplier detector. The light pulse associated with an instantaneous burst of ionization decays very rapidly, with a time constant of about 0·005 μs, so that these scintillators are useful for fast counting.

The scintillator is placed directly on the window of a photomultiplier tube (Fig. 4.15), or is coupled to it via a light guide. The window

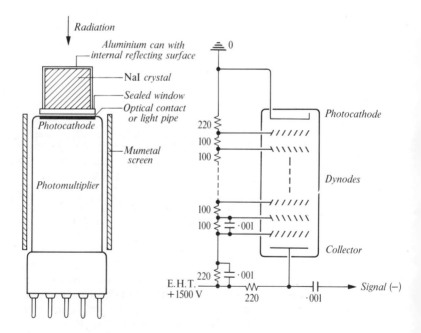

Fig. 4.15   Scintillation counter, showing circuit arrangement for supplying dynode potentials (resistances in kΩ, capacitances in μF).

is coated on the inside with a thin antimony–caesium layer from which photoelectrons are emitted with up to about 10 per cent efficiency for photons of wavelength 400–500 nm. The multiplier provides a current amplification determined by the interelectrode potential difference; external amplification may also be applied, with differentiation and integration time constants set to control the output pulse shape, essentially by defining the bandwidth of the amplifier. This should be suitable for transmitting pulses with a rise time characterized by the decay time of the luminescence centres in the scintillator. It is also necessary to minimize pulse pile-up and to discriminate as far as possible against circuit noise. In practice, the differentiation and integration time constants are usually made equal.

In the detection of γ-radiation by a sodium iodide scintillator an incident photon of energy, say 1 MeV, interacts with an atom or an electron of the scintillator and produces photoelectrons or Compton electrons together with lower-energy radiation which may itself interact similarly in the crystal. This primary process, converting photon energy into electron energy, takes place in $\approx 10^{-10}$ s, with an efficiency, as far as photon detection is concerned, determined by the absorption coefficient of the material, e.g. about 30 per cent for a cylindrical crystal of 25 mm × 40 mm diameter. The primary electrons excite radiation from the luminescence centres and the fluorescent radiations eject photoelectrons from the photocathode of the multiplier. When all efficiency factors are included, the 1-MeV incident photon may produce only about 2000 photoelectrons, with an associated statistical fluctuation of $\sqrt{2000}$, i.e. 2 per cent. The resulting spread of the final output pulse-height distribution, assuming that all radiations are completely absorbed in the scintillating crystal, is further increased by variations in the dynode multiplication factor and by variations in the efficiency of light collection over the detector volume. In practice, the pulse-height distribution for a homogeneous radiation of energy $E_\gamma = 1$ MeV gives a resolution of about 10 per cent. The contribution to the resolution due to statistical fluctuations is expected to vary as $E_\gamma^{-1/2}$ and the mean pulse height is closely proportional to the energy E_γ. The spectra observed with a pulse-height analyser for radiations from ^{57}Co, ^{137}Cs and ^{22}Na detected by a sodium iodide crystal of dimensions 25 × 40 mm are shown in Fig. 4.16. Their characteristic features are determined by the relative probabilities of the basic photon interaction processes in the crystal. For ^{57}Co, $E_\gamma = 122$ keV, there is a full energy peak due to the production of photoelectrons and the capture of all radiations resulting from the atomic vacancy produced. A much weaker, lower energy 'escape peak', indicating that sometimes the atomic K X-ray (28·6 keV for iodine) leaves the crystal, may also be resolved.

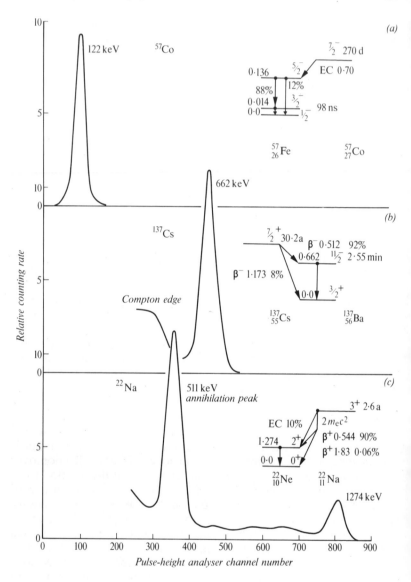

Fig. 4.16 Scintillation counter: pulse-height distribution for radiations from (*a*) ⁵⁷Co, (*b*) ¹³⁷Cs, and (*c*) ²²Na detected in a 25×40 mm NaI(Tl) crystal. In (*a*) the weak 136 keV transition is not resolved; in (*b*) internal conversion electrons accompany the 0·662 MeV transition but the Ba X-rays of energy 32 keV are not shown. In all spectra, peaks due to external back-scattering of forward-going photons and due to the escape of the 28·6-keV X-rays of iodine are omitted. The lifetimes shown are *halflives*, $t_{1/2}$.

For ^{137}Cs, with $E_\gamma = 662$ keV, a similar full energy peak appears, but there is a contribution to this from Compton scattering and below the main peak there is a 'Compton distribution' due to recoil electrons for which the Compton-scattered photon leaves the crystal.

With ^{22}Na, $E_\gamma = 1\cdot 27$ MeV, both photoelectric and Compton effects, and especially the latter, take place and in addition pair production ($\gamma \rightarrow e^+ e^-$) with a threshold of $1\cdot 02$ MeV. The positron produced in this last effect will ultimately annihilate either in the crystal or near it, with the production (generally) of two 511-keV annihilation quanta. If both of these are fully detected by the crystal, a pulse with height corresponding to E_γ is produced, since the annihilation energy adds to the kinetic energy of the $e^+ e^-$ pair. If one annihilation quantum escapes, there is a pulse at energy $E_\gamma - m_e c^2$ and if both escape, one at $E_\gamma - 2m_e c^2$. The resulting spectrum is therefore somewhat complex even for a homogeneous photon energy. The width of the full energy peaks is seen to be ≈ 10 per cent.

The main advantage of the scintillator detector is that a high efficiency coupled with moderate resolution for gamma-ray detection may be obtained with NaI(Tl) in a reasonably small size and at a reasonably low cost. Organic scintillators provide excellent timing signals and can be built into the very large and complex arrays necessary for particle location and identification in high-energy physics.

4.5.2 Semiconductor detectors

Solid-state detectors operate essentially through the promotion of electrons from the valence band to the conduction band of a solid as a result of the entry of the particle or photon to be detected into the solid. Since the band gap in some solids is only about 1 eV, production of an electron–hole pair as part of the general energy-loss by collision may require only 3–4 eV on average. The stopping of a particle of energy 1 MeV may thus produce 3×10^5 electron-hole pairs and the associated statistical fluctuation would be only $0\cdot 2$ per cent which offers a great advantage in basic resolution over the scintillation detector, and indeed over the simple gas-filled ionization chamber for which an energy expenditure of 30 eV per ion pair is required.

The difficulty in realizing the attractive features of solid-state detectors in practice has been to obtain materials in which residual conductivity is sufficiently low to permit conduction pulses due to single particles to be distinguished above background and in which the charged carriers are not rapidly 'trapped' by impurities. This has, however, been achieved for certain *semiconductors* in: (*i*) junction detectors, and (*ii*) gamma-ray detectors.

(i) *Junction detectors for charged particles.* In an n-type semiconductor, conduction is due to the motion of electrons in the conduction band, and in p-type material the process is effectively a motion of positive holes resulting from the rearrangement of electrons between atoms in the crystal. The two materials are prepared from an intrinsic semiconductor such as silicon by the controlled addition of electron-donating or electron-accepting elements (Fig. 4.17*a*). In the case of intrinsic silicon these elements could be phosphorus and indium respectively and the effect of their existence within surroundings of silicon is to produce hydrogen-like structures with their own set of energy levels, known as impurity levels. These are, again respectively, slightly below the conduction band and slightly above the valence band, so that thermal excitation of electrons to the former and from the latter is readily possible. The corresponding conduction process is then by electrons (n-type) or by holes (p-type); it is assumed that all impurities are ionized.

If a junction between p- and n-type regions is formed in a crystal, then conduction electrons predominate in the n-region but a few will diffuse into the p-region (Fig. 4.17*b*) where the conduction electron density is low. Similarly, holes from the p-region will

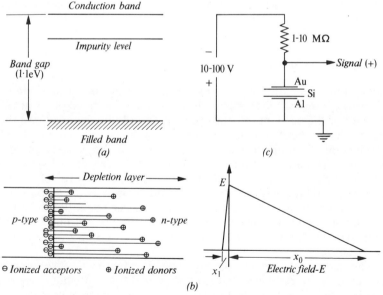

Fig. 4.17   Semiconductor counter: (*a*) Bands and impurity level in semiconductor. (*b*) Charge and electric field distributions in the depletion layer of a p–n junction. (*c*) Circuit arrangement (Dearnaley, G., *Contemp. Phys.*, **8**, 607, 1967).

diffuse into the n-region and the joint effect of these motions is to leave a positive space charge in the n- and a negative space charge in the p-region, near the boundary. The electric field resulting from this double layer generates a potential difference between the n- and the p-regions and this causes a drift current of thermally excited electrons or holes in the opposite direction to the diffusion current so that a dynamic equilibrium exists in which there is no net transfer of charge. The region of the crystal between the space charges has lost charge carriers in comparison with the rest of the solid and is able to sustain an electric field; it is known as the *depletion layer*.

The depth z_0 of the depletion layer depends on the density of impurity centres and on any applied potential difference V (reverse bias making the n-region more positive with respect to the p-region). If ρ is the resistivity of the n-type region, then

$$z_0 \approx (\rho V)^{1/2}$$

For silicon of the highest practicable resistivity a depletion depth of about 5 mm may be obtained.

If a charged particle enters the depletion layer the electron–hole pairs produced are swept away by the field existing across the layer and a pulse may be detected in an external circuit. Because of the low value of the energy required to form an electron–hole pair, and because of the existence of another favourable factor (Fano factor F) which recognizes that energy losses for a particle of finite energy are not wholly independent, a resolution as good as 0·25 per cent is obtainable at 5 MeV. Figure 4.18 shows the spectrum of α-particles from ^{212}Bi observed with a silicon detector of resistivity 27 Ωm.

Two types of p–n detector have been developed:

(a) *the diffused junction* detector in which a donor impurity, usually phosphorus, is introduced into a p-type (boron doped) silicon single crystal to form a depletion layer at the diffusion depth; and

(b) *the surface barrier* detector in which a p-type layer is formed on the surface of n-type silicon by oxidation. Contact is made to the detectors through a layer of gold. The counter base is often a layer of aluminium. Signals are derived from the surface barrier detector by the connections shown in Fig. 4.17c. Since the depth of the depletion layer, and consequently the inter-electrode capacitance, depends on the applied voltage, a charge-sensitive rather than voltage-sensitive pre-amplifier is used. Semiconductor counters are not themselves amplifying, and a typical output voltage might be 20 mV for a particle of energy 5 MeV.

The advantages of the silicon p–n junction as a detector of heavy particles are its excellent resolution and linearity, its small size and

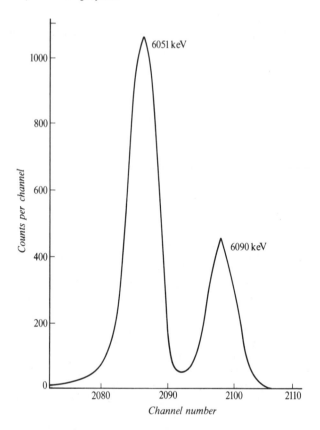

Fig. 4.18 Semiconductor counter: pulse-height distribution for 6-MeV α-particles from ^{212}Bi (ThC) detected in a silicon surface-barrier counter 134 μm thick. The groups are of width 16 keV at half maximum intensity (England, J. B. A.).

consequent fast response time permitting a high counting rate, and the fairly simple nature of the necessary electronic circuits. The solid-state counter is not actually a windowless counter because of the existence of *p*- or *n*-type insensitive layers of thickness 0·1–0·5 μm in surface barrier and diffused junction detectors respectively. This is not a serious drawback although it adds to the line width and its effect is outweighed by the overall simplicity of the counters, permitting the use of multidetector arrays. The handling of data from such complex systems increasingly demands the availability of an on-line computer for storage of information and data processing. A special facility of high value in the use of solid-state detectors is the easy construction of *particle identification* telescopes based on energy-loss × energy product signals (Sect. 3.2.5). Counters of a length of a few centimetres may also be used as *position sensitive* indicators in the focal plane of a spectrometer, since it can be arranged that output pulse height depends on the distance of entry of the incident particle from the collecting electrode. A disadvantage of silicon detectors for use with heavy ions and fission fragments has already been mentioned (Sect. 3.2.3); not only does nuclear scattering by the silicon nuclei worsen resolution, but it actually gives rise to an energy defect because of the occurrence of non-ionizing transitions in the solid lattice. For this reason, gas-filled detectors can give a better resolution in this application.

Increased sensitive volumes of semiconductor detectors can be obtained by compensating the acceptor centres in a p-type semiconductor by the controlled addition of donor impurities. The most suitable donor is lithium because this atom has a very low ionization potential in a semiconductor and also a high mobility. The lithium is 'drifted' into the bulk material from a surface layer by raising the temperature to about 420 K in the presence of an electric field. Compensated depths of about 15 mm and active volumes (in germanium) up to $100\ \mathrm{cm}^3$ can be made. Large-volume detectors may also be prepared from highly purified germanium (impurity concentration about 1 part in 10^{13}). The Ge(Li) detectors must be kept at a temperature below 150 K to prevent the lithium drifting away from the counting region.

Solid-state particle detectors may show channelling effects (Sect. 3.2.4) and these must be avoided by suitable orientation of the crystal.

(ii) *Gamma-ray detectors.* The major application of the lithium-drift technique is to the germanium counter (GeLi), which has transformed the subject of gamma-ray spectroscopy. Germanium has a band gap of only $0·67\ \mathrm{eV}$ and must be cooled to liquid nitrogen temperature when used as a counter in order to minimize thermal excitation of electrons to the conduction band. This is an inconvenience which results in a preference for silicon counters (although these are also improved by cooling) for charged-particle spectrometry. For gamma-ray detection, however, the higher atomic number of germanium ($Z_{\mathrm{Ge}} = 32$, $Z_{\mathrm{Si}} = 14$) leads to a marked improvement in efficiency as may be seen from the expressions for the basic processes of photoelectric, Compton, and pair-production absorption given in Chapter 3. The difference between Ge(Li) and Si(Li) counters is especially marked in the energy ranges in which the photoeffect and pair production predominate because of the particular Z-dependence. The Ge(Li) detector cannot yet be made as large as a sodium iodide scintillator and is less suitable than the latter when the highest detection efficiency is required, but its resolution is better by more than an order of magnitude, and line widths of less than 1 keV have been reported. These are narrower than those observed for charged particles because there is no counter 'window' and energy loss in atomic collisions is not important.

Figure 4.19 shows a spectrum of ^{24}Na γ-rays taken with a Ge(Li) detector compared with one obtained with a scintillation counter. Low-energy radiations, e.g. electrons of ≈ 10 keV or characteristic X-rays, may be recorded with high resolution by Si(Li) counters.

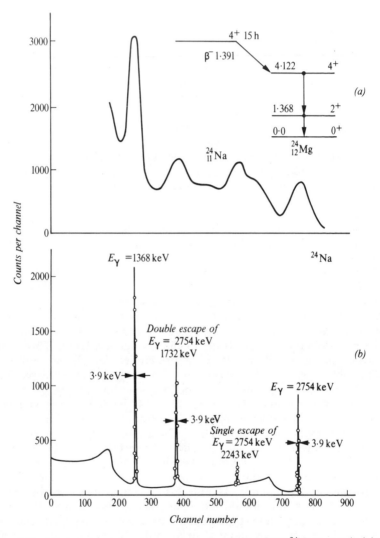

Fig. 4.19 Semiconductor counter: spectrum of radiations from ^{24}Na as seen in (*a*) a 25 mm × 40 mm NaI(Tl) crystal, and (*b*) in a 30-cm^3 Ge(Li) detector (adapted from Orphan, V. J. and Rasmussen, N. C., *Nucl. Instrum. Meth.*, **48,** 282, 1967).

4.5.3 Bubble chambers (Ref. 4.5)

The bubble chamber is a direct descendant of the original expansion chamber of C. T. R. Wilson and like that instrument in the early days of nuclear physics, now lends aesthetic quality as well as analytical power to the events that it depicts. Following its invention in 1952 by Glaser, it has become a major detector in particle physics, offering high spatial resolution, large solid angle of detection, momentum determination if operated in a magnetic field, and often particle identification by bubble or gap density. Its time resolution, however, is not better than about 1 ms. An example of a bubble chamber picture is given in Fig. 4.20.

In the operation of a bubble chamber, a volume of nearly-boiling liquid is rapidly expanded just before the entry of a particle beam pulse from an accelerator. Boiling takes place along particle tracks, stimulated by the ionization, the bubbles reach visible size in about 1 ms and are then flash-photographed. By this time the original centres on which the bubbles formed have disappeared (so that the chamber is not adaptable to counter control) and the chamber is recompressed as quickly as possible, a complete cycle taking only about 20 ms.

The only limit to the size of bubble chambers is their cost and a 2·5-m deep liquid hydrogen chamber 3·75 m in diameter is working at the Argonne Laboratory, USA. In addition to hydrogen, deuterium and helium, bubble chamber liquids include propane and freon, which are useful for gamma-ray and neutrino detection. A hybrid chamber, known as the *track-sensitive target chamber*, contains a volume of hydrogen within a neon–hydrogen mixture. Events originating within the hydrogen may be identified without ambiguity while the Ne–H mixture provides higher density for electromagnetic conversion processes.

Although a bubble chamber cannot be triggered, it is possible to trigger the light flash in response to a signal from associated counters, so that only certain events are photographed. This principle is used in the *vertex detector* in which the role of the chamber is to examine the interaction vertex corresponding to selected types of event.

The measurement of events recorded on bubble chamber film is now a highly specialized activity based on automatic machines under computer control. The accuracy of momentum determination from track curvature is about 0·5 per cent. For particles of momentum 5 GeV/c, multiple-scattering errors in liquid hydrogen exceed this value. From the observed momentum and angles of all visible tracks, e.g. in Fig. 4.20, energy and momentum conservation can be tested in accordance with hypotheses about the nature of the primary event. This may suggest the emission of neutral particles in

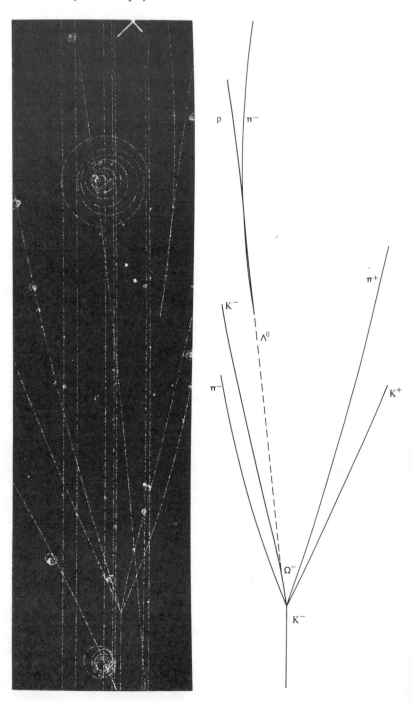

the interaction. The mass spectrum for particular groups of particles may demonstrate the formation of resonant states.

4.5.4 *Electronic instruments for particle physics*

The bubble chamber offers excellent spatial resolution and very full information on multiparticle events, but its time resolution is poor, and, moreover, not more than about 10 charged particles per accelerator pulse can usually be handled. Electronic detectors, though offering less complete spatial information (except with great complexity), can provide a much greater data-taking capacity, approaching perhaps 10^6 particles per second, together with some positional and ionization density information. The advantages of visual and electronic detectors are combined in the *spark chamber* (Ref. 4.5), in which a counter signal triggers the application of a high-voltage pulse in about 1 μs to a bank of thin plates between which discharges then take place near the particle tracks. The pulse can be applied before the primary centres disappear, so that counter control is possible, in contrast with the bubble chamber. Readout from triggered chambers, which usually operate in a magnetic field, may be visual (on film), visual (using a TV camera), acoustic (using transducers) or electrical (using wires to fill chamber planes). Spark chambers have dead times of about 10 ms and may be triggered perhaps 20 times during an accelerator beam pulse.

A popular instrument of high data-gathering power is the *multi-wire proportional counter* (MWPC), which is an array of anode wires 20 μm in diameter with 1–2 mm spacing placed between high-voltage conducting cathode planes in a gas at low pressure. The operation of an individual wire/plane is exactly as for a simple proportional counter (Ref. 4.4) and offers a time resolution of 20 ns with a dead time of less than 1 μs, compared with perhaps 10 ms for the much heavier discharge used in the spark chamber. Such chambers are operated continuously but the signals are small and amplification is required for each individual wire. In the *drift chamber*, additional spatial resolution is obtained by measuring the time that electrons from the particle track and subsequent multiplication take to travel to the nearest positively charged wire. The cathode planes are replaced by negative wires, and the total number of wires can be reduced with respect to the MWPC and higher positional accuracy obtained.

Fig. 4.20 Bubble chamber picture of the production and decay of the strange baryon Ω^- in the kaon–proton reaction

$$K^-p \rightarrow \Omega^- \pi^- \pi^+ K^+ K^0 \ (K^0 \text{ not seen})$$

$$\Omega^- \rightarrow K^- \Lambda^0; \ \Lambda^0 \rightarrow p\pi^-$$

(Birmingham University Bubble Chamber Group.)

Electronic methods of velocity (β) measurement based on the Cherenkov effect have already been mentioned (Sect. 3.1.3). For a particle of known momentum p, these immediately give mass information, since (with $c = 1$) $p = \beta\gamma M = \beta M/(1 - \beta^2)^{1/2}$. Simpler still is the time-of-flight method using fast counters that is widely used for neutron energy measurement in all ranges of velocity.

4.6 Summary

This chapter surveys some of the developments in nuclear technology that have taken place since 1930. Accelerators for charged particles have developed from the voltage-multiplier of Cockcroft and Walton, the drift-tube accelerator of Wideroe and the cyclotron of Lawrence to the modern alternating-gradient proton synchrotron and the electron linear accelerator, each able to feed storage-ring systems. Brief mention is made of the beam-handling equipment that conveys accelerated particles to a target, in which nuclear transmutations, including elementary particle reactions, may take place. The study of these events requires a detector system that is matched to the performance of the accelerator with which it is to be used. The main features of a simple magnetic spectrometer, and of scintillation and solid-state particle and photon counters are described. The significance of track chambers and proportional counters for particle physics is noted.

Examples 4

4.1 Calculate the velocity of a proton of energy 10 MeV as a fraction of the velocity of light. [0·15]

How long would a proton take to move from the ion source to the target in a uniform 10 MV accelerating tube 3 m long? [$1·4 \times 10^{-7}$ s]

4.2 A belt system of total width 0·3 m charges the electrode of an electrostatic generator at a speed of 20 m s^{-1}. If the breakdown strength of the gas surrounding the belts is 3 MV m^{-1} calculate:

(*a*) the maximum charging current, [320 μA]

(*b*) the maximum rate of rise of electrode potential, assuming a capacitance of 111 pF and no load current. [2·9 MV s^{-1}]

4.3 A magnetic field is being explored by the 'floating wire' method. If the current in the wire is 2 amperes and the tension is set at 0·5 kg weight, what is the energy of the proton whose trajectory is followed by the wire? [252 MeV]

4.4 In high-energy physics, it is customary to measure momenta in the unit MeV/c. Using the relativistic formula (4.14) find the momentum (in MeV/c) of:

(*a*) a 1000-MeV proton, (1696]

(*b*) a 1-MeV proton, [43·4]

(*c*) a 10-MeV photon, [10]

(*d*) a 100-MeV electron. [101]

4.5 Calculate the value of β ($= v/c$) for protons of the following energies: 100 MeV, 500 MeV, 1000 MeV. [0·43, 0·76, 0·87]

4.6 The peak potential difference between the dees of a cyclotron is 25 000 V and the magnetic field is 1·6 T. If the maximum radius is 0·3 m, find the energy acquired by a proton in electron volts and the number of revolutions in its path to the extreme radius. [11·0 MeV, 220]

4.7 Deuterons of energy 15 MeV are extracted from a cyclotron at a radius of 0·51 m by applying an electric field of 6 MV m^{-1} over an orbit arc of 90°. Calculate the equivalent reduction of magnetic field and the resulting increase in orbit radius Δr. [0·16 T, 0·053 m]

4.8 Calculate the cyclotron frequency for non-relativistic deuterons in a field of 1 T. [7·6 MHz]

4.9 A pulse of 10^{10} particles of single charge is injected into a cyclic accelerator and is kept circulating in a stable orbit by the application of a radiofrequency field. What is the mean current when the radiofrequency is 7 MHz? [11·2 mA]

4.10 Calculate the energy between the photopeak and the high-energy edge of the Compton electron distribution in the pulse-height spectrum from a scintillator detecting γ-radiation of energy $m_e c^2$. What would the energy of the back-scattered peak be? [170 keV]

4.11 Verify that the ratio of peak heights in Fig. 4.16c is reasonable, using the mass attenuation coefficients given in Reference 2.1, Fig. 1.6, p. 717.

4.12 In the time-of-flight method for determining particle energies (Sect. 4.1) particles of rest energy E_0 and kinetic energy T are observed over a flight path L. Show that the time of flight is $t = L/c \cdot (1 - E_0^2/(E_0 + T)^2)^{-1/2}$.

4.13* A heavy ion of mass number A_1 bombards a target atom of mass number A_2. If each nucleus behaves as a sphere of charge of radius $r_0 A^{1/3} (r_0 = 1·2 \text{ fm})$ show that the Coulomb potential barrier is

$$V_C = \frac{(A_1 + A_2) Z_1 Z_2 e^2}{4\pi\varepsilon_0 A_2 r_0 (A_1^{1/3} + A_2^{1/3})}$$

where Z_1, Z_2 are the atomic numbers. Evaluate this for $A_1 = A_2 = 125$, $Z_1 = Z_2 = 52$. [648 MeV]

4.14* In the Stanford linear accelerator, electrons reach an energy of 20 GeV in a distance of 3 km. Assuming that increments of energy are uniform and that the injection energy is 10 MeV, find the effective length of the accelerator to an electron.

4.15* Taking the magnetic potential in a quadrupole field to be $V = Gxy$, deduce the x, y components of the force on a charged particle moving with uniform velocity v along the z-axis and show that they are proportional (a) to the field gradient, and (b) to the component displacement.

5 The two-body system

The simplest complex nucleus is the deuteron (^{2}H), the bound state of the neutron–proton system. Because of its structural simplicity, it provides little guidance to the understanding of the average nucleus of, say, $A = 100$. The two-body system, however, does give information on the nucleon–nucleon force, which is an example of the strong (hadronic) interaction, and this force must provide the major stabilizing influence in nuclear matter. The properties of infinite nuclear matter, from which optimistically one may hope to understand some of those of finite nuclei, will be outlined in Chapter 6; the present chapter is concerned only with information available from a study of the neutron–proton and proton–proton interactions (Ref. 5.1).

5.1 General nature of the force between nucleons

From the fact that complex nuclei exist, the force must be *attractive*, and sufficiently strong to overcome the repulsion of the Z protons contained within the nuclear volume (according to the neutron–proton model). The *range* was indicated crudely by the early α-particle scattering experiments, which showed deviations of the force from pure Coulomb repulsion for light nuclei at distances of approach of the order of a nuclear radius, say, 4×10^{-15} m. A closer estimate was obtained by Wigner from a consideration of the stability of the simple nuclei ^{2}H, ^{3}H and ^{4}He.

Extension of Wigner's calculations led to the difficulty that an assembly of nucleons interacting attractively by a short-range force would collapse to a size of the order of the force range. The total potential energy of a nucleus containing A nucleons would then increase roughly as the number of interacting pairs, i.e. proportionally to A^2, whereas the evidence of mass measurements (Ch. 6) is that the total binding energy varies only as A.

It therefore becomes necessary to postulate a force that provides *saturation*, i.e. that limits the number of attractive interactions within the nucleus.

One way of ensuring saturation is to assume that the potential energy V of a pair of nucleons has the form shown in Fig. 5.1, analogous in radial variation to that describing van der Waals' forces between molecules. Such a potential indicates an attractive force $F = -\partial V/\partial r$ at extreme range and a repulsive force at very short distances. If the interparticle distance at the potential minimum $V = -V_0$ is r, then according to the uncertainty principle the particles must have a relative momentum of the order of $\hbar/r$. The corresponding kinetic energy is $\frac{1}{2}(\hbar^2/\mu r^2)$ where μ is the reduced mass of the system, assuming non-relativistic motion. If

$$V_0 > \tfrac{1}{2}(\hbar^2/\mu r^2) \tag{5.1a}$$

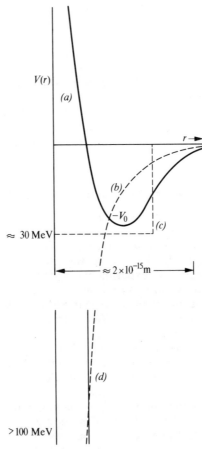

Fig. 5.1 The internucleon potential: (*a*) Schematic, showing repulsive core. (*b*) Form of Yukawa potential. (*c*) Square-well potential. (*d*) Well depth with sharp hard core.

a bound state may be formed, with a negative discrete energy. If, however,

$$V_0 < \tfrac{1}{2}(\hbar^2/\mu r^2) \tag{5.1b}$$

the total energy of the system lies in the continuum and the interaction is a scattering process.

The existence of a repulsive core creates difficulties in nuclear matter calculations. It is therefore usual, at least for problems concerned only with low relative momentum, to approximate the potential by simple spherically symmetrical forms with a suitable long-range behaviour, e.g. the *square-well potential*:

$$V = -V_0 \quad \text{for} \quad r < R \tag{5.2}$$

or the *Yukawa potential*:

$$V = -V_0 \exp{(-r/R)}/(r/R) \tag{5.3}$$

or the oscillator potential (Sect. 1.4.2). The potentials (5.2) and (5.3) are sketched in Fig. 5.1.

The depth of the potential well depends on the nature of the repulsive core. For nucleons interacting in free space through a Yukawa-type potential with a sharp hard-core radius of 0·5 fm the well depth obtained phenomenologically varies between 100 and 1000 MeV for the main attractive central part of the potential. In complex nuclei, however, observed nuclear densities (Ch. 6) show that the average spacing between nucleons is about 1·8 fm and at this separation the force is much weaker. In discussions of the nucleon–nucleon system at low energies and of the average potential in nuclei (Ch. 7) it will be convenient to use a simple square-well approximation, with a range of 1·5–2 fm and a depth of 30–60 MeV.

5.2 The scattering of neutrons by protons

5.2.1 Experimental

Because of the absence of Coulomb scattering the neutron–proton nuclear interaction can be studied with reactor neutron beams at sub-thermal energies. It can also be followed through the MeV region to the highest energies available for secondary particles from proton accelerators. Throughout this extensive range of energies, total cross-sections (Sect. 1.2.5) are observable by simple attenuation methods using hydrogen targets and the results obtained are shown in Fig. 5.2. The rise in cross-section below the energy 1 eV occurs because at these energies the two protons in the hydrogen molecule can no longer be treated as free, or independent. When these effects are corrected for it is found that the neutron-free proton total cross-section is approximately constant at about 20·4 b over a range of several hundred electron volts in energy.

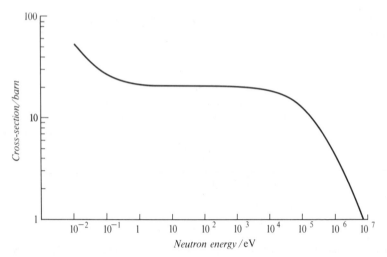

Fig. 5.2 Total cross-section for neutron–proton interaction, measured by attenuation methods, as a function of laboratory neutron energy.

In accordance with the arguments presented in Section 1.2.6, a principal effect of increasing the energy of neutrons interacting with protons is to introduce additional partial waves. Although this alters the integrated cross-section it is more directly observable in the angular distribution of scattering, at least at energies for which more than s-waves ($l = 0$) are involved. Figure 5.3 shows the centre-of-mass angular distribution of neutrons of a number of energies scattered by protons.

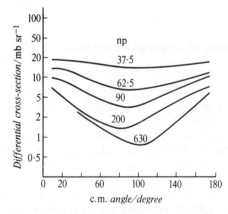

Fig. 5.3 Differential cross-section for neutron–proton scattering. The laboratory neutron energies in MeV are shown on the curves. At large angles in the c.m. system the proton takes most of the incident neutron energy and the process is then described as *charge-exchange scattering* (Lock, W. O. and Measday, D. F., *Intermediate Energy Nuclear Physics*, Methuen, 1970).

150

From the angular distribution and total cross-section data, phase shifts for the different partial waves as a function of energy may be extracted. These form both a convenient representation of the experimental results and a starting point for a theoretical analysis in terms of potentials. Figure 5.4 gives the phase shift curve for the neutron–proton system in the states $^1S_0$ (spins of neutron and proton opposed) and $^3S_1$ (spins parallel). The way in which these two states of motion are distinguished experimentally will now be described. A brief discussion of the high-energy data will be found in Section 5.6.2.

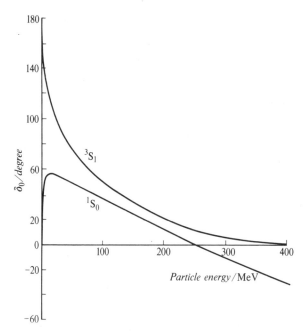

Fig. 5.4   The s-wave phase shifts δ_0 for the neutron–proton interaction (schematic). The curves embody observations on proton–proton scattering.

5.2.2 Description of low-energy (np) scattering

Low-energy scattering will be taken to refer to energies for which only s-waves, with zero orbital angular momentum in the c.m. system, need be considered and non-relativistic formulae are usually adequate. If the neutron and proton are assumed to have equal mass M, then the reduced mass of the interacting system (for the equivalent single-particle problem) is

$$\mu = M/2 \tag{5.4}$$

151

and the wavenumber k of the relative motion, for a laboratory neutron energy T_n is given by setting the c.m. energy $\frac{1}{2}T_n$ equal to $\hbar^2 k^2 / 2\mu$, the non-relativistic expression for momentum k. This gives

$$k^2 = (2\mu/\hbar^2)\tfrac{1}{2}T_n \tag{5.5}$$

For neutrons of energy 1 MeV the reduced de Broglie wavelength is $\lambdabar = 1/k = 9{\cdot}1 \times 10^{-15}$ m and as this is very much greater than the expected range of nuclear forces the assumption of s-wave interaction is justified for this energy, and will be assumed valid up to, say, $T_n = 10$ MeV.

The only inelastic reaction possible between neutrons of energy $1-10$ MeV and free protons is the capture reaction

$$n + p \rightarrow d + \gamma \tag{5.6}$$

and experimentally this has a cross-section of less than 1 per cent of the scattering cross-section, so that it may be disregarded for present purposes. The total cross-section is then equal to the integrated elastic scattering cross-section.

Following now the general treatment of elastic scattering given in Section 1.2.7, we represent the region of interaction by a spherically symmetrical potential well of depth V_0 and radius R. The incident neutron wave is an s-state solution of the free-space Schrödinger equation and this, from equations (1.69) and (1.71), has the form (valid for all kr)

$$u_i(r) = r\psi_i(r) = (A/k) \sin kr \tag{5.7}$$

Similarly, the final wave may be taken to be the s-wave part of equation (1.72), with $\eta = 1$ for elastic scattering, namely

$$u_f(r) = r\psi_f(r) = (A \exp(i\delta_0)/k) \sin(kr + \delta_0) \tag{5.8}$$

which shows the phase shift δ_0 due to the interaction. This is positive for an attractive potential (Sect. 1.2.7). The integrated elastic scattering cross-section is given by

$$\sigma_{el}^0 = (4\pi/k^2) \sin^2 \delta_0 \tag{5.9}$$

from equation (1.80), and a measurement of σ_{el}^0 gives δ_0.

To relate δ_0 to the parameters of the potential well we impose the condition of continuity of the final wavefunction and its derivative at the boundary $r = R$ at which the wavenumber changes since the Schrödinger equation for $r < R$ contains the potential V_0. In this region, the $l = 0$ solution is

$$u = r\psi(r) = (B/K) \sin Kr \tag{5.10}$$

where

$$K^2 = (2\mu/\hbar^2)(V_0 + \tfrac{1}{2}T_n) = 2\mu V_0/\hbar^2 + k^2 \tag{5.11}$$

The continuity condition, expressed in terms of a useful dimensionless parameter ρ, then gives

$$\rho = \left(\frac{r}{u}\frac{du}{dr}\right)_{r=R} = KR \cot KR = kR \cot (kR + \delta_0) \qquad (5.12)$$

If $k \ll K$, which means physically that the incident energy T_n is much less than the depth of the potential well, this equation shows that the phase shift is determined by the well parameters.

It is instructive to proceed with the evaluation of δ_0 in the *low-energy limit* $k \to 0$, assuming that the cross-section remains constant, so that equation (5.9) requires that in the limit $\sin \delta_0 / k$ shall remain finite. From equation (5.12)

$$(1/k) \tan (kR + \delta_0) = R/\rho \qquad (5.13)$$

so that as $k \to 0$

$$R + \tan \delta_0 / k = R/\rho \qquad (5.14)$$

or

$$\tan \delta_0 / k = -R(1 - 1/\rho) = -a(k) \qquad (5.15)$$

In the limit $k = 0$, $a(k) = a$ is defined to be the *scattering length*; it is also equal in absolute magnitude, for $\delta_0 \approx \tan \delta_0$, to the s-wave *scattering amplitude* given by equation (1.77) so that it then has the physical interpretation that it is the amplitude of the wave *scattered* by the potential well for unit incident amplitude. The scattering length, as defined by equation (5.15), is shown in Fig. 5.5a in the case that the internal wavefunction $\sin Kr$ just turns over within the distance R, i.e. $KR \geqslant \pi/2$. It will be seen in Section 5.2.3 that this is also the condition for the existence of a *bound state*, in fact the deuteron. If the neutron and proton collide with parallel spins, the scattering state will be a state of the same potential well as binds the deuteron, i.e. a ${}^3S_1$ (triplet) state orthogonal to the ${}^3S$ bound state. The figure makes it clear that as $k \to 0$ the phase shift δ_0 given by

$$\tan \delta_0 = -ka \qquad (5.16)$$

tends to π.

There is, of course, a further scattering state, the ${}^1S_0$ (singlet) state, in which the neutron and proton collide with spins opposed. This is not known as a stable state of the deuteron and will involve a different potential well. The scattering length in this case is shown in Fig. 5.5b, which indicates a wavefunction *rising* at the potential boundary so that only scattering states exist. The internal wavefunction does *not* turn over in the radial distance R and $KR \leqslant \pi/2$, owing to a shallower singlet potential well. The scattering length is negative and the phase shift, given by equation (5.16), tends to zero as $k \to 0$.

(a)

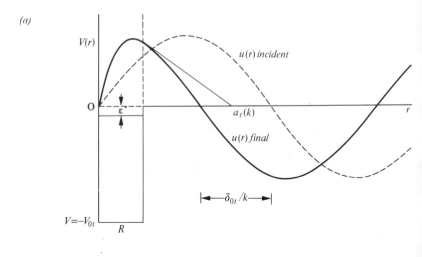

(b)

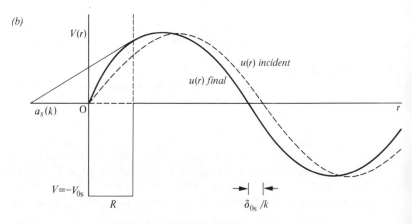

Fig. 5.5 Neutron–proton scattering at low energy, showing wave amplitudes, phase shifts and scattering lengths in relation to the square-well potential for (a) triplet (^{3}S) state, (b) singlet (^{1}S) state. In the triplet case the binding energy ε of the deuteron is indicated.

The special case of a bound state at zero energy is found for $KR = \pi/2$, or $(2n+1)\pi/2$, and gives $\rho = 0$. From Fig. 5.5, the scattering length then becomes infinite and $\delta_0 = (n+\frac{1}{2})\pi$. This is an example of *resonance scattering*.

In all three cases the scattering cross-section at zero energy in the present approximation is given by equations (5.9) and (5.15) as $\sigma_{el}^0 = 4\pi a^2$.

In a collision between unpolarized neutrons and protons, there are four possible combinations of the two intrinsic spins of $\frac{1}{2}\hbar$; of these, three belong to the triplet state $^3S_1$ and one to the singlet $^1S_0$. The relative weights of these two states are, therefore $\frac{3}{4}$ and $\frac{1}{4}$. Using these factors and denoting the triplet and singlet scattering lengths by a_t, a_s the cross-section for elastic scattering may be written, using (5.9) and (5.15),

$$\sigma_{el}^0 = (4\pi/k^2)[\tfrac{3}{4}\sin^2\delta_{0t} + \tfrac{1}{4}\sin^2\delta_{0s}]$$

$$= (4\pi/k^2)\left[\frac{3}{4(1+\cot^2\delta_{0t})} + \frac{1}{4(1+\cot^2\delta_{0s})}\right]$$

$$= \pi[3/(k^2 + 1/a_t^2) + 1/(k^2 + 1/a_s^2)] \tag{5.17}$$

For $k \approx 0$ the cross-section becomes

$$\sigma_{el}^0 = \pi(3a_t^2 + a_s^2) \tag{5.18}$$

and this near-zero-energy cross-section of 20·4 b is one of the basic observables for the two-body system.

For higher energies there is some energy dependence in the cross-section and it is useful to expand the inverse scattering length $1/a$ as a power series in k^2, which gives

$$k\cot\delta_0 = -1/a + \tfrac{1}{2}r_0 k^2 \cdots \tag{5.19}$$

where the coefficient r_0 is known as the *effective range* and corresponds physically (Appendix 3) with an average distance of interaction between the neutron and proton. It also depends on the well depth, but not on its shape, and formula (5.19) taken as far as the k^2 term applies in the *shape-independent approximation*. It is the validity of this approximation (Appendix 3) that justifies the use of the simple square well for discussion of the s-wave (np) interaction. The cross-section in this approximation becomes

$$\sigma_{el}^0 = \pi\{3/[k^2 + (1/a_t - \tfrac{1}{2}r_{0t}k^2)^2] + 1/[k^2 + (1/a_s - \tfrac{1}{2}r_{0s}k^2)^2]\} \tag{5.20}$$

and this formula has been applied to analyse scattering data up to about 10 MeV incident neutron energy.

The necessity for both singlet and triplet scattering amplitudes in low-energy (np) scattering was first pointed out by Wigner, when it had become clear that the triplet amplitude alone, as predicted from the binding energy of the deuteron (Sect. 5.2.3) led to a cross-section of only about 2 b compared with the observed 20·4 b. This is clear evidence for the *spin-dependence* of the interaction potential.

5.2.3 *The bound state (the deuteron)*

The deuteron has a binding energy of $2·2245 \pm 0·0002$ MeV and no stable excited states. It has an angular momentum of $1\hbar$ and both a magnetic dipole moment μ_d and an electric quadrupole moment. The magnetic moment is not equal to the algebraic sum of the magnetic moments of the proton μ_p and neutron μ_n, but it is sufficiently near to preclude the possibility of relative orbital motion of the two particles in a first approximation. The simplest structure to assume for the ground state of the deuteron is, therefore, a neutron and a proton in an S-state of orbital motion with parallel spins, i.e. a $^3S_1$ state, with even parity. The $^1S_0$ state of the deuteron, in which the proton and neutron have antiparallel spins, is unbound.

It has already been noted that $^3S_1$ and $^1S_0$ states are needed to describe the scattering problem, and the wavefunction for the deuteron (^3S) may easily be obtained by applying the formalism of Section 5.2.2 to the case of a total energy that is negative and equal numerically to the binding energy ε (Fig. 5.6). From the free-space Schrödinger equation for $r > R$,

$$\frac{d^2u}{dr^2} + \left(\frac{2\mu}{\hbar^2}\right)(-\varepsilon)u = 0 \tag{5.21}$$

the solution

$$u(r) = C \exp(-\alpha r) \tag{5.22}$$

where

$$\alpha^2 = 2\mu\varepsilon/\hbar^2 = (0\cdot232)^2 \text{ fm}^{-2} \tag{5.23}$$

is obtained after the additional solution $e^{\alpha r}$ has been excluded because of its unphysical behaviour for large r. The quantity $1/\alpha = 4\cdot31$ fm is a 'size' parameter for the deuteron, measuring the radial extent of the wavefunction (5.22); it is determined by the binding energy ε.

The external wavefunction must join smoothly with the wavefunction for $r < R$, which is a solution of the Schrödinger equation

$$d^2u/dr^2 + (2\mu/\hbar^2)(-\varepsilon + V_0)u = 0 \tag{5.24}$$

and has the form of (5.10), i.e.

$$u(r) = (B/K_d) \sin K_d r \tag{5.25}$$

with

$$K_d^2 = (2\mu/\hbar^2)(V_0 - \varepsilon) \tag{5.26}$$

The necessary boundary condition is

$$\rho/R = (1/u \cdot du/dr)_{r=R} = K_d \cot K_d R = -\alpha \tag{5.27}$$

from (5.25) and (5.22), and it follows that

$$\cot K_d R = -\alpha/K_d = -[\varepsilon/(V_0 - \varepsilon)]^{1/2} \tag{5.28}$$

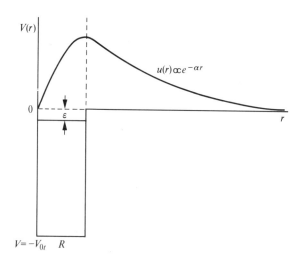

Fig. 5.6   The deuteron wavefunction in relation to the square-well potential.

This is a small quantity so that $K_d R \approx \pi/2$. It could also be $\approx 3\pi/2$, $5\pi/2$, etc., but these wavefunctions would have radial nodes and would not give the state of lowest energy.

Assuming as a first approximation that $K_d R = \pi/2$ it follows that

$$2\mu V_0 R^2/\hbar^2 = \pi^2/4$$

and

$$V_0 R^2 = \pi^2 \hbar^2/8\mu = 10^{-28}\ \text{MeV} \times \text{m}^2 \qquad (5.29)$$

which is a relation between the *range of the force and the depth of the triplet well.* Equation (5.29) implies that no bound state is possible in the well unless $V_0 R^2 \geqslant \pi^2 \hbar^2/8\mu$. The deuteron binding does not appear in (5.29); it is sufficient for the conclusion that a bound deuteron, of small binding energy, exists. Furthermore, so long as $|\varepsilon| \ll V_0$ a relation of the form (5.29) will be obtained whatever the shape of the potential, as already indicated generally in equation (5.1a). Moreover, it may now be seen that for $k = 0$ the unbound triplet wavefunction for $r < R$ will be very similar to that of the deuteron (Fig. 5.6) because K becomes equal to K_d if both k and ε are zero.

The triplet scattering length may be connected with the properties of the deuteron by using equations (5.15) and (5.27). These give

$$a_t = R + 1/\alpha \qquad (5.30)$$

so that if R is known, a_t determines α, i.e. the binding energy, and conversely. If R is assumed negligible compared with a_t (*zero-range*

157

approximation) we obtain from (5.17) a scattering cross-section, assuming only the triplet state,

$$\sigma_{el}^0 = 4\pi/(k^2 + 1/a_t^2) = 4\pi/(k^2 + \alpha^2)$$

$$= \frac{2\pi\hbar^2}{\mu} \Big/ (\tfrac{1}{2}T_n + \varepsilon) = 5 \cdot 2/(\tfrac{1}{2}T_n + \varepsilon) \times 10^{-28} \, m^2 \quad (5.31)$$

where T_n and ε are measured in MeV. This is much too small as already observed, and it is not brought into agreement with experiment by using, instead of (5.30), the effective-range formula for a_t, namely (see Ex. (5.4)):

$$1/a_t = \alpha - \tfrac{1}{2}r_{0t}\alpha^2 \quad (5.32)$$

Only the addition of singlet scattering correctly predicts σ_{el}^0.

The triplet scattering length given by the zero-range approximation $a_t = 1/\alpha = \hbar/(2\mu\varepsilon)^{1/2}$ is 4·3 fm. From the total cross-section at zero energy the singlet scattering length is then obtained using equation (5.18) as $|a_s| = 24$ fm. Working back at the present very crude level of approximation it follows that the singlet state of the deuteron should have an energy of $2 \cdot 2 \times (a_t/a_s)^2 \, MeV = 70 \, keV$ although it is not determined whether the state is virtual (unbound) or not. It is likely, however, as shown in Figs. 5.5a and 5.5b, that the singlet well is shallower than the triplet well. A set of figures, consistent with the evidence so far available, is as follows:

State	Range of potential	Depth of potential	VR^2
$^3S_1$	$R_t = 1 \cdot 93$ fm	$V_{0t} = 38 \cdot 5$ MeV	$1 \cdot 43$ MeV × b
$^1S_0$	$R_s = 2 \cdot 50$ fm	$V_{0s} = 14 \cdot 3$ MeV	$0 \cdot 89$ MeV × b

The ranges R are both less than $1/\alpha = 4 \cdot 3$ fm so that there is a high probability of the neutron and proton being separated by more than the range of the force; the 'size' of the deuteron is determined by its binding energy and not by the range. The square-well ranges R are comparable with the reduced Compton wavelength of a pion, $\hbar/m_\pi c = 1 \cdot 42$ fm and with the effective ranges r_0 (Table 5.1).

We now consider the additional information that enables the parameters a_t, a_s, r_{0t}, r_{0s} to be most accurately determined.

5.2.4 Coherent scattering and the (np) parameters

Further relations between the singlet and triplet scattering parameters and observable cross-sections may be obtained from a study of the interference effects found in the scattering of very slow neutrons by the nuclei of hydrogen molecules or by nuclei bound in solids. For coherent scattering by hydrogen the neutron wavelength must exceed the intermolecular distance 0·078 nm and in fact it must be greater than $0 \cdot 2$ nm (corresponding to $T = 90$ K) if inelastic effects resulting from the conversion of parahydrogen (nuclear spins opposed) to orthohydrogen (spins parallel) are to be avoided. Because of the established spin dependence of the (np) interaction the elastic scattering of neutrons from ortho- and parahydrogen under these long wavelength conditions will differ.

Let the spin vectors divided by $\hbar$ for the neutron and *one* of the hydrogen nuclei of the molecule be s_n, s_p. Then the scattering length for both singlet and triplet collisions may be written

$$a = \tfrac{1}{4}(3a_t + a_s) + (a_t - a_s)s_n \cdot s_p \qquad (5.33)$$

because for a triplet collision $s_n \cdot s_p = \tfrac{1}{4}$ and for a singlet collision $s_n \cdot s_p = -\tfrac{3}{4}$ as may be seen by inserting eigenvalues in the equation

$$(s_n + s_p)^2 = s_n^2 + s_p^2 + 2s_n \cdot s_p \qquad (5.34)$$

For scattering from the *two* protons of the molecule the scattering length is thus

$$a_H = \tfrac{1}{2}(3a_t + a_s) + (a_t - a_s)(s_n \cdot S_H) \qquad (5.35)$$

where $S_H = s_{p1} + s_{p2}$ and the assumption of the same phase for the scattering from the two protons is made. The cross-section for elastic scattering is then $\sigma = 4\pi a_H^2$ where

$$a_H^2 = \tfrac{1}{4}(3a_t + a_s)^2 + (3a_t + a_s)(a_t - a_s)s_n \cdot S_H + (a_t - a_s)^2(s_n \cdot S_H)^2 \qquad (5.36)$$

The middle term is zero on average if the neutron spin is unaligned and for a similar reason cross-terms in the final bracket vanish, leaving

$$(s_n \cdot S_H)^2 = s_{nx}^2 S_{Hx}^2 + s_{ny}^2 S_{Hy}^2 + s_{nz}^2 S_{Hz}^2$$
$$= \tfrac{1}{4}S_H^2 \text{ (since } s_{nx} = \tfrac{1}{2}) = \tfrac{1}{4}S_H(S_H + 1) \qquad (5.37)$$

and this has the value 0 for a singlet and $\tfrac{1}{2}$ for a triplet state. The parahydrogen cross-section is, therefore

$$\sigma_{\text{para}} = \pi(3a_t + a_s)^2 \qquad (5.38)$$

and the orthohydrogen cross-section

$$\sigma_{\text{ortho}} = \pi(3a_t + a_s)^2 + 2\pi(a_t - a_s)^2 \qquad (5.39)$$

Experimentally, $\sigma_{\text{para}} \approx 4$ b and $\sigma_{\text{ortho}} \approx 130$ b for neutrons of energy about 1 meV so that it is immediately apparent that a_s must be large and negative, as has already been indicated for an unbound state, i.e. the singlet state of the deuteron.

The parahydrogen scattering length

$$\tfrac{1}{2}(3a_t + a_s) = 2(\tfrac{3}{4}a_t + \tfrac{1}{4}a_s) \qquad (5.40)$$

159

is double the scattering length $\frac{3}{4}a_t + \frac{1}{4}a_s$ that would be defined for an encounter between a low-energy neutron and a free proton. It is, however, *equal* to the coherent scattering length a_H applicable to the scattering of neutrons by protons bound in crystals. This quantity may be determined accurately by crystal diffraction methods and especially by reflection of thermal neutrons from liquid hydrocarbon mirrors, from which

$$a_H = \tfrac{1}{2}(3a_t + a_s) = -3 \cdot 707 \pm 0 \cdot 008 \text{ fm} \qquad (5.41)$$

If this value is combined with the most accurate zero-energy total cross-section

$$\sigma(0) = \pi(3a_t^2 + a_s^2) = 20 \cdot 44 \pm 0 \cdot 23 \text{ b} \qquad (5.42)$$

then values for a_t and a_s may be found. The effective-range formula (5.32) and the deuterium binding energy then give r_{0t} and the higher-energy cross-section formula (5.20) then yields r_{0s}. The values of the (np) scattering parameters so found are included in Table 5.1.

5.3 The scattering of protons by protons

5.3.1 Experimental

Because of the Coulomb force between two protons, Rutherford scattering is the predominant interaction at low energies, and the transmission methods used for measuring neutron–proton cross-sections cannot be simply applied, at least at energies of a few MeV or less. On the other hand, the availability of intense beams of particles from accelerators and the convenience of detection methods for charged particles has permitted the accumulation of a very large amount of precise data at all energies up to the limit of accelerator performance. Both total and differential cross-sections are available over the greater part of the energy range since for high energies at least Coulomb effects are chiefly apparent at very small laboratory angles and can be corrected for under simplifying assumptions. The results for cross-section as a function of energy are shown in Fig. 5.7 and some c.m. angular distributions are given in Fig. 5.8.

As with the neutron–proton data, phase shifts may be derived from the differential cross-sections. The Pauli principle eliminates half the states found in the (np) system, e.g. the $^3S_1$ state is forbidden. The phase shifts for the allowed states, e.g. $^1S_0$, are determined to high accuracy for energies for which elastic scattering is the main process of interaction. Above the pion production threshold at about 300 MeV, inelastic reactions are possible and the phase shifts are complex, and less accurate. The $^1S_0$ phase shift has

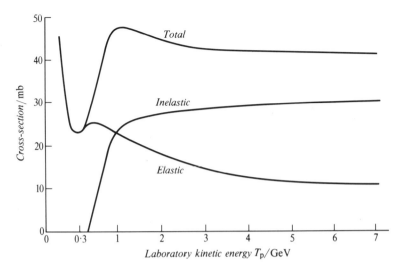

Fig. 5.7 Cross-sections for the proton–proton interaction as a function of energy (plotted on a scale that is linear in momentum). The inelastic cross-section above 0·3 GeV represents meson production.

in fact been presented in Fig. 5.4, since it applies equally to the neutron–proton system (Sect. 5.2) and provides the best singlet scattering information for that system at high energies.

5.3.2 Description of low-energy (pp) scattering

The Coulomb scattering of identical particles was calculated wave mechanically by Mott. Experimental results confirm the predicted interference at a given angle between waves representing elastically scattered protons and identical recoil particles. This scattering, resulting from a long-range force, involves both singlet and triplet states (i.e. those allowed by the Pauli principle) and all relative orbital momenta. For energies up to approximately 10 MeV, only $^1S_0$ states are allowed for interactions involving the short-range nuclear force although again account must be taken of the identity of the particles.

At energies and angles for which the scattering amplitudes for the Coulomb and nuclear force are comparable, interference effects may be seen, e.g. by the appearance of a minimum in the differential cross-section (Fig. 5.8). The cross-section for scattering involving Coulomb forces and the nuclear s-wave interaction may be written in terms of the corresponding phase shift δ_{0c} as

$$\sigma(\theta) = \sigma_{\text{Mott}}(\theta) - A(\theta, \delta_{0c}) \sin \delta_{0c} + \sin^2 \delta_{0c}/k^2 \qquad (5.43)$$

161

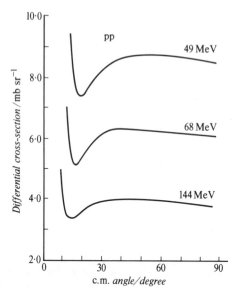

Fig. 5.8 Differential cross-sections for proton–proton scattering showing Coulomb–nuclear interference (Lock, W. O. and Measday, D. F., *Intermediate Energy Nuclear Physics*, Methuen, 1970).

where k is the wavenumber for the relative motion (see eqn (5.5) with $T_n = T_p$) and $A(\theta, \delta_{0c})$ is a calculable coefficient. The last term is the nuclear scattering and the second term is due to Mott–nuclear interference. This term permits the sign as well as the magnitude of δ_{0c} to be found and it is positive, corresponding with an attractive nuclear force.

The effective-range theory used to discuss the (np) interaction has been adapted to describe the variation of the $^1S_0$ phase shift for (pp) scattering with energy. In the shape-independent approximation the formula corresponding to (5.19) is

$$C^2 k \cot \delta_{0c} + (m_p e^2/4\pi\varepsilon_0\hbar^2)h(\eta) = -1/a_s + \tfrac{1}{2}r_{0s}k^2 \quad (5.44)$$

where

$$C^2 = 2\pi\eta/(\exp 2\pi\eta - 1) \quad (5.45)$$

is the Coulomb penetration factor and $h(\eta)$ is a calculable function of the *Coulomb parameter* $\eta = e^2/4\pi\varepsilon_0\hbar v$. The factor C^2 appears when a fit is required between internal and external (Coulomb) wavefunctions at the nuclear boundary. In the case of neutron–proton scattering the s-wave phase shift is derived directly from the observed scattering cross-section, but for the proton–proton system δ_{0c} is obtained by analysis of the angular distribution at a number of

162

energies; (5.44) then yields values for the singlet parameters a_s and r_{0s}. The (pp) scattering length a_s is found to be negative, corresponding with an unbound state for the two-proton system, in analogy with the singlet state of the (np) interaction (Fig. 5.5b). Recent results are shown in Table 5.1. For comparison with the (np) data, correction must be made for all electromagnetic effects, and a change in the scattering length results, whereas r_{0s} changes only by about 1 per cent.

5.4 The scattering of neutrons by neutrons

The large electromagnetic correction necessary in obtaining a nuclear scattering length from (pp) data justifies consideration of the charge-symmetric (nn) singlet system. Until experiments with colliding beams of neutrons become possible, this requires the study of processes in which two neutrons are left in a final state, e.g.

$$\pi^- + d \rightarrow \gamma + 2n \qquad (5.46a)$$

with stopping pions in which the detailed shape of the energy spectrum of both photons and neutrons depends on the $^1S_0$ (nn) scattering length. Other reactions that have been used are:

$$\left.\begin{array}{l} n + d \rightarrow p + 2n \\ ^3H + {}^3H \rightarrow {}^4He + 2n \end{array}\right\} \qquad (5.46b)$$

From these, the values for a_s and r_{0s} shown in Table 5.1 have been obtained.

5.5 Comparison of low-energy parameters

Table 5.1 assembles recent results for the reactions so far discussed in the energy region up to 10 MeV. Allowing for the large errors in the (nn) data, and for some uncertainty in corrections to the (pp) scattering length, there is support for the hypothesis of *charge symmetry* between (nn) and (pp) forces in the $^1S_0$ interaction. The (np) singlet scattering length differs substantially from the other two, but this can be accounted for by a very slight difference in the strength of the (np) force, since the $^1S_0$ state is near to zero binding energy and is sensitively dependent on the precise potential that exists (cf. Fig. 5.5b). Disregarding this difference, which may arise because charged pions as well as the lighter neutrals can mediate the (np) force (Fig. 2.8) but only neutrals the (nn) and (pp) interaction, it is plausible to assert the *charge independence* of nuclear forces for nucleons in the same state of relative motion. This leads immediately to the significance of a new quantity known as *isobaric spin*.

163

TABLE 5.1 Parameters of the two-nucleon system

Parameter	Value/10^{-15} m			Method of determination
	(np)	(pp)	(nn)	
Triplet scattering length a_t	$5 \cdot 425 \pm 0 \cdot 0014$	—	—	Low-energy (np) scattering Coherent scattering
Singlet scattering length a_s	$-23 \cdot 714 \pm 0 \cdot 013$	$-7 \cdot 821 \pm 0 \cdot 004^*$	$-17 \cdot 4 \pm 1 \cdot 8$	Low-energy (np) and (pp) scattering Coherent scattering Final state interactions
Triplet effective range r_{0t}	$1 \cdot 749 \pm 0 \cdot 008$	—	—	Binding energy of deuteron Photodisintegration
Singlet effective range r_{0s}	$2 \cdot 73 \pm 0 \cdot 03$	$2 \cdot 830 \pm 0 \cdot 017$	$2 \cdot 4 \pm 1 \cdot 5$	Low-energy (np) and (pp) scattering Capture of slow neutrons by protons Final state interactions

* The proton–proton scattering length becomes about −17 fm when correction is made for the Coulomb forces.

5.6 Isobaric (isotopic) spin (Ref. 1.1a, p. 87)

5.6.1 Formalism

The neutron and proton have a similar mass and identical spin and, as has just been seen, apparently participate similarly in mutual interactions so long as Coulomb effects are disregarded and so long as the pair of interacting particles is in the same quantum state. It was therefore proposed as early as the 1930s and soon after the discovery of the neutron, that the two particles might be regarded as two states of a single particle, the *nucleon*, distinguished simply by the label of charge. Since electromagnetic forces are weak compared with nuclear forces within the range of the nuclear interaction, the disturbance of this simplifying assumption by charge-dependent effects should be small.

To describe this in symbols suitable for quantum calculations it is useful to start by drawing an analogy with the quantum theory of angular momentum. It is well known that a state with angular momentum J has $2J+1$ substates J_z which become distinct in a finite magnetic field. Similarly, an intrinsic spin has $2s+1$ orientations with respect to an axis of quantization, and for $s = \frac{1}{2}$ this indicates two states. The two states of the nucleon can, therefore, be described by assigning a fictitious vector T with quantum number $T = \frac{1}{2}$ to the basic particle and by regarding the proton and neutron as substates with $T_z = +\frac{1}{2}$ and $T_z = -\frac{1}{2}$ respectively. These substates become distinct once the underlying symmetry is broken by Coulomb effects, i.e. by the recognition of charge. Because of the close analogy with intrinsic spin the new quantity is known as *isobaric* or *isotopic spin* (*i*-spin) although it has nothing to do with ordinary mechanical variables. Like intrinsic spin, it exists only in virtue of the assumption of a symbolic space in which it can assume permitted orientations. In each orientation the charge state is specified by the component T_z. The value of this formal description is that it may easily be extended to systems of particles and that it facilitates the expression of a new conservation law.

The discussion of conserved quantities in Section 1.3 emphasizes their connection with symmetries and with the fact that the corresponding operator commutes with the Hamiltonian operator H. Thus, for ordinary angular momentum J the Hamiltonian is invariant under rotations of the coordinate system in ordinary space and

$$[H, J] = 0 \qquad (5.47)$$

so that J is conserved in an isolated system. Convenient commuting operators are then J^2 and J_z with eigenvalues given by

$$\left. \begin{array}{l} J^2 \psi = J(J+1)\hbar^2 \psi \\ J_z \psi = J_z \psi = M\hbar\psi \quad \text{with} \quad |M| \leqslant J \end{array} \right\} \qquad (5.48)$$

where ψ is a wavefunction in ordinary space. And for isobaric spin, we now postulate that the Hamiltonian operator is invariant under rotations in the fictitious *i*-spin space. In analogy with (5.47) and (5.48) we then have, omitting the angular momentum $\hbar$,

$$[H, \boldsymbol{T}] = 0 \tag{5.49}$$

$$\left.\begin{array}{l} \boldsymbol{T}^2\phi = T(T+1)\phi \\ \boldsymbol{T}_z\phi = T_z\phi \qquad |T_z| \leqslant T \end{array}\right\} \tag{5.50}$$

where ϕ is a wavefunction in *i*-spin space and T is the isobaric spin quantum number. The inference from (5.49) that T_z is a constant of the motion expresses the *conservation of charge*, which is a well-known experimental fact, but the conclusion that $\boldsymbol{T}^2$ is constant, i.e. that *total isobaric spin is conserved* is new and is a generalization of the charge independence hypothesis. It means that the strong interaction between particles depends on $\boldsymbol{T}$ but not on T_z.

A simple illustration of the isobaric spin formalism in classifying particle states is found in the two-nucleon system discussed in this chapter. This system comprises the three possibilities (pp), (np) and (nn). By the rules for addition of quantum vectors, the total isobaric spin quantum number may be $T = T_1 + T_2 = 1$ or 0. The third component, measuring total charge, must be $T_z = 1, 0, -1$ in the three cases, and each of these is a substate of the state with $T = 1$. Moreover, the (np) system, with $T_z = 0$, is also a substate of total *i*-spin $T = 0$. It can now be seen immediately that the two-nucleon singlet states have $T = 1$, i.e.

$$^1S_0 \qquad \text{pp (np)}_{\text{singlet}} \text{ (nn)} \qquad T = 1, T_z = 1, 0, -1 \tag{5.51a}$$

and the two-nucleon triplet state occurs only with $T = 0$, i.e.

$$^3S_1 \qquad \text{(np)}_{\text{triplet}} \qquad T = 0, T_z = 0 \tag{5.51b}$$

In this particular case of two similar particles such as nucleons, with isospin $\frac{1}{2}$, there is no mixing of isospin states for a given state of motion, e.g. $^1S_0$ or $^3S_1$. The validity of total isospin as a good quantum number is then directly equivalent to charge independence of the nuclear force, i.e. that the interaction is the same in all similar states of motion (e.g. $^1S_0$) with the same T (e.g. 1) but different T_z and is different from that in states of other T (e.g. 0). A more restrictive possibility is that the interaction is the same for the same T and $|T_z|$, i.e. for (pp) and (nn), but different for (np) which has $T_z = 0$. This would be the result of charge symmetry of the nuclear force.

In a state of $T = 1$, but not of $T = 0$, a neutron or proton may be changed to the other particle without offending the exclusion principle. The *i*-spin concept thus permits a useful generalization of the Pauli principle; permitted states of two nucleons are those which have antisymmetrical wavefunctions for exchange of *all* coordinates, i.e. isobaric spin, ordinary spin and position. Thus, the two-nucleon state $T = 1$, $^1S_0$ is symmetrical in i-spin, antisymmetric in ordinary spin and symmetrical in space coordinates.

We may extend this discussion to the levels of complex nuclei, under the reasonable assumption that these are determined by interactions for which charge independence is valid. For the nucleons, we note that the particle charge Q may be expressed in units of $|e|$ as

$$Q = (T_z + \tfrac{1}{2}) \tag{5.52}$$

and for a nucleus containing N neutrons and Z protons with $N + Z = A$, the third component of total i-spin is

$$T_z = (Z - N)\tfrac{1}{2} = (A - 2N)\tfrac{1}{2} = (2Z - A)\tfrac{1}{2} \tag{5.53}$$

whence

$$Q = Z = (T_z + A/2) \tag{5.54a}$$

In elementary-particle language this would be written

$$Q = (T_z + B/2) \tag{5.54b}$$

using B the baryon number in place of A. This expression, and also (5.52), relates to non-strange particles and may be generalized to include both strangeness and charm as shown in Table 2.2.

Since from (5.50) $|T_z| \leq T$, the total isobaric spin for a nuclear state in a nucleus with known T_z (i.e. known charge Z) must have $T \geq |T_z|$. A level of known T, e.g. 1, will therefore be expected to occur in $(2T + 1)$ isobaric nuclei, though not at the same excitation in all because of the corrections necessary for neutron–proton mass difference and Coulomb repulsion. Figure 5.9 shows an isobaric triad for $A = 10$. (Odd-mass mirror nuclei such as ^{13}C, ^{13}N provide many examples of isobaric doublets.)

In Fig. 5.9*b* the $(0^+, T = 1)$ level of ^{10}B is the isobaric analogue of the ground state of ^{10}Be. Such states are also found in much heavier nuclei, and at much greater excitation because of the increasing Coulomb energy of the proton-richer isobar. The analogue state may then, in many cases, be excited directly by a proton-induced nuclear reaction, and the reaction yield as a function of proton energy shows a characteristic feature known as an *isobaric analogue resonance*.

Isobaric spin conservation is found experimentally to be valid generally for the strong interaction (Ch. 11) apart from violations due to electromagnetic forces. Two illustrations are given in the following section.

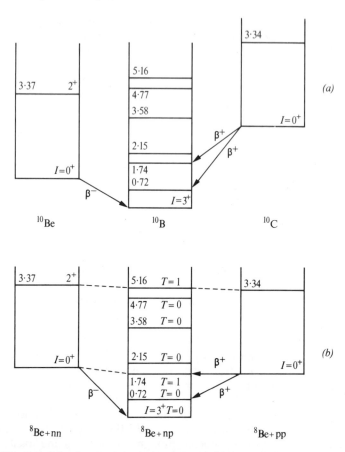

Fig. 5.9 Low-lying levels of the isobars of mass 10. Energies are marked in MeV and spins and parities are indicated by the symbols 0^+, 1^+, etc. (*a*) Masses of bare nuclei. (*b*) Nuclear masses corrected for Coulomb energy and for neutron–proton mass difference, showing isobaric triplet levels.

5.6.2 Application to scattering problems

(i) *Nucleon–nucleon scattering.* Proton–proton collisions take place in a state of pure isobaric spin $T = 1$ because $T = T_z = \frac{1}{2}$ for each particle. Only the partial waves for the states ^1S, ^3P, ^1D, ... are allowed to contribute to the scattering amplitude as noted in Section 5.3.1.

The s-wave neutron–proton collision has been analysed so far in terms of singlet (^1S) and triplet (^3S) spin states with relative weights of 1 and 3 (eqn (5.17)). From the discussion in Section 5.6.1 it follows that these spin states are pure states of $T = 1$ and $T = 0$

respectively. If the scattering cross-section is to be analysed in terms of isobaric spin eigenstates, however, it must be remembered that only one of the $T = 1$ triplet of states is accessible to the (np) system, namely that with $T_z = 0$. This system must, therefore, be considered to be equally in the $T = 1$ and $T = 0$ states. Charge independence apart from Coulomb effects exists between the $T = 1$ substates, but not between the (pp) state and the complete mixed (np) system.

Figure 5.7 gives the variation of the proton–proton cross-sections over a range of energies for which several partial waves are important, although their number is limited by centrifugal barrier penetration effects. From this information and from similar data for the (np) system the variation of the $T = 1$ and $T = 0$ cross-sections over the same energy range can be extracted. Differential cross-sections for the (pp) (Fig. 5.8) and (np) (Fig. 5.3) interactions have also been obtained over a wide range of energies, and together with the elastic cross-sections and some observations of spin-polarization effects provide the data from which phase-shift analyses up to about 400 MeV laboratory energy may be made; beyond this extra parameters are needed to deal with the inelastic channels.

The phase shifts, two of which are plotted in Fig. 5.4, join smoothly with the low-energy values δ_{0s}, δ_{0t} discussed in Section 5.2.2. The $^1S_0$ phase shift ($T = 1$ state) becomes negative at about 250 MeV and this is evidence for a repulsive core in the nucleon–nucleon interaction at a radius of about 0.5 fm, as assumed in Section 5.1. Altogether, the phase shifts summarize a large body of experimental data and can be used for testing potential models of the nucleon–nucleon interaction.

The (np) differential cross-section at energies of about 100 MeV is nearly symmetric about 90° c.m. and the depth of the 90° dip increases with energy. The high intensity of neutrons apparently scattered through 180° (c.m.) cannot be understood if the (np) well depth is only about 30 MeV as indicated by the low-energy experiments, because the whole of the neutron energy has been transferred to the proton. If, however, the proton and neutron can change their identity in the collision, then a backward neutron peak can occur for low momentum transfer. This phenomenon is known as *charge exchange scattering* and is evidence for an *exchange behaviour* in the nucleon–nucleon force. The forward peak in the scattering cross-section is due to the ordinary, non-exchange scattering.

(ii) *Pion–proton scattering.* The pion (Sect. 2.2.1) is an isobaric triplet, that is to say it has $T = 1$ and three charge states with $T_z = 1$, 0, -1 corresponding with π^+, π^0 and π^- particles. The total isobaric spin number for the (πp) system is, therefore, $T = \frac{3}{2}$ or $\frac{1}{2}$ with $T_z = \frac{3}{2}$, $\frac{1}{2}$, or $-\frac{1}{2}$, of which the first is found only in the $T = \frac{3}{2}$ state. Experimentally, differential cross-sections have been measured for

169

the elastic scattering processes

$$\pi^+ + p \rightarrow \pi^+ + p \tag{5.55}$$

$$\pi^- + p \rightarrow \pi^- + p \tag{5.56}$$

and for the charge-exchange process

$$\pi^- + p \rightarrow \pi^0 + n \tag{5.57}$$

For process (5.55) the pion–nucleon system is an eigenstate of isobaric spin with $T = \frac{3}{2}$, $T_z = \frac{3}{2}$ and if charge independence is valid should have the same cross-section as $(\pi^- n)$ scattering, with $T = \frac{3}{2}$, $T_z = -\frac{3}{2}$. These cross-sections would also be equal to those for other pion–nucleon states such as $(\pi^- p)$, $(\pi^+ n)$, $(\pi^0 p)$ and $(\pi^0 n)$ if these belonged wholly to the $T = \frac{3}{2}$ multiplet. This, however, is not so, because the T_z values in these cases permit a mixture of $T = \frac{1}{2}$ states and the pion–nucleon states must be represented by the proper superpositions of eigenstates of T, T_z. This situation differs from that discussed for the two-nucleon system because the interacting particles are now not identical and the different T values do not necessarily imply different space–spin states (e.g. ^3S, ^1S).

The expansion of a product wavefunction $\phi(\text{pion}) \times \phi(\text{nucleon})$ in terms of eigenstates of T, T_z may be obtained directly from the table of Clebsch–Gordan or vector coupling coefficients for the case labelled $j_1 = 1$, $j_2 = \frac{1}{2}$ given in Appendix 4. The m-values in the table are the T_z-values for the pion and nucleon concerned. The same table enables an eigenstate of T, T_z (labelled J, M) to be decomposed into a sum of products of pion and nucleon states. Thus, for the $(\pi^- p)$ system with a T_z of $-1 + \frac{1}{2} = -\frac{1}{2}$ we have

$$\phi(\pi^- p) = \phi_1(\pi^-)\phi_2(p) = \sqrt{\tfrac{1}{3}}\psi(\tfrac{3}{2}, -\tfrac{1}{2}) - \sqrt{\tfrac{2}{3}}\psi(\tfrac{1}{2}, -\tfrac{1}{2}) \tag{5.58}$$

Also, the eigenstate $\psi(\frac{3}{2}, -\frac{1}{2})$ may be related to the pion–nucleon systems with the same T_z of $-\frac{1}{2}$, namely, $(\pi^- p)$ and $(\pi^0 n)$:

$$\psi(\tfrac{3}{2}, -\tfrac{1}{2}) = \sqrt{\tfrac{1}{3}}\phi(\pi^- p) + \sqrt{\tfrac{2}{3}}\phi(\pi^0 n) \tag{5.59}$$

These and similar formulae permit the hypothesis of conservation of isobaric spin in the pion–nucleon interaction to be tested by predicting the ratio of cross-sections for the processes (5.55), (5.56) and (5.57).

Let the (angle-dependent) scattering amplitude be $f_{3/2}$ for the $T = \frac{3}{2}$ state and $f_{1/2}$ for the $\frac{1}{2}$ state, each independent of T_z in accordance with charge independence within a T-multiplet. Then in the $(\pi^+ p)$ scattering the incident state $\phi(\pi^+ p)$ is the i-spin state $\psi(\frac{3}{2}, \frac{3}{2})$ and leads simply to a final state with the same (conserved) i-spin but including a factor $f_{3/2}$.

For $(\pi^- p)$ scattering, however, the initial state is mixed in i-spin, as shown in (5.58) above and an i-spin conserving interaction leads to a final state containing the wavefunction

$$\sqrt{\tfrac{1}{3}}f_{3/2}\psi(\tfrac{3}{2}, -\tfrac{1}{2}) - \sqrt{\tfrac{2}{3}}f_{1/2}\psi(\tfrac{1}{2}, -\tfrac{1}{2}) \tag{5.60}$$

This can be put back into pion × nucleon states using expressions such as (5.59) and the result is

$$(\tfrac{1}{3}f_{3/2} + \tfrac{2}{3}f_{1/2})\phi(\pi^- p) + \left(\frac{\sqrt{2}}{3}f_{3/2} - \frac{\sqrt{2}}{3}f_{1/2}\right)\phi(\pi^\circ n) \tag{5.61}$$

from which we conclude that the elastic scattering amplitude, process (5.56), is $(\tfrac{1}{3}f_{3/2} + \tfrac{2}{3}f_{1/2})$ and the charge exchange amplitude, process (5.57), is $\sqrt{2}/3(f_{3/2} - f_{1/2})$.

This is, so far, only descriptive, but if now it is assumed that for a particular energy the amplitude $f_{1/2}$ happens to be small (e.g. in the neighbourhood of the $\Delta(1236)$ resonance which has $T = \tfrac{3}{2}$), then the differential cross-sections for the three reactions at any angle should stand in the ratio $(\tfrac{1}{3})^2 : (\sqrt{2}/3)^2 : 1$, i.e. $1 : 2 : 9$. This is, in fact, verified to an encouraging degree of precision, and the result lends important support to the hypothesis of the conservation of isobaric spin in this hadronic interaction.

More detailed examination of the extensive body of data that now exists permits the scattering amplitudes $f_{3/2}$ and $f_{1/2}$ to be obtained over a wide range of energy involving a series of partial waves. Their variation as a function of c.m. energy displays peaks at the mass values of nucleon isobars, e.g. the $\Delta(1236)$. They are normally expressed in terms of relativistically invariant variables, and may be analysed to yield a value for the pion–nucleon coupling constant (Sect. 2.3).

5.7 Electromagnetic properties of the deuteron; non-central forces

5.7.1 Neutron–proton capture and photodisintegration

The processes of thermal neutron capture

$$n + p \rightarrow d + \gamma \tag{5.62}$$

and its inverse, photodisintegration

$$\gamma + d \rightarrow n + p \tag{5.63}$$

exchange energy between the neutron–proton system and an electromagnetic field. The minimum exchange is the binding energy of the deuteron and process (5.62) indeed yields the most accurate value of this quantity:

$$\varepsilon = 2 \cdot 2245 \pm 0 \cdot 0002 \text{ MeV}$$

The spins of the particles concerned (n, p $= \frac{1}{2}$ and d $= 1$) allow each of the processes to occur by a magnetic dipole interaction, e.g. $^1S \rightarrow {}^3S$ for capture and $^3S \rightarrow {}^1S$ for photoeffect. The intrinsic spins 'flip' in this process and the angular momentum change $1\hbar$ is conveyed to or from the photon. The coupling between the nucleons and the electromagnetic field is via the intrinsic magnetic moments and the cross-sections in the zero-range approximation include a factor $(\mu_p - \mu_n)^2$. There is also a factor $(1 - \alpha a_s)^2$, where α is the deuteron size parameter and a_s is the singlet scattering length, and this makes the cross-section sensitive to the sign of this latter quantity. The observed value of the capture cross-section confirms that the ^1S state of the deuteron is unbound (a_s negative).

The photodisintegration process (5.63) may also take place through the electric dipole transition $^3S \rightarrow {}^3P$ and this is responsible for the major part of the total cross-section except very near threshold, where the magnetic transition dominates because the emission of p-wave particles is retarded by a centrifugal barrier-penetration factor. The photoelectric cross-section reaches a maximum at a photon energy $h\nu \approx 2\varepsilon$ and is then of the order of magnitude of the 'area' of the deuteron $\pi/4\alpha^2$ multiplied by the fine structure constant, which represents the coupling to the electromagnetic field. The cross-section in zero-range approximation can readily be corrected for finite range and then contains a factor $(1 - \alpha r_{0t})^{-1}$ from which the triplet effective range may be extracted.

5.7.2 Magnetic and electric moments; tensor and spin-orbit forces

The magnetic moment of the deuteron in units of the nuclear magneton $\mu_N = e\hbar/2m_p$ is less than the sum of the intrinsic magnetic moments of the neutron and proton, taken antiparallel for the ^3S configuration:

$$\mu_d = 0 \cdot 857\,411 \pm 0 \cdot 000\,019 \qquad \mu_n = -1 \cdot 913\,15 \pm 0 \cdot 000\,07$$
$$\mu_p = 2 \cdot 792\,71 \pm 0 \cdot 000\,02$$
$$\mu_n + \mu_p = 0 \cdot 879\,56 \pm 0.000\,07$$

The closeness of the two values confirms that the magnetic moment of the neutron is negative, but it is not possible to explain the difference if the two particles are in a pure s-state of relative motion. Moreover, such a state is spherically symmetrical and has no electric moments, whereas the deuteron is known to have a positive electric quadrupole moment (Sect. 3.4.1) of $0 \cdot 29 \, \text{fm}^2$ corresponding with an elongation of the density distribution along the spin axis.

These facts can be explained if the deuteron wavefunction contains a d-state component, indicating some motion of the two particles with a relative *orbital* angular momentum quantum number of 2. The D-state consistent with the observed spin of the deuteron is $^3D_1$ and the ground state wavefunction would then be written

$$\psi(\text{SD}) = a\psi(^3S) + b\psi(^3D) \tag{5.64}$$

where the mixing is to be produced by an appropriate property of the internucleon potential. If the mixing coefficient is expressed by the perturbation theory formula

$$b/a = \int (\psi_D^* |V| \psi_S) \, d^3r/(E_D - E_S) \tag{5.65}$$

where E_D and E_S are the energies of pure D- and pure S-states, then it can immediately be seen that if V is of a purely radial nature, i.e. if the force is purely central, the matrix element vanishes because of orthogonality of ψ_D and ψ_S. In other words, central forces do not mix states of different angular momentum, whereas equation (5.65) requires a force for which L^2 is *not* a constant of the motion if b is to be finite.

A suitable non-central interaction is provided by adding a *tensor* component V_T to the potential. This may be written

$$V_T = V_T(r)S_{12} \tag{5.66}$$

where

$$S_{12} = 3(s_p \cdot r_p)(s_n \cdot r_n)/r^2 - (s_p \cdot s_n) \tag{5.67}$$

and s_p, s_n are the spin vectors divided by $\hbar$. This gives a velocity-independent force depending on the relative orientation of the spin vectors and the line joining the particles (Fig. 5.10); it resembles the classical interaction between magnetic dipoles. Detailed calculations show that if about 7 per cent of the ground state wavefunction is D-state and if $V_T < 0$ then the electromagnetic moments of the deuteron are correctly predicted.

Tensor forces are not effective in singlet states because there is no preferred spin axis. It might, therefore, be possible to ascribe the whole of the spin dependence of nuclear forces to the existence of non-central effects in the triplet state, and a tensor-type interaction is indeed suggested by meson-exchange theories. The existence of spin polarization effects in nucleon scattering and the theory of nuclear shell structure (Ch. 7), however, jointly require the existence of a coupling between spin and orbital motion of a single nucleon in a potential field. Such a force in a complex nucleus may derive from a two-body *spin-orbit* force between a pair of nucleons with a potential

$$V_{LS} = V_{LS}(r)L \cdot S \tag{5.68}$$

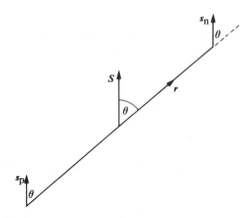

Fig. 5.10 Spin vectors in the (triplet) ground state of the deuteron (spin S). The minimum energy for the tensor force is when $\theta = 0$ or $180°$. The deuteron is then cigar-shaped, with a positive quadrupole moment.

Such a force does not mix states of different L (because $\boldsymbol{L} \cdot \boldsymbol{S}/\hbar^2$ is a number (cf. Ex. (7.4)) and cannot account for the properties of the deuteron.

5.8 Summary and theoretical description

From the evidence presented in this chapter it appears that the force between nucleons has the following properties:

(a) short range (≈ 2 fm);
(b) spin dependence;
(c) charge symmetry and most probably charge independence to a good approximation except for electromagnetic effects;
(d) an exchange behaviour;.
(e) a repulsive core of radius ≈ 0.5 fm in the ^1S state and probably also in other states of motion, and an attractive potential outside the core in the ^1S and ^3S states at least;
(f) a tensor and a spin-orbit term in addition to the central forces.

The potential energy function for the central part of the force is typically as shown in Fig. 5.1.

The meeting-point between experiment and theory in the discussion of the nuclear force is the comparison between observed and calculated two-nucleon phase shifts. Many *phenomenological potentials* (Hamada–Johnston, Yale University, Tabakin) that predict the phase shifts have been proposed. In addition to conforming with the requirements summarized above, they are constrained by the general invariance principles relating to space translation or rotation, parity and time reversal (Sect. 1.3). A general type of potential may be written

$$V = V_C + V_{CS} + V_{LS} + V_T + V_{QLS} \qquad (5.69)$$

where V_C is the spin-independent central potential;

V_{CS} is the spin-dependent central potential;

V_{LS} is the two-body spin-orbit potential;

V_T is the tensor potential;

V_{QLS} is a quadratic spin-orbit potential (which will not be discussed).

Each potential term in (5.69) has a radial dependence determined by the range of the force, and contains operators for spin, isospin, position and momentum which give the necessary properties (cf. (5.67) and (5.68)). Thus, the isospin operators give a potential depending on total isospin rather than its third component (charge independence) and can be used to describe a force arising from the exchange of charge between a neutron and proton. Similarly, the spin operators can be arranged to describe forces arising from the exchange of spin, and the spin–isospin operators jointly describe forces connected with change of position.

The several types of exchange force that are formally described in this way were originally introduced (by Heisenberg) to account for the saturation of the force (Sect. 5.1) because some of the exchange forces have opposite signs in even and odd angular momentum states. A repulsive core, of course, also contributes to a saturation requirement, but the evidence of (np) scattering (Sect. 5.6.2) makes it clear that exchange characteristics must be included in the general potential.

The phenomenological potentials leading to attractive forces have well depths of ≈ 1000 MeV at the hard-core radius. Since the deuteron binding energy is only 2·2 MeV, it follows that the strength of the forces is only just sufficient to produce a bound state, and that this arises from the long-range part of the force.

Both ordinary and exchange forces are predicted if a meson is literally *exchanged* between the interacting particles, as shown in Fig. 2.8. As pointed out by Yukawa, the short range of the force is a consequence of the finite mass of the exchanged particle, and the range $\hbar/\mu c$, with $\mu = m_\pi$ thus indicated (Sect. 2.2.1) agrees well with observation. The attractive one-pion exchange potential (OPEP) calculated in this way has several of the features of the phenomenological potential (5.69), particularly charge independence, tensor and exchange properties. The OPEP deals only with the 'long-range' part of the force; to include the shorter distances and especially the repulsive core, more complex exchanges are required in addition, especially of two pions and of ω, ϕ and ρ particles.

   Realistic potentials should be used in nuclear matter and nuclear structure calculations but formidable difficulties arise because of the singular nature of the 'hard' repulsive core. One way in which these difficulties have been circumvented, in the case of nuclear matter, is outlined in Section 6.4. For nuclear structure, simplified forms of the potential may be used with considerable success in predicting the sequence of low-lying excited states once the basic single-particle states have been established. An example will be given in Section 8.6.

Examples 5

5.1 Using the tabulated values of the magnetic moment of the neutron and proton, calculate the force between these two particles in a triplet state at a separation of 3×10^{-15} m, and calculate the work required, on account of this force, to bring the neutron from infinity to this distance from the proton. Assume that the spins always point along the line joining the particles. [1 N, 6267 eV]

5.2* Estimate the range of the repulsive core of the nucleon–nucleon potential given that the s-wave ($^1S_0$) phase shift vanishes for a laboratory energy of 250 MeV.

5.3 Using $r_{0t} = 1.75$ fm, find the correction to the triplet scattering length introduced by the effective range term in equation (5.32). [1.1 fm]

5.4 Using equation (5.19) and writing $k = i\alpha$ for the bound state, deduce the relation (5.32). You may assume that $i\delta_0$ is a large quantity, which is required if the external wavefunction is to be well behaved at infinity (Ref. 6.4).

5.5* Assuming that the wavefunction $u(r) = r\psi(r) = Ce^{-\alpha r}$ is valid for the deuteron from $r = 0$ to $r = \infty$, obtain the value of the normalization constant C.
   If $\alpha = 0.232$ fm^{-1} find the probability that the separation of the two particles in the deuteron exceeds a value of 2 fm.
   Find also the average distance of interaction for this wavefunction.

5.6 Suppose that the deuteron is represented by a square-well potential of depth 20 MeV and radius 2.5 fm. Plot the effective potential for $l = 1$ and $l = 2$ states and comment on the possibility of the formation of a bound state.

5.7 A particle of kinetic energy T and mass M is scattered by a square-well potential of radius a and depth b. Show that the s-wave phase shift for $k \to 0$ is given by

$$\tan \delta_0 = k \tan k'a/k' - \tan ka$$

where

$$k^2\hbar^2 = 2MT, \qquad (k')^2\hbar^2 = 2M(T + b)$$

5.8 The triplet and singlet scattering lengths for the neutron–proton system are 5.4 fm and 23.7 fm respectively. Calculate the elastic scattering cross-section expected for neutrons of energy 1 eV. [20.3 b]

5.9 Using the principle of isospin conservation, predict the ratio of the total cross-sections for the reactions $p+p \to d+\pi^+$ and $n+p \to d+\pi^0$ as far as isospin factors are concerned. [2/1]

5.10 Using the 1, $\frac{1}{2}$ table of Appendix 4 write down a complete list of pion–nucleon states in terms of eigenstates of isospin. Write down also the expression of the isospin eigenstates in terms of pion–nucleon states.

5.11 The total cross-section for (Λp) scattering at $T = 10$ MeV (c.m.) is 100 mb. Assuming that the singlet and triplet scattering lengths are each -1.7 fm, find

the effective range, also assumed the same in singlet and triplet states.
[3·09 fm]

What is the physical meaning of a negative scattering length in both states?

5.12* Using equation (1.77), write down the elastic scattering amplitude for the (np) system at an energy at which only s- and p-waves are effective and inelastic processes may be disregarded. Neglecting terms of second order in δ_1, find the differential cross-section and show that if δ_1 is positive $\sigma(180°) < \sigma(0°)$ whereas for δ_1 negative $\sigma(180°) > \sigma(0°)$. (A repulsive force in odd-parity states is characteristic of the *Majorana* type of exchange force.)

5.13* By expressing two-nucleon states in terms of states of pure isospin, show that the integrated elastic cross-section in the state $T = 0$ at a given energy is $\sigma_{T=0} = 2\sigma_{np} - \sigma_{pp}$.

6 Nuclear mass, nuclear size and nuclear matter

The nucleus was identified in the α-particle scattering experiments of Rutherford and Geiger because of the concentration of the positive charge Ze of the nuclear protons and of nearly the whole mass of the atom within a volume of radial extent less than about 10^{-14} m. Following the recognition of the fundamental neutron–proton structure of the nucleus, interest in the charge as an observable property now centres on its radial and angular distribution, on the interpretation of this distribution in terms of model wavefunctions, and on the effects of the electromagnetic interaction on nuclear structure. A complete theory of the charge density must also allow for the part played by exchanged particles such as pions and for the formation of nucleon isobars. The nuclear mass is a more fundamental property in the sense that it embodies not only the baryon number $B = N + Z = A$ but also energies derived from the basic nucleon–nucleon interaction, modified, if need be, by the inclusion of many-body interactions which cannot be sensed by the two-body scattering experiments described in Chapter 5. Because of the strong interaction between nucleons, the spatial distributions of charge and mass must be similar.

The calculation of the properties of finite nuclei from an assumed internucleon potential is an undertaking of considerable technical difficulty. It is useful, however, as an intermediate step to consider the nature of an infinite nuclear medium to see whether any general trends in nuclear mass values can be predicted. An immediate result of such a consideration is an understanding of the reasons why many nuclear properties, as well as mass, show the persistence of single-particle motion in a many-body system. The resulting *shell structure* is considered later (Ch. 7); the present chapter is primarily concerned with bulk properties.

6.1 The mass tables; binding energy

Two scales for mass measurements are now recognized:

(*a*) *The absolute scale*, related to the kilogram.

(b) *The atomic mass scale,* defined by setting the mass of one atom of the nuclide ^{12}C equal to $12 \cdot 000 \ldots$ atomic mass units (a.m.u. or u).

The absolute value of the atomic mass unit is obtained by noting that for 1 mole of ^{12}C $(0 \cdot 012 \text{ kg})$:

$$0 \cdot 012 \text{ kg} = N_A \times 12 \text{ a.m.u.}$$

where N_A is Avogadro's number, i.e. the number of atoms in one mole on the physical scale. This gives

$$1 \text{ a.m.u.} = 0 \cdot 001 / N_A \text{ kg} = 1 \cdot 661 \times 10^{-27} \text{ kg} = 931 \cdot 481 \text{ MeV}/c^2$$

Both N_A and the a.m.u. depend on the standard substance, and were slightly different when masses were referred to ^{16}O.

Any particle mass may be specified in any of these units. Usually for general calculations kilograms are used, for particle physics kinematics, MeV/c^2, and for nuclear structure studies, a.m.u. Mass tables, such as those of Reference 6.1, give the masses of *neutral atoms* in a.m.u. rather than those of bare nuclei since the former are obtained as relative values directly by mass spectrometry. The atomic mass $M(A, Z)$ of a nuclide, i.e. a specific nucleus of mass number A, is related to the *nuclear mass* M_N by the equation

$$M(A, Z) = M_N + Zm_e - B(Z)/c^2 \tag{6.1}$$

where m_e is the mass of the electron and $B(Z)$ the total electron binding energy. The binding energy term can be estimated from known X-ray energies not to exceed about 10^{-3} per cent of the atomic mass $M(A, Z)$ even for the heaviest elements and it may usually be neglected. The expression 'nuclear mass' is often loosely used to mean the neutral-atom mass in a.m.u.

The relative masses of all stable isotopes have now been determined with a precision of the order of 1 part in 10^6 by the refined techniques of mass spectrometry. Similar techniques, based on the deflection of charged particles in electric and magnetic fields, are used in the most precise determinations of nuclear reaction energy releases, which are related to mass changes through the equation

$$\Delta E = c^2 \, \Delta M \tag{6.2}$$

in which, of course, only changes of binding energy and not changes of the mass of the nucleons are involved. Combination of the data obtained in these ways leads to the present mass tables (Ref. 6.1), of which a small section is given in Table 6.1; the accuracy claimed in the relative atomic masses is better than 1 part in 10^6 for masses of the order of 30 a.m.u., corresponding with a determination of reaction energy released to a precision of a few keV or better.

TABLE 6.1 Atomic mass table (Ref. 6.1)

Nuclide ^A_Z	Atomic mass a.m.u.	$B(A, Z)$ MeV	B/A MeV/*nucleon*	$S_n(A, Z)$ MeV	$P_n(A, Z)$ MeV
$^{131}_{54}$Xe	$130·905\,08 \pm 0·3^*$	1103·55	8·41	6·6	—
$^{132}_{54}$Xe	$131·904\,16 \pm 0·3$	1112·49	8·42	8·9	1·2
$^{133}_{54}$Xe	$132·905\,89 \pm 0·9$	1118·94	8·41	6·5	1·1
$^{134}_{54}$Xe	$133·905\,39 \pm 0·5$	1127·48	8·41	8·5	1·0
$^{135}_{54}$Xe	$134·907\,14 \pm 1·2$	1133·93	8·40	6·45	0·9
$^{136}_{54}$Xe	$135·907\,22 \pm 0·6$	1141·92	8·40	8·0	1·4
$^{137}_{54}$Xe	$136·911\,75 \pm 2·2$	1145·78	8·36	3·9	—

* The errors in the mass values (a.m.u.) refer to the last significant figure.

The mass tables show that nuclidic masses are sufficiently near to multiples of the hydrogen-atom mass to suggest that nuclei are built up of particles of mass comparable with that of the proton, and this is consistent with a neutron–proton model. If this model is assumed the *total binding energy* $B(A, Z)$ of the nucleus (A, Z), which is the work required to break down the nucleus into Z protons and N ($= A - Z$) neutrons, is given by the equation

$$B(A, Z) = ZM_H + NM_n - M(A, Z) \qquad (6.3)$$

This expresses $B(A, Z)$ in a.m.u., but conversion to MeV is straightforward; typical values are given in Table 6.1 together with the average binding energy per nucleon B/A. The definition (6.3) implies that a stable nucleus has a positive binding energy, though a *negative* total energy with respect to that of its uncombined constituents at infinite separation.

The *separation energy* $S(A, Z)$ of a *specific particle*, e.g. a neutron, in the nucleus (A, Z) is the work required to remove that particle from the nucleus. For a neutron this converts the nucleus (A, Z) to the nucleus $(A - 1, Z)$ so that, for the least-bound particle,

$$\begin{aligned} S_n(A, Z) &= B(A, Z) - B(A - 1, Z) \\ &= M(A - 1, Z) - M(A, Z) + M_n \end{aligned} \qquad (6.4a)$$

The S_n values calculated from this expression are greater for even N than for odd N (cf. Fig. 7.2) and a similar staggering effect is seen in the proton separation energies. This is due to attractive forces between identical pairs of nucleons with opposite spins in similar states of motion. If the average trend of separation energies is required, an alternative definition

$$S_n(A, Z) = \tfrac{1}{2}[B(A + 1, Z) - B(A - 1, Z)] \qquad (6.4b)$$

may be used. To exhibit the pairing effect, however, it is useful to define a *pairing energy* P by the expression

$$P_n(A, Z) = (-1)^N \tfrac{1}{4}\{2S_n(A, Z) - S_n(A - 1, Z) - S_n(A + 1, Z)\}$$
$$(6.5)$$

for neutron pairing and similarly an energy $P_p(A, Z)$ for proton pairing. Values of P_n derived from S_n-values given by equation (6.4a) are shown in Table 6.1.

Separation energies are often determined from the Q-values for transfer or capture reactions (see Ex. 6.4).

6.2 Nuclear size

If a nucleus is held together by short-range forces it is expected to have a constant density (liquid-drop model). If also it is spherical, with a sharp edge, then the radius R for mass number A is easily seen to be

$$R = r_0 A^{1/3} \qquad (6.6a)$$

For such an idealized, uniform object the potential energy of a hypothetical strongly-interacting positive charge $Z_1 e$ as a function of distance from the nuclear centre would be approximately as shown in Fig. 6.1, in which the Coulomb repulsion is artificially

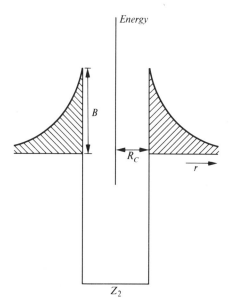

Fig. 6.1 Idealized potential well of radius R_C with barrier of height B. Spherical symmetry is assumed.

181

terminated at a radius R_C where it creates a *barrier* of height B given by

$$B = Z_1 Z_2 e^2 / 4\pi\varepsilon_0 R_C \qquad (6.6b)$$

if the nuclear charge is $Z_2 e$. For $Z_1 = 1$, $Z_2 = 92$ and $R_C = 7$ fm, $B = 19$ MeV. At smaller distances the Coulomb potential is replaced by the attractive nuclear potential; for a hypothetical point-neutron there is simply a uniform well of radius R_C. In both cases centrifugal potentials are disregarded.

This uniform model is adequate, with a suitable choice of r_0 in equation (6.6a), for the calculation of the Coulomb term in the mass formula, equation (6.10), and for estimating penetrabilities and geometrical cross-sections (Ch. 11) but it is a highly inadequate representation of the density of nuclear matter within a nucleus. To make the concept of nuclear size more realistic we now consider exploration of the interaction region shown in Fig. 1.1 using (*a*) an electromagnetic probe, specifically high-energy electrons, and (*b*) a hadronic probe, specifically nucleons.

(i) *Electron scattering.* If the particle approaching the nucleus is an electron, there is no attractive nuclear force, but the Coulomb force itself becomes attractive. The scattering can reveal detail of the charge distribution if the reduced de Broglie wavelength of the incident electrons is less than a typical nuclear dimension, say $\lambda \approx 1$ fm, $E \approx 200$ MeV. The electron is not known to have any structure and, therefore, behaves like a point charge. The effect of finite size of the target is then seen directly as a reduction of the scattering at a given angle below that expected for relativistic point-charge scattering as calculated by Mott. In the pioneer experiments of Hofstadter and his collaborators at Stanford, using linear accelerators providing electrons with energy up to 550 MeV, the ratio of observed and calculated intensities at a given angle was expressed as the square of a form factor F. Since the electromagnetic interaction is completely known F can be given directly in terms of the nuclear charge density $\rho(r)$ on the assumption that the scattering can reasonably be described by the Born approximation of quantum mechanics. In terms of the momentum transfer $\boldsymbol{q}$ to the nucleus ($Z_2 = Z$) as defined, for instance, in equation (1.6)

$$F(q) = (4\pi\hbar/qZe) \int_0^\infty \rho(r) \sin(\boldsymbol{q} \cdot \boldsymbol{r}/\hbar) r \, dr \qquad (6.7a)$$

where $q = |\boldsymbol{q}| = (2E/c) \sin \theta/2$, E is the total energy of the electron and θ the angle of scattering, both in the c.m. system. To simplify this and similar formulae it is usual to write the momentum transfer as $\hbar\boldsymbol{q}$ so that q becomes a reciprocal length and (6.7a) reads

$$F(q) = (4\pi/qZe) \int_0^\infty \rho(r) \sin(\boldsymbol{q} \cdot \boldsymbol{r}) r \, dr \qquad (6.7b)$$

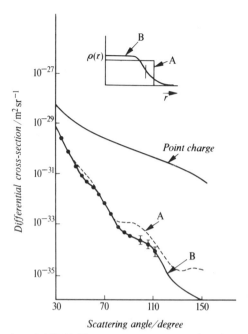

Fig. 6.2   Scattering of 153-MeV electrons by gold with (inset) charge distributions that may be used to fit the data points; the 'half-way' radius is marked. Reproduced from *Nuclear Structure* Vol. I, 1969, by Aage Bohr and Ben R. Mottleson with permission of the publishers, Addison-Wesley/W. A. Benjamin Inc.

Experimental data from which values of $F(q)$ may be derived are shown in Fig. 6.2. If a sufficiently large range of q-values can be covered then $\rho(r)$ can be obtained by making a Fourier transform of $F(q)$ with respect to q. In general, however, this is not possible and a density distribution must be assumed e.g.

$$\rho(r) = \rho_0/[1 + \exp{(r - R_{1/2})/a}] \qquad (6.8a)$$

Comparison with experiment then gives best-fit values for the half-way radius $R_{1/2}$ and the surface thickness parameter a. In the inset to Fig. 6.2 such a distribution is shown, together with the uniform distribution of radius R_C which best fits the data points. The mean-square radius for a given $\rho(r)$ is defined as

$$\langle r^2 \rangle = (1/Ze) \int_0^\infty r^2 \rho(r) 4\pi r^2 \, dr \qquad (6.8b)$$

and for the density distribution (6.8a) we have $\langle r^2 \rangle^{1/2} < R_{1/2} < R_C$. For a wide range of nuclei with $A > 20$, an $A^{1/3}$ dependence of $R_{1/2}$ is found.

It should be noted that this very direct determination of charge density does not give precisely the distribution of nuclear protons because the proton itself has a charge structure which must be folded in with the spatial density to give the observed $\rho(r)$. It should also be observed that the use of the q-dependence of elastic scattering in the present way is an example of a general method of *structure* determination that is used in other fields of physics. We shall encounter later (Sect. 9.4.3) a similar use of the q-dependence of *inelastic* form-factors but the information then obtained is mainly *spectroscopic*.

Other methods of determining the mean-square radius of the charge distribution, or its variation from one nucleus to another, are observations of:

(*a*) energy displacements in the spectra of muonic atoms (Sect. 3.3);
(*b*) the Coulomb energy of mirror nuclei such as ^{13}C, ^{13}N which have the same structure except that Z and N are interchanged;
(*c*) optical and X-ray isotope shifts (Sect. 8.1.2).

The results of all these methods are consistent with those of the electron-scattering determinations.

(ii) *Nucleon scattering.* The density distribution of nuclear matter $\rho_m(r)$, as distinct from that of nuclear protons $\rho_p(r)$, is best studied by proton–nucleus scattering at energies in excess of the Coulomb barrier. Such experiments are described in Section 7.2.4 in connection with the shell-model potential, and in Section 11.4 in connection with the optical model for nuclear reactions. Here, it will only be noted that despite the facts that the probe is not a point particle and that the interaction is not known, it is indeed possible to obtain potential parameters. The radial variation of the interaction potential is usually taken to have the form used for the charge density in equation (6.8a) with an overall negative sign but as it is not certain that the half-way radius means exactly the same thing we shall denote it by R_V. (It is, in principle, possible to predict R_V from $R_{1/2}$ (matter) by folding the nucleon–nucleon interaction with a density distribution.)

Experimental results confirm that R_V has the expected $A^{1/3}$ dependence and suggest that the radial dependence of the matter density is very similar to that of the charge density. Only for the heaviest nuclei is there some indication of an increased radius for the neutron distribution.

(iii) *Summary.* Inserting the best average values for r_0, nuclear size may be specified by the following radii for a spherical shape:

$$R_{1/2} = 1 \cdot 1 A^{1/3} \text{ fm} \qquad \text{(matter distribution)}$$
$$R_C = (1 \cdot 2 - 1 \cdot 3) A^{1/3} \text{ fm} \qquad \text{(equivalent square well for}$$
$$\text{charge distribution)} \qquad (6.9a)$$
$$R_V = 1 \cdot 25 A^{1/3} \text{ fm} \qquad \text{(optical potential)}$$

Other radii, usually less well defined, will be used in this book, especially in connection with heavy-ion collisions (Ch. 11). In such cases the crude picture of Fig. 6.1 is very far from the truth because nuclear interactions set in at an interparticle distance considerably greater than $1 \cdot 1(A_1^{1/3} + A_2^{1/3})$ fm. The *strong-absorption radius*, at which such processes are first seen, is approximately

$$R_{SA} = (1 \cdot 4 - 1 \cdot 5)(A_1^{1/3} + A_2^{1/3}) \text{ fm} \qquad (6.9b)$$

6.3 The semi-empirical mass formula

This formula connects the experimental information on masses with predictions from the theory of nuclear matter. A simple spherical liquid drop of uniform density and radius given by equation (6.6a) is assumed, although independent particle motion is also invoked, as will be seen.

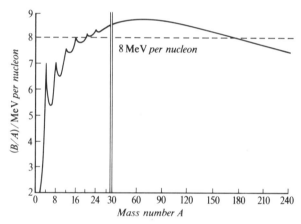

Fig. 6.3 Average binding energy per nucleon of the stable nuclei as a function of mass number (Ref. 2.1).

Figure 6.3 shows the dependence of the observed binding energy per nucleon B/A on the mass number A. The most important features of this diagram are:

(*a*) The appearance of approximately the same B/A ($7 \cdot 3$ to $8 \cdot 7$ MeV) for all nuclei with $A > 16$ so that for all but the lightest nuclei it is a good approximation to assume that B is proportional to A. If all pairs of nucleons in the nucleus interacted attractively, B would be more nearly proportional to A^2. It therefore follows that attractive interactions are, in effect, confined to near neighbours and that the total *volume energy* of the nucleus may be expressed in the form αA where α is a constant to be determined from the masses and to be predicted theoretically. A simple expression of this form,

185

however, overlooks the fact that nucleons near the surface of the nucleus will probably be less effective in producing binding than those in the interior. From equation (6.6a) the nuclear surface area varies as $A^{2/3}$ and the *surface energy* correction may be written $\beta A^{2/3}$.

(b) The general upward trend of B/A for the light nuclei. In a liquid-drop model, this trend is explained by the increasing ratio of the volume energy (αA) to the surface energy $(\beta A^{2/3})$ as A increases. There are marked increases of stability for $A = 4n$, $Z = N = 2n$ corresponding to equal numbers of neutrons and protons, which are not predicted by the liquid-drop model and must indicate a structure in which pairs of like particles can exist close together under attractive forces. If this effect is a general one there will be a correction when Z differs from N, since disregarding other effects the isobaric nucleus with $Z = N = A/2$ would be more stable. The resulting *asymmetry energy* may be estimated (see Ex. 6.1) from a model in which the nucleons fill the states of a Fermi distribution in accordance with the exclusion principle; it has the form $\gamma(A - 2Z)^2/A$ or $4\gamma T_z^2/A$ where T_z is the isobaric spin component (Sect. 5.6).

(c) The decrease of binding energy per nucleon from a maximum value of 8·7 MeV at $A = 60$, to 7·3 MeV at $A = 238$. This is due to the mutual electrostatic repulsion of the nuclear protons which does operate between all pairs because of the long-range nature of the force. In the liquid-drop model the corresponding *Coulomb energy* is simply the potential energy of a sphere of charge Ze whose radius is $\approx 1·25A^{1/3}$ fm (Sect. 6.2); it is written $\varepsilon Z^2/A^{1/3}$ (sometimes $Z(Z-1)$ rather than Z^2).

In heavy nuclei the disruptive effect of the protons is compensated by the attractive bonds associated with an increased number of neutrons. The stability of such nuclei will be affected by the nature of the spatial states occupied by the nucleons and the importance of shell-structure corrections is now established (Sect. 11.7).

(d) The existence of the pairing energy $P(A, Z)$ given by equation (6.5). If the general trend of masses is that of the odd-A nuclei, then for the even-A nuclei those with Z, N even are more stable, and those with Z, N odd are less stable, than the average. This correction is additional to the asymmetry energy and is written $\delta(A, Z)$ where $\delta \approx P(A, Z)$ and explicitly (Ref. 6.2):

$\delta = 0$ for A odd

$\delta = +12A^{-1/2}$ MeV for A even, Z odd, sometimes given as $33·5A^{-3/4}$ MeV

$\delta = -12A^{-1/2}$ MeV for A even, Z even, sometimes given as $-33·5A^{-3/4}$ MeV

If now all these energies are expressed in the same units, e.g. a.m.u. or MeV, the mass of the neutral atom may be written

$$M(A, Z) = ZM_H + NM_n - \alpha A + \beta A^{2/3} + \gamma (A - 2Z)^2/A$$
$$+ \varepsilon Z^2/A^{1/3} + \delta(A, Z) \quad (6.10)$$

where $\delta(A, Z)$ is negative for A even, Z even, zero for A odd, and positive for A even, Z odd. This is the *semi-empirical mass formula* giving the mass $M(A, Z)$ or the binding energy $B(A, Z)$ from equation (6.3) as a smoothly varying function of A and Z which may be fitted to the curve shown in Fig. 6.3. No corrections so far appear for shell structure.

Several sets of constants α, β, γ, ε have been proposed. With one set, due to A. E. S. Green, the values in MeV are $\alpha = 15 \cdot 8$, $\beta = 17 \cdot 8$, $\gamma = 23 \cdot 75$ and $\varepsilon = 0 \cdot 71$. The total binding energy of a nucleus of $Z = 100$, $A = 250$ would be

$$B = ZM_H + NM_n - M(A, Z)$$
$$= A[15 \cdot 8 - 2 \cdot 8 - 0 \cdot 95 - 4 \cdot 51 - \delta A^{-1}]$$
$$= A[7 \cdot 5 - \delta A^{-1}] \, \text{MeV}$$

showing that the corrections to the volume energy are substantial.

The use of the semi-empirical mass formula in indicating stability limits and in demonstrating the importance of shell structure will be mentioned in Sections 6.5 and 7.1.1. The next section will be devoted to a consideration of the origin of the volume energy term αA of the formula, since this must be a property of nuclear matter as well as of finite nuclei.

6.4 Nuclear matter

6.4.1 Characteristics

Nuclear matter is the result of an extrapolation of a typical heavy nucleus to a uniform infinite medium in which the numbers of neutrons and protons are equal, and no Coulomb forces exist. In such a system there is no preferred centre, and particle motions should be described by plane waves.

The physical properties of nuclear matter are summarized by stating the binding energy per particle E/A and the equilibrium density ρ, for which it is reasonable to assume that the binding energy reaches a maximum value. Numerical values of E/A are obtained by extrapolation from the semi-empirical mass formula (6.10) to the nuclear matter limit $A = \infty$, $N = Z$ which gives, with constants specified in Section 6.3,

$$E/A = (B/A)_\infty = \alpha = 15 \cdot 8 \, \text{MeV} \quad (6.11)$$

The density ρ is found from electron-scattering determinations of the central charge-density of heavy nuclei and is usually taken to be

$$\rho = 3/4\pi r_0^3 = 0\cdot17 \text{ nucleons fm}^{-3} \tag{6.12}$$

using a radius parameter $r_0 = 1\cdot1$ fm.

Although nuclear forces are strong, they are insufficiently strong to confer the properties of a solid on nuclear matter, and this substance will be regarded as a degenerate Fermi gas containing two varieties of fermion. It is then possible to give an alternative expression for the density, exhibiting the maximum momentum $p_F = k_F\hbar$ belonging to such a system; k_F is the wavenumber. In terms of k_F

$$\rho = 2k_F^3/3\pi^2 \tag{6.13}$$

as in the electron theory of metals, remembering that for a given type of nucleon, the density is $\rho/2$. For nuclear matter, equations (6.12) and (6.13) give $k_F = 1\cdot38 \text{ fm}^{-1}$.

In such a Fermi gas the average kinetic energy of a particle is

$$\langle T\rangle = \tfrac{3}{5}(\hbar^2 k_F^2/2m_p) = 23 \text{ MeV} \tag{6.14}$$

for $k_F = 1\cdot38 \text{ fm}^{-1}$ so that with a volume energy term of $15\cdot8$ MeV the attractive field in nuclear matter must provide a binding energy of $38\cdot8$ MeV. It is the task of nuclear-matter theory to relate this binding energy to the properties of the internucleon force.

6.4.2 Theory of the binding energy

The first important fact about nuclear matter as a Fermi gas is that the Pauli exclusion principle prevents the close approach of like nucleons in the same state of motion, so that the effective interaction is less than would be expected for particles obeying classical statistics. The Pauli principle has the further effect that an interaction between two nucleons, which we regard as a scattering (Fig. 6.4a), is only possible if both nucleons after the interaction occupy vacant momentum states. Since all states are filled up to a momentum $p_F = \hbar k_F$, such a scattering process involves a high momentum transfer, and is suppressed, at least for particles well down in the Fermi sea. It is for this reason that single-particle motion can be envisaged in nuclear matter and in finite nuclei where it allows the persistence of shell-model orbital motion (Ch. 7). The scattering process shown in Fig. 6.4 is virtual and energy conservation must be restored by further processes that return the interacting particles to their original states.

The Pauli exclusion principle operates for like fermions whatever the form of interaction between them. Nucleons both like and

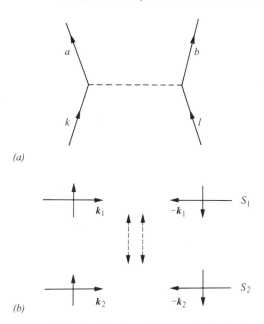

Fig. 6.4 (*a*) Particles k, l in the Fermi sea interact, leaving holes in the sea and becoming particles a, b above the sea. This is a virtual process contributing to the energy of nuclear matter. (*b*) Head-on collision of two nucleons with equal and opposite momenta and intrinsic spin.

unlike are also kept apart by the probable existence of a repulsive core of short range in their mutual interaction. This core may be very strong but the product of the core potential v and the actual wavefunction for two interacting particles ψ_{kl} will be finite if an interaction exists. We do not know what the actual wavefunction for the interacting pair is, but we may use a *model* wavefunction ϕ_{kl} and relate this to ψ_{kl} by the equation

$$G\phi_{kl} = v\psi_{kl} \qquad (6.15)$$

where G is an *effective potential* which, with suitable choice of ϕ_{kl} may be less singular than the actual potential v. For nuclear matter the model wavefunctions ϕ_{kl} will be constructed from plane waves; for finite nuclei, oscillator functions are more appropriate.

The use of the effective potential in nuclear matter and nuclear model calculations is discussed in References 6.3 and 6.4. The key techniques in these calculations were developed by Brueckner, Bethe and Goldstone, who set up an operator equation for the effective potential G in the form

$$G\phi_k\phi_l = v\phi_k\phi_l + [vQ_{kl}/(e_{kl} - H_0)]G\phi_k\phi_l \qquad (6.16)$$

189

where ϕ_k, ϕ_l are single-particle wavefunctions in a potential well U that represents the average field in nuclear matter, $H_0 = H_0(1) + H_0(2)$ is the unperturbed Hamiltonian operator, giving the energy of the two nucleons in the absence of correlations induced by their interaction, e_{kl} is the perturbed energy of the pair, and Q_{kl} is the Pauli factor that ensures that the particles occupy states permitted by the exclusion principle. If G is calculated from (6.16) generalized to a system of A particles, and used in the Schrödinger equation with any suitable H_0 and U, the energy of nuclear matter then involves the matrix elements of G between uncorrelated states, which are known, rather than the matrix elements of v between actual correlated states, which are unknown.

The details of these calculations will not be given here, since they are algebraically complex. It is found that the internucleon force introduces most perturbation in the s-waves, but even so the correlation disappears within a distance of about 2 fm. The tensor component of the effective internucleon force and the existence of exchange components in the central force jointly produce the nuclear matter saturation effect at the postulated equilibrium density. Typical results of matter calculations, for the so-called Reid soft-core potential, are given in Table 6.2, which shows the effect of introducing three-body and four-body effects. The uncertainties in some of the contributions to binding energy are such that the differences from the experimental value of 15·8 MeV is not unexpected. It may be concluded that the energy and density of nuclear matter are consistent with the properties of actual forces. It will also be noted from the table that the effective interaction in nuclear matter is density-dependent.

TABLE 6.2 Energy of nuclear matter (MeV) (from Ref. 6.4, p. 297)

k_F fm^{-1}	1·2	1·36	1·6
Two body	−27·71	−34·02	−42·05
Three body	−1·6	−1·59	−1·1
Four body	−0·9	−1·1	−1·5
Kinetic energy	17·92	22·97	31·85
E/A after corrections	−13·5	−14·1	−10·4

If nuclear matter with an unequal number of neutrons and protons is analysed in the same way, the coefficient of the asymmetry term in the semi-empirical mass formula may be found; it is also consistent with the experimental value (Sect. 6.3).

6.4.3 The pairing energy

On the assumption that the methods outlined in the previous section permit the definition of a short-range effective interaction in nuclear matter, despite the existence of a repulsive core, one may further enquire whether any effects remain that may be attributed to the longer-range attractive parts of the internucleon force that are not included in the previous calculation. One such effect gives rise to the *pairing energy* $\delta(A, Z)$ that appears in the semi-empirical mass formula (6.10). This has little to do with the main distribution of nucleons in nuclear matter, but is of special significance in finite nuclei. It arises in a particular type of collision between like nucleons (Fig. 6.4b) in which the colliding pair has zero resultant linear momentum (in nuclear matter) or angular momentum (in a nucleus) and oppositely directed spins. The collision transfers the pair from a state S_1 with momentum vectors k_1 to a state S_2 with momentum vectors k_2, conserving momentum; further collisions return the pair to S_1, conserving energy. If the force between the particles is attractive, then it is found, as in the case of a superconducting metal, that a state containing such pairs is lowered in energy.

To see the origin of such an effect we consider for simplicity two like nucleons outside the Fermi sphere and we specify a small number of plane-wave two-particle states $S_1, S_2, \ldots, S_n$ which they may occupy. The two-particle wavefunctions, in the absence of interaction, will be written $\phi_1, \ldots, \phi_n$ and the *single*-particle kinetic energies will be taken as those corresponding with plane waves of momentum vector k, i.e. $e_1 = \hbar^2 k_1^2 / 2m$, etc. The wavefunctions ϕ are solutions of the field-free Schrödinger equation

$$H_0 \phi = E_0 \phi \tag{6.17a}$$

and $E_0 = 2e_1$, etc. in this simple case.

When an interaction V exists, i.e. when collisions take place, the Schrödinger equation becomes

$$(H_0 + V)\psi = E\psi \tag{6.17b}$$

where the perturbed wavefunction ψ may be expanded in terms of the ϕ-functions in the form

$$\psi = \sum_{i=1}^{n} c_i \phi_i \tag{6.18}$$

The perturbed energies E are then obtained by substituting in (6.17b), multiplying through by a chosen ϕ_k^*, and taking an integral over all space using the orthogonality property of the ϕ's. This gives, using also equation (6.17a) for the unperturbed energies $2e_k$,

$$(2e_k - E)c_k + \sum_{.1}^{n} c_i \langle \phi_k^* | \, V \, | \phi_i \rangle = 0 \tag{6.19}$$

where

$$\langle \phi_k^* | \, V \, | \phi_i \rangle = \int \phi_k^* V \phi_i \, \mathrm{d}^3 r \tag{6.20a}$$

191

We now assume that the interaction is confined to states with momenta very close to the Fermi momentum and that for all such states

$$\langle \phi_k^* | V | \phi_i \rangle = -F \tag{6.20b}$$

and is zero otherwise. We then have

$$c_k = CF/(2e_k - E) \tag{6.21}$$

where $C = \sum_1^n c_i$ and an eigenvalue equation is obtained in the form

$$1/F = \sum_1^n 1/(2e_k - E) \tag{6.22}$$

This may be solved graphically (Fig. 6.5) and gives the actual energies E in relation to the unperturbed energies $2e_k$. The figure shows that the lowest eigenvalue is shifted downwards more than the others, because of the form of equation (6.22), and if these latter states are close together the downward shift causes a gap in the sequence.

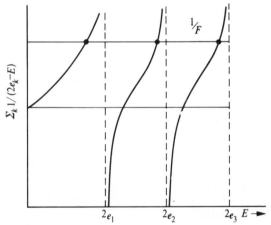

Fig. 6.5 Lowering of 'ground state' energy of two nucleons by the pairing interaction. The figure is a graphical solution of equation (6.22), in which the dotted lines represent unperturbed energy eigenvalues. The dots show the actual eigenvalues and indicate an increased displacement for the ground state (adapted from Ref. 6.4).

This treatment may be generalized and extended to the many-body problem with similar results. It is shown in Reference 6.4 that if V has the simple form of a delta-function interaction, the matrix element (6.20) indeed has a constant value and all pairs contribute in phase to a lowered energy.

In what follows we consider explicitly the case of finite nuclei, because the spectrum of basic states near the Fermi energy accentuates pairing effects in comparison with those same effects in nuclear matter. The main consequences of pairing are that:

(*a*) The Fermi surface is no longer sharp and both holes below it and particles above it can co-exist. (Excitations that are mixtures of a particle state and a hole state are known as *quasiparticles*.)

(*b*) An *energy gap*, usually given as 2Δ, arises between the ground state and first excited state of many even–even nuclei (though not those that will be found in Ch. 7 to have closed major shells). Such gaps are known experimentally, and their appearance led A. Bohr, Mottelson and Pines originally to suggest that the ground state might exhibit superfluid properties. The difference between this state and other states with a similar constitution in terms of basic states is simply that there is a greater degree of coherence in the nuclear motion.

An odd–odd nucleus has two unpaired and dissimilar nucleons and its ground state may be expected to be displaced by an energy of the order of 2Δ above that of the isobaric even–even nucleus for this reason. From the semi-empirical mass formula, equation (6.10), we see that this difference is expressed by the quantity $2\delta(A, Z)$ so that δ is equal to half the energy gap. The pairing energy $P(A, Z)$ defined in equation (6.5) is then also approximately equal to Δ which is seen from Table 6.1 to be about 1 MeV at $A = 134$.

6.5 Nuclear stability

If nuclei are arranged in order of increasing N and Z on a rectangular grid, all nuclides of the same N (isotones) appear in, for example, horizontal lines and all nuclides of the same Z (isotopes) in vertical lines. This neutron–proton diagram, sometimes known as the Segrè chart, is shown for the naturally occurring nuclides in Fig. 2.1, p. 51. Stable nuclei of low mass number have $N = Z = A/2$ but medium-weight and heavy nuclei have many excess neutrons. The unstable nuclides mainly form a fringe to the band of stability but, for some elements, intermingle with the stable isotopes.

The semi-empirical mass formula is, of course, constructed to ensure that the stability of known nuclei is correctly predicted. For odd-A nuclei for instance, for which δ is zero, the most stable nucleus is that with an atomic number Z given by the setting $\partial/\partial Z[M(A, Z)] = 0$. From equation (6.10), remembering that $N = A - Z$, and using the constants due to Green

$$Z = A/(1 \cdot 98 + 0 \cdot 015 A^{2/3}) \qquad (6.23)$$

which shows that Z falls short of $A/2$ as A increases. Physically, this is due to the fact that Coulomb repulsion is a long-range force and that it becomes increasingly important with respect to short-range nuclear forces as the nuclear charge increases. The repulsive effect has to be balanced by the presence of extra neutrons. These provide extra volume energy, but they must also conform with the Pauli principle and the extra kinetic energy (a disruptive effect) is represented by the asymmetry term. If, in fact, the variation of $B(A, Z)$ with A for constant Z, i.e. the dependence of B on neutron number is examined, equation (6.10) shows that $[\partial B/\partial A]_Z$ vanishes for $Z/A \approx \frac{1}{3}$, i.e. when there are twice as many neutrons as protons, additional neutrons do not contribute extra binding. If, however, the nucleus becomes very large indeed, so that gravitational forces can play a part, stability will again result in a body consisting mainly of neutrons; this is a *neutron star*.

If a third coordinate representing the atomic mass $M(A, Z)$ is added to the neutron–proton diagram, a three-dimensional *mass surface* is defined. Stable nuclei group round the bottom of a valley in this surface. Intersections of the mass surface by planes of constant A pick out groups of isobars comprising both stable and unstable nuclei. Formula (6.10) shows that for A constant, $M(A, Z)$ is a parabolic function of Z (Fig. 6.6). For odd A, one parabola is obtained but for even A there are two, separated by an energy $2\delta(A, Z)$.

From Fig. 6.6 there is just one stable isobar for A odd but for even A there may be two or three stable species. Stability is reached in an isobaric sequence by electron (β^-) or positron (β^+) emission or

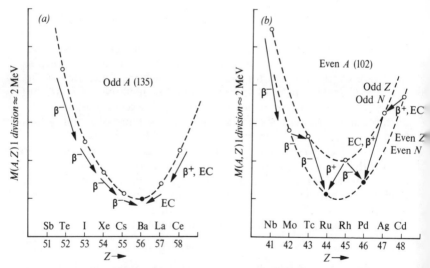

Fig. 6.6 Stability of isobars, showing atomic mass M plotted against atomic number Z. (*a*) Odd A. (*b*) Even A. Electron capture (EC) is in principle also possible whenever β^+-emission may take place (Ref. 2.1).

electron capture (EC) as indicated in the figure. Such radioactive decay processes are possible for a nuclide of atomic mass $M(A, Z)$ if this quantity is related to the mass $M(A, Z \pm 1)$ of the product nuclide by the following inequalities:

For electron decay to $M(A, Z+1)$ $M(A, Z) > M(A, Z+1)$

For positron decay to $M(A, Z-1)$ $M(A, Z) > M(A, Z-1) + 2m_e$
where m_e is the electron mass in a.m.u.

For atomic electron capture to $M(A, Z-1)$ $M(A, Z) > M(A, Z-1) + B_e$
where B_e is the ionization energy (in a.m.u.) for the electron captured in the neutral atom of charge $Z-1$.

$$(6.24)$$

These relations assume that the mass of the neutrino that is known to be emitted in such semi-leptonic processes (Sect. 2.1.6) is zero. If $\Delta M = [M(A, Z) - M(A, Z \pm 1)]\,\text{MeV}/c^2$ then the Q-values in MeV for the processes (6.24), defined in accordance with equations (1.7) and (1.20) as the kinetic energy available for all product particles (or quanta), are:

$$Q_{\beta^-} = \Delta M$$

$Q_{\beta^+} = \Delta M - 2m_e c^2$ (because the change of Z frees an atomic electron in addition to the decay positron) (6.25)

$$Q_{EC} = \Delta M - B_e$$

Sometimes, however, the Q-value for a positron decay process is quoted as $Q = \Delta M$, and then a $2m_e c^2$ adjustment, namely $1 \cdot 02\,\text{MeV}$, must be made in calculating the maximum *positron* energy. In each case, if the decay process leads to a state of excitation E_γ in the final nucleus, the Q-values are reduced by E_γ. The actual observable kinetic energy for a given Q is subject to a correction for c.m. motion, but this is very small for leptonic processes (see Ex. 10.3).

The width of the valley in the mass surface in which stable nuclei are found is determined by the criteria of β-stability (6.24). Other types of spontaneous decay such as α-particle emission may be energetically possible, but are often highly retarded by barrier effects. For $Z > 83$, however, the β-stable nuclei are actually α-active.

6.6 Nuclidic abundance (Ref. 6.5)

The abundance of the chemical *elements* in nature does not differ greatly, on present evidence, between the solar system and the rest of the universe. The standard solar-system number abundance is conventionally related to the number of silicon atoms ($Si \equiv 10^6$) and is derived in large part from observations on meteorites, since terrestrial samples have been subject to chemical dispersion during geological time. For the stars and for interstellar space, spectroscopic and cosmic-ray observations are used. The interest in this data is its connection with the processes of element building through nuclear reactions that have taken place in stars. The end-products of this nucleosynthesis also contain information relevant to nuclear stability rules and to nuclear structure.

Such information is best sought in the abundance figures for specific *nuclides* (N, Z), which can be obtained from the elemental data using isotope ratios measured in both meteoritic and terrestrial samples. The main conclusions from both the elemental and nuclidic abundances are that:

(*a*) The elements of highest abundance in the universe are hydrogen and helium.

(*b*) There is a deficiency of the elements Li, Be and B in solar-system abundances presumably because the corresponding nuclides are readily consumed in nucleosynthesis. The cosmic rays, however, contain larger relative quantities of these nuclei, because they can be produced by interactions in interstellar space.

(*c*) Beyond the light elements a marked peak in the abundance of the even-Z, even-N nuclei occurs around $A = 56$ (Fig. 6.7) which correlates with the maximum binding energy shown for this A-value in Fig. 6.3. These are the most stable elements, and would be formed as end-products of an element-building stage in which thermal equilibrium was reached.

(*d*) Beyond the iron group, distinct peaks are seen at A-values corresponding with neutron numbers $N = 50$, 82 and 126. If these nuclides are built from lighter species by successive neutron capture and β^--decay then such peaks will be expected when a shell closes because the next neutron is less tightly bound (Ch. 7). Arguments can be given to relate the additional peaks at $A = 80$, 130 and 194 to the same shell structure.

Further evidence for shell structure in nuclei from abundance data comes from an examination of the distribution of *isotopes* (Z constant) and *isotones* (N constant) with respect to the line of stability shown by Fig. 2.1, p. 51. It is found that

(*e*) The number of stable and long-lived isotopes is greater for $Z = 20$, 28, 50 and 82 than for nearby elements.

(*f*) The number of stable and long-lived isotones is greater for $N = 20$, 28, 50, 82 and 126 than for nearby N-values.

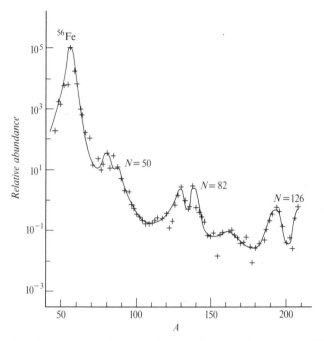

Fig. 6.7   Relative abundance (number of atoms) for even Z, even N nuclides as a function of mass number A for $A > 50$ (Ref. 6.2).

The distribution of stable nuclides according to individual neutron and proton numbers is given in Table 6.3 (taken from Ref. 2.1). The figures in the table illustrate the effect of the general properties of the nucleon–nucleon interaction, operating in nuclear matter. Thus, the approximate equality of number of odd-Z – even-N and odd-N – even-Z nuclides suggest the equivalence of neutrons and protons in element-building reactions (charge symmetry). The preponderance of even-Z – even-N nuclides is a direct consequence of the pairing force. The presence of two *unpaired* nucleons, as in nuclides of odd Z and odd N, results in instability in all but the lightest nuclei ^{2}H, ^{6}Li, ^{10}B and ^{14}N.

TABLE 6.3   Distribution of stable nuclides (Ref. 2.1)

A	Z	N	Number of cases
odd	odd	even	50
odd	even	odd	55
even	odd	odd	4
even	even	even	165

197

6.7 Summary

The most important 'bulk' properties of a nucleus are its mass and its size, to which all constituent nucleons contribute. Definitions are given of the atomic mass unit and of binding energy, separation energy and pairing energy. The binding energy is described by a semi-empirical formula and some understanding of the origin of its main attractive term has been reached by nuclear matter calculations. The particular role of pairing effects in the low-lying states of finite nuclei is emphasized.

Examples 6

6.1* Write down the total kinetic energy of a gas containing N neutrons and Z protons, regarding these as independent particles. By comparing this energy with that for the case $Z = N = A/2$, deduce the form of the asymmetry energy.

6.2* The scattering of α-particles through $60°$ by lead ($Z = 82$) begins to deviate appreciably from the predicted Coulomb value at an energy of 27 MeV. Estimate a radius for the target nuclei.

6.3 What is the mean-square radius of a square-well charge distribution of radius R? $[\frac{3}{5}R^2]$

6.4 The energy release in the reaction $A(d, p)B$ is Q. Show that the neutron separation energy for nucleus B is $S_n = Q + \varepsilon$ where ε is the binding energy of the deuteron.

6.5* Calculate the level shift of the electron in a hydrogen atom due to the finite size of the nucleus, assuming a proton radius $R = 0.6 \times 10^{-15} \text{ m}$. (Use first-order perturbation theory, and take the hydrogenic wavefunction to be $(1/\pi a_0^3)^{1/2} \exp(-r/a_0)$.)

6.6 Oxygen has isotopes of mass numbers 16, 17, 18 with relative abundance 99.758, 0.0373 and 0.2039. If Avogadro's number on the physical scale with $^{16}O = 16$ is 6.0249×10^{23} calculate its value on the chemical scale ($O = 16$). Find also the value of this constant on the physical scale with $^{12}C = 12$ if the mass of ^{12}C on the ^{16}O scale is 12.0039. $[6.0233 \times 10^{23}, 6.0229 \times 10^{23}]$

6.7 Calculate the binding energy B, the binding energy per nucleon B/A, and the neutron and proton separation energies S_n, S_p for ^{114}Cd given that the mass excesses (i.e. the difference between the mass in a.m.u. and the mass number A) are $p = 7.29 \text{ MeV}$, $n = 8.07 \text{ MeV}$, $^{114}Cd = -90.01 \text{ MeV}$, $^{113}Cd = -89.04 \text{ MeV}$, $^{113}Ag = -87.04 \text{ MeV}$. $[972.55, 8.53, 9.04, 10.26 \text{ MeV}]$
Compare your results with prediction from the semi-empirical mass formula with the constants given in the text.

6.8 Consider a large spherical nucleus of radius R containing A nucleons. At what total mass would the gravitational energy per nucleon equal a nuclear kinetic energy $\hbar^2/2m_p r_0^2$ with $r_0 = 1.2 \text{ fm}$? $[\approx 10^{29} \text{ kg}]$

6.9 Show that in a Fermi gas of neutrons the number of degenerate states with energy less than $\hbar^2 k^2/2m_n$ is $(2/9\pi)(kR)^3$ where R is the linear dimension of the containing volume.

6.10 No decay process linking the nuclei $^{150}_{60}Nd$ and $^{150}_{61}Pm$ is known. The (atomic) mass difference between the nuclei $^{150}_{60}Nd$ and $^{150}_{62}Sm$ has, however, been measured as 3633 ± 4 micromass units on the carbon scale and the accepted decay energy for the beta process $^{150}_{61}Pm \rightarrow ^{150}_{62}Sm$ is $3.46 \pm 0.03 \text{ MeV}$. Determine the relative stability of ^{150}Nd and ^{150}Pm and the decay energy. Suggest a possible mode of decay. $[76 \pm 30 \text{ keV}, \text{EC}]$

6.11 Given that the mass defect curve falls from $+0.14$ mass units for uranium to

198

−0·06 mass units in the middle of the periodic table, estimate in kWh the energy which could theoretically be obtained from 1 kg of ^{235}U. $[2·7 \times 10^7]$

6.12 Assuming that the binding energy of an even-A nucleus may be written

$$-B(A, Z) = f(A) + \frac{0·083}{A}(A/2 - Z)^2 + 0·000\,627 Z^2/A^{1/3} \pm 0·036 A^{-3/4} \text{ a.m.u.}$$

(+ for Z odd, − for Z even) determine the number of stable nuclides of mass $A = 36$. Take $A^{1/3} = 3·3$, $A^{1/4} = 2·45$ and confine your attention to the range $13 \leqslant Z \leqslant 20$. [2]

6.13 In the carbon atom the K-edge is at 300 eV and the L_I-, L_{II}-edges at 60 eV. The atomic mass is 12·0000. Calculate:
   (a) the mass of the carbon nucleus in a.m.u. neglecting electron binding [11·996 71]
   (b) the percentage correction introduced when electron binding is allowed for. $[10^{-5}]$

6.14 The tabulated masses of $^9_3$Li, $^9_4$Be and $^9_5$B are 9·0268, 9·0122 and 9·0133 a.m.u. Calculate the value of the asymmetry coefficient in the semi-empirical mass formula, equation (6.10), ($M_H = 1·007\,83$, $M_n = 1·008\,67$ a.m.u.). [16 MeV]

6.15 From observation of the decay fragments of the hypernucleus ^{11}B it is found that the binding energy of the Λ-particle in this nucleus is 10·24 MeV. If the mass of ^{10}B is 10·0129 a.m.u., what is the mass of $^{11}_\Lambda$B? ($m_\Lambda = 1115$ MeV/c^2). [11·1995 a.m.u.]

6.16 Estimate the energy of the μK X-ray for a phosphorus nucleus ($Z = 15$), given that the ionization potential for hydrogen is 13·5 eV. [472 keV]

6.17 For any given muonic atom, calculate the principal quantum number of the muonic orbit that is just inside the electronic K-shell. [14]

6.18* For a uniform-density nucleus of (sharp) radius R show that the form factor for electron scattering with momentum transfer $q\hbar$ is, apart from a normalizing factor, $F(q) = [\sin qR - qR \cos qR]/q^3$. Compute the quantity $|F(q)|^2$ for points in the range $0·1 < qR < 5·0$.

7 The single-particle shell model and nuclear moments

In the previous chapter it has been seen that despite the strength of the interaction between nucleons, it is possible to envisage independent particle motion within nuclear matter. Obviously, the effect of surrounding nucleons cannot be completely forgotten and a realistic discussion of nuclear properties must certainly take account of the pairing interaction at least. Despite this, a simple nuclear model based on the concept of the motion of non-interacting fermions in a spherically symmetrical potential well with a mean free path greater than the radius of the system, has been extremely useful in coordinating a wide range of nuclear phenomena. The model was developed from earlier proposals by Mayer and by Haxel, Jensen and Suess in 1949; since for a spherical system the basic states have wavefunctions that are eigenfunctions of the angular momentum operator, both orbital and total angular momentum are of crucial significance and lead to the definition of the quasi-atomic shells and subshells from which the model takes its name.

In this chapter the empirical evidence that strongly suggests the relevance of shell structure in nuclei will be reviewed, and the simplest interpretation in terms of a potential model will be presented. Some inadequacies of the single-particle model will be mentioned as an introduction to the extended models discussed in Chapter 8. The importance of the single-particle shell model in defining the nucleon configurations from which realistic wavefunctions of nuclear states are constructed will be indicated.

7.1 Empirical evidence for the regularity of nuclear properties

7.1.1 Nuclear mass and binding energy

Although the total binding energy of nuclei is not greatly affected by the character of the motion of the last few ('valence') nucleons, discontinuities are seen when accurately measured masses are compared with the predictions of a smoothly varying mass formula, e.g. equation (6.10). Figure 7.1 shows a plot of this difference against

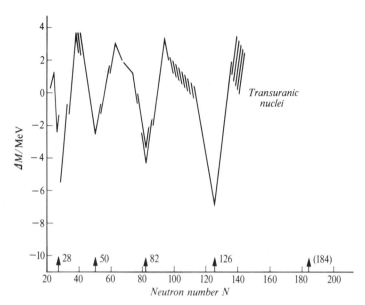

Fig. 7.1   Periodicity of the nuclidic masses. ΔM is the observed mass less that predicted by a smoothly varying formula (adapted from Kümmel, H. *et al.*, *Nuclear Physics*, **81,** 129, 1966).

neutron number and discontinuities are seen at $N = 28$, 50, 82 and 126. This type of effect is reminiscent of the periodicity seen in plots of atomic ionization potentials against atomic number, and like that effect, is interpreted as due to the formation of closed shells of particles. A similar periodicity at the same nucleon numbers up to 82 is seen if nuclidic mass differences are displayed as a function of atomic number Z, so protons and neutrons behave similarly in this respect.

The same discontinuities are seen, without reference to the empirical mass formula, in the nucleon separation energies S_n, S_p. The neutron separation energy S_n near $N = 82$ is shown in Fig. 7.2; for N values just above 82 the separation energies are lower, allowing for pairing effects, than for N just less than 82.

Abundance data, reflecting nuclear stability, have already been mentioned (Sect. 6.6). There is correlation with neutron numbers $N = 50$, 82 and 126 in the sense that nuclides with slightly greater N values are relatively less abundant (Fig. 6.7) and this is again a manifestation of shell closure. The number of isotopes and isotones as a function of Z or N gives similar evidence and additionally indicates that the potential wells are similar for neutrons and protons since the favoured nucleon numbers are the same in each case.

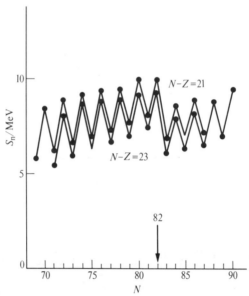

Fig. 7.2 Neutron separation energies near $N = 82$. The lines join nuclides of constant neutron excess $N - Z$. The values are calculated from equation (6.4a) and show odd–even staggering. Reproduced from *Nuclear Structure*, Vol. I, 1969, by Aage Bohr and Ben R. Mottleson with permission of the publishers, Addison-Wesley/W. A. Benjamin Inc.

The semi-empirical mass formula predicts the energy release in α- and β-decay to be a smooth function of A. In fact, these energies show non-monotonic variations, although this is essentially the same effect as that shown in Fig. 7.1.

The striking effects of corrections that must be ascribed to shell structure on the barriers impeding the fission of heavy nuclei are recent and strong evidence for the shell model. The double-humped barrier, leading to dynamical effects sometimes known as fission isomerism, will be described later (Sect. 11.7).

7.1.2 Nuclear levels and moments

Even–even nuclei, in which nucleons are paired, are best for displaying systematic effects. Their ground states without exception have spin zero and even parity ($I^\pi = 0^+$) and in most examples the first excited state has $I^\pi = 2^+$. The excitation energy of these 2^+ states gradually decreases with increasing A but also shows maxima at shell closures.

Another special class of nuclear levels showing periodicity is that of the isomeric states (Sect. 9.2.2). These are β- or γ-decaying states for which the radiative transition is of long life because of low energy or large spin change. The states occur in well-defined groups below the 'magic numbers' (Fig. 9.6).

At higher excitations, the density of nuclear levels also shows shell effects, closed-shell nuclei having a much smaller density because of the extra energy required to change the state of motion of a core nucleon.

The ground state spins and parities of nuclei and their electromagnetic moments show changes that are sometimes abrupt at shell closures. This is a consequence of a basic change of wavefunction.

7.1.3 Conclusion

The evidence presented in preceding sections of the present chapter suggests that nuclei containing 20, 50, 82 or 126 neutrons or protons are particularly stable. Some phenomena also suggest the addition of the number 28. Before considering how the shell model is able to predict these 'magic numbers' we examine some direct evidence for the model that has become available in recent years.

7.1.4 Direct evidence for nuclear shells

The existence of independent particle motion in nuclei has received striking confirmation by the direct observation of proton–hole states through the (p, 2p) reaction. In this process protons of energy 50–400 MeV are used to bombard a nucleus, e.g. ^{12}C, producing a *knock-out* direct reaction

$$^{12}\text{C} + \text{p} \rightarrow ^{11}\text{B} + 2\text{p} \tag{7.1}$$

The outgoing proton energies are measured in coincidence and geometrical conditions can be imposed which minimize energy loss to the recoil nucleus ^{11}B. The overall loss of energy in the reaction then measures the binding of the knocked-out proton in the ^{12}C nucleus. Figure 7.3 shows some results obtained for this nucleus in the form of a scatter plot relating the energy of one proton to that with which it is associated in the reaction. The distinct band at 45° to the axis indicates that for these events the sum of the proton energies is constant and this means that there is a well-defined binding energy (here ≈ 16 MeV) for the struck proton; for ^{12}C this is interpreted as the energy necessary to form a hole in a p-shell ($l = 1$). The figure also shows a much broader distribution, corresponding with binding energies between 30 and 40 MeV which may be associated with the formation of holes in the s-shell ($l = 0$). The width of the distributions arises because the single-hole states are, in fact, fragmented, in the sense that their strength is spread over a range of states in the nuclear spectrum; the distribution in fact gives the *strength function* (Sect. 11.3) for the hole states.

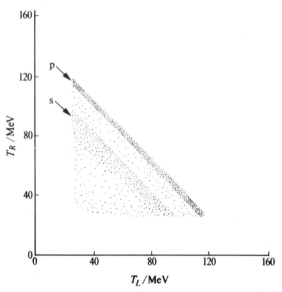

Fig. 7.3   Scatter plot of the energies T_L, T_R of proton pairs from the ^{12}C (p, 2p)^{11}B reaction observed at ±42·5° with a beam of 160-MeV protons (adapted from Gottschalk, B. *et al.*, *Nuclear Physics*, **A90**, 83, 1967).

By altering the angle with respect to the incident beam at which the two protons are observed, the momentum distribution of the struck proton may be explored and the angular momentum of the corresponding bound state may be inferred. The results are consistent with the p and s assignments for ^{12}C. Similar results have been obtained for other and heavier nuclei and with other bombarding particles, including electrons. Evidence for other angular momentum states has been obtained and altogether this type of reaction offers entirely convincing evidence for the existence of independent-particle motion within nuclei.

Single-particle states can also be investigated by the formation of hypernuclei (Sect. 2.4). If a nuclear neutron is changed to a Λ-particle by the reaction $K^-(n, \pi^-)\Lambda^0$ the energy spectrum of the outgoing π^- gives the binding energy of the Λ in the target nucleus. This spectrum has peaks corresponding with the neutron states, at least for light nuclei.

7.2 The single-particle shell model

7.2.1 Objectives

The first objective of the model is clearly to understand how the magic stability numbers of 20, 28, 50, 82 and 126 neutrons or protons arise. To these numbers it is usual to add 2 (the alpha particle) and 8 (oxygen-16) although these are numbers that arise very naturally among the light nuclides because of the simple nature of $l = 0$ and $l = 1$ levels, already known in atomic physics. If a satisfactory orbital sequence can be established then the regular filling of orbits in accordance with the Pauli principle will be expected to account for the main features of the properties discussed in Section 7.1.2.

A second objective is to ensure that the potential well adopted to account for the sequence of orbits is consistent with the observed binding energies of the single-particle levels, including virtual levels, and with the general properties of nuclear matter as determined by the internucleon interaction.

Finally, the model must indicate the directions in which it may be improved.

7.2.2 Sequence of orbits

The starting-point of all shell models is the solution of the Schrödinger equation for a particle moving in a spherically symmetrical *central* field of force, i.e. in a field in which the potential energy V of the particle with respect to the centre is a function $V(r)$ of radial distance r only. This problem has been examined generally in Section 1.4, with specific attention to the Coulomb and harmonic oscillator potentials.

Neither of these potentials produces shell closures at the magic numbers, and the same is true of other simple well shapes such as the square well. For the pure harmonic oscillator, for instance, the level order indicated in Section 1.4.2 is

$$(n, l): \text{1s; 2p; 2s, 3d; 3p, 4f;} \ldots$$

where n is the principal quantum number, or if n is redefined to be the radial quantum number n_r

$$(n, l): \text{0s; 0p; 1s, 0d; 1p, 0f;} \ldots$$

which gives shell closures at particle numbers

$$2, 8, 20, 40, 70, 112 \tag{7.2}$$

The average spacing of these levels can be obtained by choosing the basic oscillator frequency ω so that the expectation value of r^2 is equal to the mean-square nuclear radius (cf. Ex. 7.7). This gives

$$\hbar\omega \approx 40A^{-1/3} \text{ MeV} \tag{7.3}$$

and this expression is useful in tracing the systematics of single-particle states.

An essential addition to the shell-model potential, made in 1949 by Mayer and by Haxel, Jensen and Suess was a velocity-dependent term usually known as the *spin-orbit* coupling. Such an effect has already been noted as a necessary component of the two-nucleon interaction (Sect. 5.7). Within the constant-density nuclear volume a force such as this may be expected to average out, but in the nuclear surface there is a density gradient and, therefore, a direction with respect to which the orbital angular momentum $\mathbf{r} \times \mathbf{p}$ of a single nucleon may be defined. The assumption is that the spin-orbit potential depends on the relative direction of the orbital vector and the intrinsic spin vector so that it may be written

$$V_{ls}(r) = V_{ls}\left(\frac{\hbar}{m_\pi c}\right)^2 \mathbf{l} \cdot \mathbf{s}\left(\frac{1}{r} \cdot \frac{\mathrm{d}f}{\mathrm{d}r}\right) \tag{7.4}$$

where f gives the profile of the central potential (see eqn (7.12)) and $\mathbf{l}, \mathbf{s}$ are the orbital and spin momenta divided by $\hbar$. The radial factor arises naturally if the nucleon–nucleus spin-orbit potential is derived (Ref. 6.2) from the two-body spin-orbit potential V_{LS} of equation (5.68); the Compton wavelength of the pion is introduced for dimensional reasons.

The spin-orbit doublet levels for a single nucleon have total angular momentum quantum numbers given by

$$j = l \pm \tfrac{1}{2} \tag{7.5}$$

Since the observed spacing between the two components of the doublet is given (Ref. 6.2) by

$$E_{(l+1/2)} - E_{(l-1/2)} = -10(2l+1)A^{-2/3} \,\text{MeV} \tag{7.6}$$

it is necessary that the state with the larger j shall be the more stable. It is also necessary that the splitting shall increase with l (Exs. 7.4 and 7.6). Such a large displacement cannot originate in electromagnetic effects, as it does in atomic spectra.

With these specific assumptions about the spin-orbit coupling the states of the oscillator potential are enumerated as follows:

$(nl_j):$ $0s_{1/2}$; $0p_{3/2}0p_{1/2}$; $0d_{5/2}1s_{1/2}0d_{3/2}$; $0f_{7/2}$; $1p_{3/2}0f_{5/2}1p_{1/2}0g_{9/2}$; ...

where n is the radial quantum number n_r, or using the serial number $(n_r + 1)$ which indexes the order of appearance of a particular l-value in the level sequence:

$(nl_j):$ $1s_{1/2}$; $1p_{3/2}1p_{1/2}$; $1d_{5/2}2s_{1/2}1d_{3/2}$; $1f_{7/2}$; $2p_{3/2}1f_{5/2}2p_{1/2}1g_{9/2}$; ...

$$\tag{7.7}$$

This order assumes that some anharmonicity or deviation from the pure oscillator-well shape already lowers the states of high angular momentum for a given oscillator number. Each state of given j may accommodate $(2j+1)$ neutrons and $(2j+1)$ protons, although the well parameters will differ for these two particles.

The scheme (7.7), which will be used in the book, shows that the $1g_{9/2}$ level $(N=4)$ is lowered so much by the spin–orbit attractive force that it merges with the levels of oscillator number $N=3$, of opposite parity and similarly for higher N-values. Major shell closures, giving a substantial energy separation, then occur at neutron or proton numbers

$$2, 8, 20, 28, 50, 82, 126 \qquad (7.8)$$

as required by experiment. Between these magic numbers, other shells (subshells) will be completed, with smaller energy differences. The sequence of levels is shown for neutrons or protons in Fig. 7.4 for a well intermediate in shape between a harmonic oscillator and a square well. When both neutrons and protons complete a major shell we speak of a *double-closed-shell* nucleus, e.g. $^{40}_{20}\text{Ca}$, $^{208}_{82}\text{Pb}$.

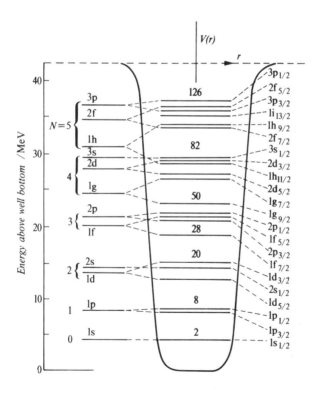

Fig. 7.4 Single-particle levels in a potential well, allowing for spin–orbit coupling. If this effect is sufficient to overcome the oscillator spacing, the magic numbers result.

Continuation of the shell structure shown in Fig. 7.4 to higher nucleon numbers suggests a double-closed shell at $Z = 126$, $N = 184$. However, more detailed considerations (Sect. 11.7.2) indicate that *superheavy elements* of long life against all forms of decay may be found in an 'island of stability' near $Z = 114$, $N = 184$. The nucleus of maximum stability ($^{294}_{110}$X) might have a lifetime comparable with the age of the Earth, and its presence in existing minerals is, therefore, not excluded.

The importance of the simple shell model, apart from giving a general description of nuclear properties in broad outline at least, is that it provides a sequence of single-particle states classified according to n (or n_r), l and j. In atomic spectroscopy, a group of x electrons with the same n- (in this case the principal quantum number) and l-values form a pure *configuration* $(nl)^x$ and the coupling together of their spins and orbital momenta in the Russell–Saunders scheme leads to a number of *terms* denoted by symbols ^{2S+1}L where S is the resultant spin and L is the resultant orbital momentum. The further coupling of spin and orbital vectors to give a resultant J leads to the possible *levels* $^{2S+1}L_J$ and if the magnetic quantum number is specified a level resolves into $2J+1$ *states* $^{2S+1}L_{J,m_J}$. The same definitions can be applied to identical nucleons, but because of the importance of spin-orbit coupling in the shell model a pure configuration specifies the j-value in the form $(nl_j)^x$ e.g. $(2d_{5/2})^3$ and there is no distinction between terms and levels, e.g. $(2d_{5/2})^3_{5/2}$ or $(2d^3_{5/2}; I = \frac{5}{2})$. A state may be written $(2d^3_{5/2}; I = \frac{5}{2}, m_I = \frac{3}{2})$. Unfortunately, it is difficult to confine the usage of the word 'state' to this precise definition; it will often imply a linear combination of terms.

The single-particle shell model asserts that when x is even, the resultant total angular momentum of the lowest term of a pure configuration is zero, because of pairing, and that when x is odd the resultant is equal to the j-value of the configuration.

The succession of nuclei in the $N-Z$ chart may be envisaged in terms of the configurations based on a single-particle potential of slowly varying depth (Sect. 7.2.4) and of a radius increasing with mass number. Neutrons and protons are added in accordance with the Pauli principle to complete the configurations and as the radius increases so it becomes possible for single-particle states of higher n and l to be bound within the potential well. The binding of a new level may produce a change in some property controlled by the single-particle motion, e.g. the ground state spin, and this accounts for the observed periodicities. Changes in scattering or reaction cross-section may also be observed for the same reason.

In an odd-A nucleus the simplest excitations (intrinsic states) are those in which the odd nucleon enters one of the series of single-particle orbits. Such states may be studied by transfer reactions

(Sect. 11.5 and Ref. 7.1) and are seen particularly clearly in nuclei such as $^{209}_{83}$Bi with one proton outside a closed shell ($Z = 82$). If also the excited states of $^{207}_{81}$Tl, with one proton hole, are measured, the whole range of proton states above and within the closed shell may be identified. They are found to follow the order shown in Fig. 7.4, with the exception of an inversion of the $f_{7/2}$ and $h_{9/2}$ states, and clearly show a major shell gap of about 4·5 MeV between the $h_{9/2}$ and $s_{1/2}$ states. The results of a similar study of the neutron states in the nucleus $^{209}_{82}$Pb ($N = 127$) are shown in Fig. 11.11.

In some odd-mass nuclei the sequence of single-particle levels is not followed faithfully because the required excitation involves passage of a particle to another major shell, e.g. $d_{3/2} \rightarrow f_{7/2}$. A more favourable arrangement may be for a pair in a lower subshell (e.g. $s_{1/2}$) to break, and a pair in an upper subshell (e.g. $d_{3/2}$) to form, leaving the configuration $s_{1/2}^{-1}$. The gain of energy from such rearrangements may sometimes be sufficient to allow passage across a shell gap (see Ex. 7.3).

7.2.3 *Parity, spin and moments of nuclear ground states*

The usefulness of a parity quantum number ($\pm$, or even/odd) is well established in atomic physics and indicates that the electromagnetic interaction is invariant under the space-inversion transformation (Sect. 1.3). In a nucleus, the strong interaction determines the energy levels and again parity is expected to be conserved. Each of these statements must be modified slightly because of a small intrusion of the parity-violating property of the weak interaction (Ch. 10). For nuclear states however, the amplitude of the impurity expected for this reason is only 10^{-6} to 10^{-7} of the regular amplitude of the wavefunction and can usually be disregarded.

The parity of a nuclear state can be measured, relatively at least, by studying reactions such as the neutron transfer collision (deuteron stripping)

$$^{16}O + d \rightarrow {}^{17}O + p \tag{7.9}$$

The theory of this process (Sect. 11.5) shows how the orbital angular momentum of the transferred neutron can be deduced from the angular distribution of the stripped protons. If this angular momentum is $l\hbar$ then the relative parity of ^{17}O and ^{16}O is $(-1)^l$ since the neutron space wavefunction has an angular dependence $P_l(\cos \theta)$, which changes sign or not under space-inversion according to the l-value. The intrinsic parity of the nucleons is even by convention. Other methods of parity determination are based on the selection rules for radiative transitions (Ch. 9).

The parity of a nucleus as a whole is the product of the parities of the neutrons and protons and is, therefore, determined by the occupation of the single-particle states predicted by the shell model. There is excellent agreement with the model, e.g.

for $A = 6$, $1s^4 1p^2$, parity even $(+)$
for $A = 7$, $1s^4 1p^3$, parity odd $(-)$
for $A = 19$, closed shell $+(2s$ or $1d)^3$, parity even $(+)$

Nuclear spins are small integral multiples of the fundamental spin $\frac{1}{2}\hbar$, which is known experimentally to be the spin of both proton and neutron. Spins are determined, often together with magnetic dipole and electric quadrupole moments, by optical hyperfine spectroscopy and by magnetic resonance experiments using both liquids and solids and also collision-free beams of atoms and molecules. The strong tendency of protons and neutrons to pair off suggests, quite independently of specific models, that *even-N–even-Z nuclei have zero ground-state spin*. Additionally, it also follows in a model-independent way, that the spin of an odd-A nucleus is equal to the total resultant angular momentum $j = l + s$ of the odd nucleon, all other pairs being assumed to couple to zero resultant. The spins of ground states of odd-A nuclei change abruptly at the magic numbers, as illustrated for some odd-proton nuclei in Table 7.1. Moreover, odd-mass nuclei with a nucleon number just below a magic value may have single-particle excited states with a spin value considerably different from that of the ground state, e.g. $I_e = \frac{9}{2}$ compared with $I_g = \frac{1}{2}$. This is the reason for isomerism (Sect. 9.2.2) in such nuclei.

Magnetic dipole moments of odd-mass nuclei which are well described by a single-particle model can be predicted by a simple calculation following the derivation of the Landé g-factor in atomic physics. The moment μ_I is given by the equation

$$\mu_I = g_I \mu_N I \tag{7.10}$$

where I is the nuclear spin quantum number, μ_N is the nuclear magneton and g_I is the resultant gyromagnetic ratio given by

$$g_I = \frac{1}{2}\left[(g_l + g_s) + (g_l - g_s)\frac{l(l+1) - \frac{3}{4}}{j(j+1)} \right] \tag{7.11}$$

In this expression g_l and g_s are the orbital and intrinsic spin gyromagnetic ratios for the odd nucleon concerned (see Appendix 6) and l and j are the orbital and total angular momentum quantum numbers, the latter being equal to the nuclear spin quantum number I for an odd-mass nucleus.

From observed values of μ_I, with knowledge of I, the l-value and hence the parity of the ground state can be predicted. In general, magnetic moments, although not quantized, change when spin values change and therefore indicate the same periodicity, illustrated in

TABLE 7.1 Properties of selected odd-proton nuclei

l_j is the state of motion of the odd proton. For the deformed nuclei l is not a good quantum number. The symbol I^π gives ground-state spin and parity.

Z	A	Atom	l_j	I^π	μ_I/μ_N	Q_I/b	Predicted configuration of protons (Fig. 7.4)		
							$1s_{1/2}$	$1p_{3/2}$	$1p_{1/2}$
1	1	H	$s_{1/2}$	$\frac{1}{2}^+$	2·793	—	1		
3	7	Li	$p_{3/2}$	$\frac{3}{2}^-$	3·256	−0·04	2	1	
7	15	N	$p_{1/2}$	$\frac{1}{2}^-$	−0·283	—	2	4	1
8									

Z	A	Atom	l_j	I^π	μ_I/μ_N	Q_I/b	$1d_{5/2}$	$2s_{1/2}$	$1d_{3/2}$
9	19	F	$s_{1/2}$	$\frac{1}{2}^+$	2·629	—	1		
11	23	Na	$(d_{5/2})^3$	$\frac{5}{2}^+$	2·218	0·14	3		
19	39	K	$d_{3/2}$	$\frac{3}{2}^+$	0·391	0·055	6	2	3
20									

Z	A	Atom	l_j	I^π	μ_I/μ_N	Q_I/b	$1f_{7/2}$
21	45	Sc	$f_{7/2}$	$\frac{7}{2}^-$	4·756	−0·22	1
25	55	Mn	$(f_{7/2})^3$	$\frac{5}{2}^-$	3·444	0·4	5
27	59	Co	$f_{7/2}$	$\frac{7}{2}^-$	4·62	0·4	7
28							

Z	A	Atom	l_j	I^π	μ_I/μ_N	Q_I/b	$2p_{3/2}$	$1f_{5/2}$	$2p_{1/2}$	$1g_{9/2}$
29	63	Cu	$p_{3/2}$	$\frac{3}{2}^-$	2·223	−0·18	1			
31	69	Ga	$p_{3/2}$	$\frac{3}{2}^-$	2·016	0·19	3			
35	79	Br	$p_{3/2}$	$\frac{3}{2}^-$	2·106	0·31	4	3		
47	107	Ag	$p_{1/2}$	$\frac{1}{2}^-$	−0·114	—	4	6	1	
49	113	In	$g_{9/2}$	$\frac{9}{2}^+$	5·523	0·82	4	6	2	9
50										

Z	A	Atom	l_j	I^π	μ_I/μ_N	Q_I/b	$1g_{7/2}$	$2d_{5/2}$	$1h_{11/2}$	$2d_{3/2}$	$3s_{1/2}$
51	121	Sb	$d_{5/2}$	$\frac{5}{2}^+$	3·359	−0·29	1				
51	123	Sb	$g_{7/2}$	$\frac{7}{2}^+$	2·547	−0·37	1				
63	151	Eu	$d_{5/2}$	$\frac{5}{2}^+$	3·464	1·1	8	5			
67	165	Ho	—	$\frac{7}{2}^-$	4·12	3·0					
71	175	Lu	—	$\frac{7}{2}^+$	2·23	5·6	Deformed nuclei				
73	181	Ta	—	$\frac{7}{2}^+$	2·36	4·2					
75	185	Re	$d_{5/2}$	$\frac{5}{2}^+$	3·172	2·7	8	6	11		
79	197	Au	$d_{3/2}$	$\frac{3}{2}^+$	0·145	0·58	8	6	12	3	
81	203	Tl	$s_{1/2}$	$\frac{1}{2}^+$	1·612	—	8	6	12	4	1
82											

Z	A	Atom	l_j	I^π	μ_I/μ_N	Q_I/b	$2f_{7/2}$	$1h_{9/2}$	$3p_{3/2}$	$2f_{5/2}$
83	209	Bi	$h_{9/2}$	$\frac{9}{2}^-$	4·080	−0·35	1			

Notes: [1] For ^{19}F, ^{79}Br, ^{121}Sb, ^{185}Re and ^{209}Bi deviations from the level order of Fig. 7.4 are seen.

[2] For ^{23}Na and ^{55}Mn the observed spin must be explained by coupling of equivalent nucleons or by deformation (Ch. 8).

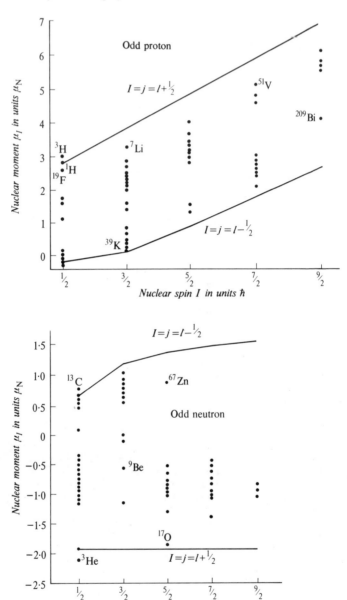

Fig. 7.5 Schmidt diagrams showing magnetic moment of odd-mass nuclei as a function of nuclear spin. The lines show the magnetic moments expected for a single nucleon at the spin values $j = l \pm \frac{1}{2}$. (Blin-Stoyle, R. J., *Theories of Nuclear Moments*, OUP, 1957).

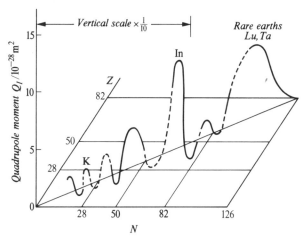

Fig. 7.6   Values of nuclear electric quadrupole moment Q_I as a function of N and Z. (Kopferman, H., *Nuclear Moments*, Academic Press, 1958).

Table 7.1. In detail, however, nuclear moments do *not* agree with simple predictions of equations (7.10) and (7.11) for odd-mass nuclei; the experimental results are given in Fig. 7.5. This displays the so-called Schmidt lines which result from the extreme single-particle prediction. Evidently, real nuclei are such that pure single-particle motion is in many cases diluted.

Electric quadrupole moments Q_I also show a periodicity, e.g. a change of sign at shell closures, as illustrated in Table 7.1 and Fig. 7.6. The outstanding features of this property, however, are:

(a) its existence at all for nuclei of the type of ^{17}O, which might be supposed to be a neutron attached to a spherical symmetrical ^{16}O core; and

(b) the very large values, in comparison with single-particle expectations, found for the group of rare earth nuclei.

These observations indicate an enhancement of single-particle motion, namely a collective effect (Ch. 8).

7.2.4 The average shell-model potential

Because of the short range of internucleon forces the radial shape of the shell-model potential may be expected to resemble the charge density distribution given by equation (6.8a). The main central nuclear potential may then be written

$$V(r) = -V_0[1 + \exp(r - R_V)/a]^{-1}$$
$$= -V_0 f(r, R_V, a)$$

(7.12)

where $f(r, R_V, a)$ is known as the Saxon–Woods *form factor*. The parameters R_V and a are not necessarily the same as those for the charge distribution because for the potential energy the range of nucleon–nucleon forces must be included (Sect. 6.2). The radial shape differs from the harmonic oscillator shape in such a way that the remaining l-degeneracy (Sect. 1.4.2) is removed and the levels of higher l for a given N are lowered in energy.

The potential well of the single-particle model may be expected to reflect some of the features apparent in the semi-empirical mass formula, to which it must contribute. The main correction to a constant well depth is an asymmetry potential proportional to $\pm T_z/A$ to represent the fact that neutrons and protons feel different forces on the average; the presence of an asymmetry term proportional to T_z^2/A in the mass formula will be recalled. There is also a Coulomb term $V_C(r)$ with the radial dependence of a point-charge field outside the nucleus, which lifts the proton well with respect to that for neutrons. Finally, a spin-orbit term $V_{ls}(r)$ as given in (7.4) must be added; it may also contain an asymmetry term.

The potential may in principle be determined by calculating the binding of the single-particle states and comparing them with experiment. Because of residual interactions so far neglected, no very precise results can be expected from this approach. This is already evident from the spread of binding energies seen in the (p, 2p) experiments. A more powerful approach is to probe the potential well through scattering experiments

The first approach of this sort was an examination of slow-neutron scattering lengths a as a function of A. If the neutron-nucleus interaction were adequately represented by a hard sphere of radius $R = r_0 A^{1/3}$ a smooth variation of a with A would be expected as shown in Fig. 7.7; in fact, discontinuities appear due to the binding of new s-orbits within a potential well at particular values of the parameter KR, where K is the internal wavenumber. It has already been noted that a bound state at zero energy implies an infinite scattering length (Sect. 5.2.2), in accordance with the trend of the results sketched in Fig. 7.7. The scattering anomalies in slow-neutron cross-sections are also in conflict with the assumption that a nucleus acts as a black, or strongly absorbing sphere. All available evidence, however, is consistent with the existence of a well of depth ≈ 50 MeV with the property of both absorbing and scattering an incident particle.

For scattering problems, involving unbound states, the shell-model potential with its absorptive and refractive properties is known as the *optical* potential. It has been extensively studied, especially in the elastic scattering of protons of energy above the Coulomb barrier by a range of nuclei. Figure 7.8 shows some of the differential cross-sections obtained, expressed as a ratio to the

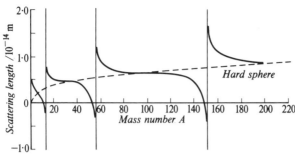

Fig. 7.7   Slow-neutron scattering length as a function of mass number, showing the expectation for hard-sphere scattering (dotted curve) and for scattering by a square-well potential (solid curves). Experimental points follow the latter.

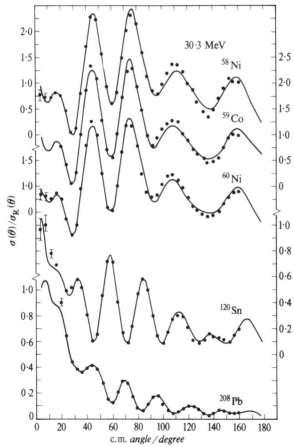

Fig. 7.8   Differential cross-section for the elastic scattering of 30·3-MeV protons by nuclei. The points are experimental and the curves are theoretical fits (Greenlees, G. W. *et al.*, *Phys. Rev.*, **171**, 1115, 1968).

215

Rutherford scattering from a point charge. The optical potential may be written in just the form of the shell-model potential, but with extra parameters to allow for the different geometrical behaviour (i.e. different form factors) of the various main terms. A frequently used form is:

$$V(r) = V_C(r) - V_0 f_1(r, R_V, a_V) - iW_0 f_2(r, R_W, a_W)$$
$$+ V_{ls}(\hbar/m_\pi c)^2 \mathbf{l} \cdot \mathbf{s} \left(\frac{1}{r} \cdot \frac{d}{dr}\right) f_3(r, R_{ls}, a_{ls}) \quad (7.13)$$

in which the main terms of the shell-model potential for bound states will be recognized, together with an imaginary term $\approx W_0$ to represent the absorptive effect of the well on the incident particles. The form factors f_1, f_2, f_3 generally have the Saxon–Woods form of equation (7.12).

The computer analysis of results such as those shown in Fig. 7.8 brings many partial waves into the scattering amplitude and yields parameters of the sort shown in Table 7.2. Usually, the radii R are taken to vary as $A^{1/3}$ and the parameter a which measures the surface thickness, is kept independent of A. The spin-orbit potential V_{ls} is found from measurements of polarization effects in the scattering.

The shell-model potential is the extrapolation of the optical potential back to an energy corresponding to a bound particle. The variation with energy of the terms V_0 and W_0 is shown (for neutrons) in Fig. 7.9. Comparison of the real central potential for neutron scattering with that for proton scattering reveals the presence of an isospin-dependent term not shown in (7.13) but already mentioned in connection with the shell-model potential.

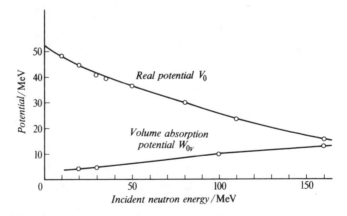

Fig. 7.9 Variation of real and imaginary parts of the optical potential with incident neutron energy.

TABLE 7.2   Optical model analysis of 30·3-MeV proton scattering for a range of nuclei; potentials are given in MeV.

	^{40}Ca	^{58}Ni	^{120}Sn	^{208}Pb
V_0	46·1	47·0	51·1	53·4
W_{0v}	0·4	3·4	1·2	4·0
W_{0s}	5·96	4·4	8·7	7·6
V_{ls}	12·0	8·8	12·0	10·2

The geometrical parameters differ for the different potentials but are approximately

$$R = r_0 A^{1/3}, r_0 = 1·25 \text{ fm}$$
$$a = 0·65 \text{ fm}$$

In this particular analysis, taken from the work of Greenlees and Pyle (*Phys. Rev.*, **149**, 836, 1966), the imaginary potential is split into a part W_{0v} representing absorption over the whole nuclear volume and a surface-peaked part W_{0s}. The value of the spin-orbit potential V_{ls} has been adjusted to take account of the use of the actual spin **s** in equation (7.13).

7.2.5   Bound states

The spherical potential defined by the scattering experiments should lead to the single-particle levels of the shell model as eigenstates. A detailed calculation (Ref. 6.2) for a potential with just central and

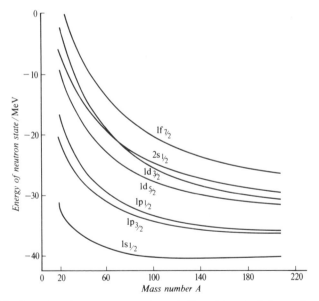

Fig. 7.10   Energies of neutron states in a shell-model potential as a function of *A* (Ref. 6.2).

(Figs 7.9 and 7.10 reproduced from *Nuclear Structure*, Vol. I, 1969, by Aage Bohr and Ben R. Mottelson with permission of the publishers, Addison-Wesley/W. A. Benjamin Inc.

spin-orbit terms gives results of the form shown in Fig. 7.10 for the energies of neutron states.

The energies calculated in this way are consistent with the general trend of observed level sequences, and with the results of (p, 2p) experiments, but are only a first approach because of the neglect of residual interactions. The eigenfunctions are also consistent with the size parameters found by electron scattering. The calculations show that a state of given (n, l) becomes more tightly bound as A increases, in conformity with the somewhat cruder prediction of single-particle level spacing (7.3).

7.3 Summary

The single-particle shell model, based upon a near-harmonic oscillator potential, gives an excellent account of many nuclear properties, and especially of the periodicity indicated by the 'magic' numbers. The success of the model results from the fact that it takes over from the theory of nuclear matter the concept of single-particle motion within a degenerate Fermi gas, with an effective interaction between nucleons that includes a two-body spin-orbit potential. In finite nuclei the residual interaction contains a significant pairing effect that exposes the last nucleon in odd-A nuclei. Its shortcomings in prediction of electric and magnetic moments arise from a neglect of the correlation between nucleons, apart from pairing.

The shell-model potential is best examined through the elastic scattering of protons and neutrons. The potential and its radial form factors can be made consistent with nucleon separation energies and with the charge distributions sensed by electron scattering experiments. It has been possible through the development of the so-called *folding models* to relate the terms of the optical potential to matter densities with an assumed nucleon–nucleon interaction (Sect. 11.4).

Examples 7

7.1 Using the relativistic relation between momentum and energy find the minimum kinetic energy of (*a*) an electron, (*b*) a proton, confined within a dimension of 7×10^{-15} m. (Assume $\Delta p \times \Delta r = \hbar$). [28, 0·42 MeV]

7.2 Write down the expected odd-particle configuration of the following nuclei:

$$^{27}_{13}\text{Al}; \, ^{29}_{14}\text{Si}; \, ^{40}_{19}\text{K}; \, ^{93}_{41}\text{Nb}; \, ^{157}_{64}\text{Gd}. \, [\text{d}_{5/2}; \text{s}_{1/2}; \text{d}_{3/2} + \text{f}_{7/2}; \text{g}_{9/2}; \text{h}_{9/2}]$$

7.3 Predict the spin-parity of the first excited state of the nuclei $^{31}_{14}\text{Si}$, $^{41}_{19}\text{K}$, $^{49}_{21}\text{Sc}$. $[\frac{7}{2}^-, \frac{7}{2}^-, \frac{3}{2}^-]$. Comment on the fact that the observed values are $\frac{1}{2}^+$, $\frac{1}{2}^+$, $\frac{3}{2}^+$.

7.4* From the equation $j = l + s$ for a single particle in an eigenstate of the operators j, l, s obtain the value of the quantity $(l \cdot s)$ and show that the energy separation of a nucleon spin-orbit doublet is proportional to $(2l + 1)$ (if the radial dependence of V_{ls} is smooth). For the case $l = 1$, $j = \frac{3}{2}$, what is the angle between l and j? [24°]

7.5   From equation (7.4), using the Saxon–Woods form-factor f given by (7.12), determine and sketch the radial variation of the spin-orbit potential $V_{ls}(r)$.

7.6*   Obtain an order-of-magnitude estimate of the spin-orbit energy that would arise due to electromagnetic effects from the motion of a proton in a p-state in a nucleus of $Z = 10$, assuming a central point charge as in the simple theory of the electronic fine structure. Compare your result with the observed spacing given by equation (7.6) for $A = 20$.

7.7*   Assuming that the expectation value of the potential energy of a harmonic oscillator state is half the total energy $(N + \frac{3}{2})\hbar\omega$, find the expectation value of the mean-square radius for a nucleon in such a state.

For the case of N large find the value of N for a nucleus containing A nucleons, assuming equal numbers of neutrons and protons.

By equating the summed $\langle r^2 \rangle$ for all the nucleons to $\frac{3}{5}AR^2$ deduce the form of equation (7.3).

7.8*   Following the method of Example 7.7, show that the spacing of hypernuclear energy levels is $\hbar\omega \approx 60A^{-2/3}$ MeV.

7.9   The low-lying levels of $^{39}_{20}$Ca have spin-parity values, starting from the ground state, of $\frac{3}{2}^+$, $\frac{1}{2}^+$, $\frac{7}{2}^-$ and $\frac{3}{2}^-$. Interpret these values on the basis of the single-particle shell model.

7.10   The single-particle potential is sometimes written with a term $(t \cdot T)/A$ where t is the isospin of a nucleon and T that of the target. Evaluate this term for a nucleus (A, N, Z) in the cases $T_> = T + \frac{1}{2}$, $T_< = T - \frac{1}{2}$.

7.11*   Calculate the ground-state magnetic moments of ^{7}Li, ^{23}Na, ^{39}K, ^{45}Sc and compare with the values shown in Table 7.1, which gives the spins of these nuclides.

8 Nuclear deformations and the unified model

The first attempt to improve the single-particle shell model (SPSM) was to consider nuclei with just two or three particles outside a closed shell and to remove the degeneracy, not by the pairing force of the SPSM, but by a more general residual interaction. The coupling of the individual spin and orbital angular momenta was treated by the methods developed by Racah for atomic spectroscopy in which either the jj or the LS (Russell–Saunders) scheme may be used. Observed nuclear spectra may require a coupling scheme intermediate between the two extremes. When both types of nucleon are involved use is generally made of the assumption of charge independence and total isobaric spin T then provides an additional quantum number. The model that thus develops from the extreme single-particle model is known as the *independent-particle model* (IPM); it regards the first few nucleons outside a closed shell as essentially *equivalent*.

IPM calculations often improve the prediction of magnetic moments, but are unable to explain the enhanced electric quadrupole moments of the rare-earth nuclei, for which a stronger cooperative effect is required. This effect must involve the nucleons not only of unfilled shells but also of the closed shells that have so far been disregarded except for their presumed role in providing the central potential with respect to which single-particle motion may be defined. The recognition of the interaction between the loose or 'valence' nucleons and the nucleons of the core in producing coherent collective nuclear motion was an achievement for which Rainwater, A. Bohr and Mottelson were awarded the Nobel Prize of 1975. Their work, dating from the early 1950s, is essentially the theory of nuclear deformations.

A deformed nucleus will generate a deformed potential well, in which neither the total nor the orbital angular momentum of a single particle is a good quantum number because of lack of spherical symmetry. Shell-model calculations should be based on this potential. The way to do this was shown by Nilsson and the model resulting, which generalizes the SPSM, is known as the *unified model.* In this chapter some of the properties of deformed nuclei will be examined and related to this model.

8.1 Early empirical evidence for collective motion and for nuclear deformation

8.1.1 The compound nucleus

The nature of the nucleon motion within a nucleus determines the general character of its excitation energies. If it is possible to describe the nucleus as a mechanical system with a small number of degrees of freedom, the energy eigenvalues will be well separated, and this is essentially the prediction of the SPSM. If, however, many degrees of freedom exist, as might be expected if there is strong coupling between individual nucleons, then the number of states per unit energy range will increase considerably.

Information on nuclear level densities at about 8 MeV excitation became available in 1935 as a result of experiments with neutrons of near-thermal energies. For such particles the (n, γ) capture reaction, e.g.

$$^{107}\text{Ag} + \text{n} \rightarrow {}^{108}\text{Ag} + \gamma \tag{8.1}$$

is easily observable, by detection of the radioactivity of the product nucleus, e.g.

$$^{108}\text{Ag} \rightarrow {}^{108}\text{Cd} + \beta^- + \bar{\nu} \tag{8.2}$$

or by a transmission experiment, and the variation of the cross-section with energy can be studied. According to the SPSM the neutron would move only briefly in the potential well provided by the target nucleus and would have a high probability of emerging (Fig. 8.1a); resonance would be observed when the incident particle created a system with an energy near one of the levels of the potential well, but the spacing of these would be $\approx \hbar^2/2MR^2$, i.e. about 1 MeV for $A \approx 100$. Also, the width would be large (≈ 1 MeV) because of the brief time of association ($\approx 10^{-21}$ s) of the neutron with the well. Variations of capture probability with energy in the thermal range are not likely to be produced by these resonances.

This picture is in sharp disagreement with observation at several points. Although the single-particle resonances are seen as broad structural features in scattering cross-sections as a function of energy for neutrons of a few MeV energy, near-thermal neutrons show small scattering and large absorption for many nuclei. In particular, the work of Moon and Tillman and of Amaldi and Fermi, using reactions such as (8.1) and (8.2), established not only the existence of thermal neutrons in systems in which slowing down

221

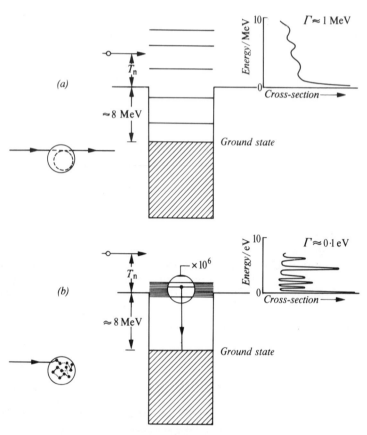

Fig. 8.1 Interaction of a neutron with a nucleus. The levels shown represent states of excitation of the system (neutron + nucleus); a neutron of zero laboratory energy T_n produces an excitation of about 8 MeV: (*a*) Potential-well model. (*b*) Compound-nucleus model. Note the different energy scales.

could take place, but also anomalies in cross-section for near-thermal energies that had to be interpreted as due to *narrow, relatively long-lived* resonant states (Fig. 8.1*b*).

In order to explain these observations, Niels Bohr in 1936 introduced the concept of the *compound nucleus*, which is a many-body system formed by the amalgamation of the incident particle with the target nucleus, e.g. reaction (8.1) proceeds in two stages:

$$\left.\begin{array}{r}{}^{107}\mathrm{Ag} + \mathrm{n} \to {}^{108}\mathrm{Ag}^* \\ {}^{108}\mathrm{Ag}^* \to \text{final state}\end{array}\right\} \tag{8.3}$$

where ${}^{108}\mathrm{Ag}^*$ represents a ${}^{108}\mathrm{Ag}$ nucleus plus the energy brought in by the incident neutron, which is the separation energy S_n plus a very small kinetic energy ε. In such a system the incident particle has a short mean free path and shares the total energy $S_n + \varepsilon$ with many of the internal degrees of freedom of the compound nucleus, so that it cannot be re-emitted until, as a result of further exchanges, sufficient energy $(>S_n)$ is again concentrated on this or a similar particle. If the incident particle is a slow neutron ($\varepsilon \approx 0$) this may be a very long time, several orders of magnitude greater than the time that a fast particle takes to cross the nucleus, and this time ($\approx 10^{-14}$ s) may be long enough to permit the electromagnetic coupling to generate the emission of a photon so that (8.3) is completed by the process

$$^{108}\text{Ag}^* \rightarrow {}^{108}\text{Ag} + \gamma \tag{8.4}$$

as shown in Fig. 8.1*b*.

The clear experimental evidence for the compound nucleus in slow-neutron capture reactions indicates the closely coupled nature of nuclear motion, and consequently the possibility of collective modes. In the early theories of N. Bohr and Kalckar, the dynamical situation was likened to that of a classical liquid drop and this concept was fruitful in the evolution of the semi-empirical mass formula (Sect. 6.3) and in the theory of fission (Sect. 11.7.2). It is now more useful, in the light of present knowledge of nuclear densities and Fermi momenta, to speak of a Fermi gas rather than a Fermi liquid, but the feature of collective motion applies equally to each. The application of statistical methods to the compound nucleus level system at high excitations has also shown agreement with many features of nuclear reactions and confirms the concept of correlated motion.

8.1.2 Radius anomalies

Electron scattering experiments (Sect. 6.2) determine the mean-square radius of the nuclear charge distribution and show that it varies with mass number in accordance with the formula

$$\langle r^2 \rangle \propto A^{2/3} \tag{8.5}$$

for the natural sequence of nuclei in which roughly equal numbers of protons and neutrons are added in passing from one nucleus to another. If only neutrons are added, the change in $\langle r^2 \rangle$ should be given, according to (8.5), by

$$\delta\langle r^2 \rangle / \langle r^2 \rangle = 2\delta A / 3A \tag{8.6}$$

Such a change in nuclear dimensions will cause an *isotope shift* (volume or field effect) in the lines of optical, X-ray and muonic atom spectra, additional to the shift expected from the mass change. Optical spectroscopic observations available in 1958 for pairs of

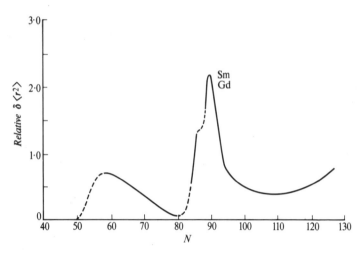

Fig. 8.2   Relative $\delta\langle r^2\rangle$ from isotope shifts as a function of neutron number N. The line represents the trend of observed shifts (Ref. 6.2).

even–even nuclei differing in neutron number by two (N and $N-2$) indicated changes of $\langle r^2\rangle$, relative to that predicted by equation (8.6), as shown in Fig. 8.2. The jump in $\delta\langle r^2\rangle$ between $N = 88$ and $N = 90$ is entirely unconnected with any features of the SPSM and represents the onset of a deviation of the nuclear shape from spherical symmetry. Later and more comprehensive experimental evidence on deformation shows that it vanishes for closed-shell nuclei.

8.1.3 Origin of nuclear deformation

The original proposal by Rainwater, elaborated and developed by Bohr and Mottelson, was based on the liquid-drop model of the nucleus, as already used in setting up the semi-empirical mass formula (Sect. 6.3). This formula indicates that since for a given volume a spherical shape has the minimum surface area, deformed nuclei should have an energy increasing with the square of a suitable parameter, such as the change of major axis, representing the deformation. This situation is represented in Fig. 8.3, curve (*a*).

If the motion of the individual nucleons is considered, the spherical equilibrium shape for nuclei with a few nucleons outside closed shells or subshells may be ascribed to the pairing force. The collective motion in this case is a vibration about the spherical shape. As the number of 'valence' nucleons increases, however, distorting forces due to the quadrupole component of the internucleon force may become important. The frequency of the collective vibration

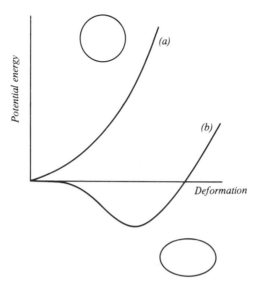

Fig. 8.3   Potential energy of a nucleus as a function of deformation. (*a*) Spherical shape. (*b*) Deformed shape.

diminishes and finally the spherical shape becomes unstable and the nucleus acquires a *permanent* deformation. The origin of this may be understood by considering the actual orbital motion of the nucleons; if a deformation of the core occurs the outer nucleons move to 'orbits' of larger radius and if the angular momentum stays constant, the kinetic energy *decreases* as the radius parameter increases. Equilibrium is reached at the stable deformation (Fig. 8.3, curve (*b*)).

A few nucleons of one kind and of given l outside a closed shell, with the nucleons of the other kind paired off, would tend to occupy equatorial orbits of high m-values. Interaction of these nucleons with the core would produce an equatorial bulge, which corresponds with a negative quadrupole moment (Fig. 8.4*a*). If, however, there are *holes* in the closed shell the opposite behaviour is favoured and a positive quadrupole moment develops when there are sufficient holes (Fig. 8.4*c*). At the closed-shell nucleon number the moment is zero (Fig. 8.4*b*). This variation of quadrupole moment with nucleon number is also expected according to the single-particle model (Table 7.1), but the present picture shows how the moments may be enhanced, and indeed very much enhanced if a stable deformation exists.

If there is such a deformation, then as pointed out by A. Bohr vibrational and rotational motion of the deformed nucleus may co-exist with the intrinsic motion of the individual nucleons that

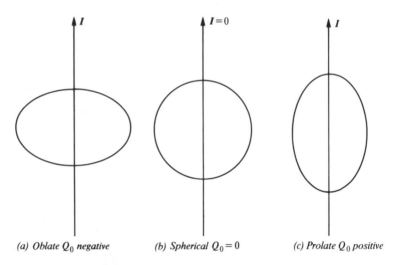

Fig. 8.4 Intrinsic quadrupole moments Q_0.

creates the deformed shape. It is necessary for a simple development of the theory of this coupled collective motion that the
single-particle energies associated with the shell-model nucleon
states should be large compared with the vibrational and rotational
energies so that the period of the collective motions is relatively
long and it is sensible, for instance, to define a moment of inertia.
The collective excitations then provide a fine structure in the overall
pattern of intrinsic, e.g. single-particle, states. At the same time the
deformation of the nucleus has a profound effect not only on the
static electromagnetic moments but also on radiative transition
rates, which are often enhanced above single-particle values.

The simplest type of deformation is ellipsoidal, with an axis of
symmetry Oz', as shown in Fig. 8.5. The axis Oz' is fixed in the
nucleus but may orient with respect to a space-fixed axis Oz. The
magnitude of this type of deformation may be specified by a
parameter δ given by

$$\delta = \Delta R / R_0 \tag{8.7}$$

where R_0 is the average nuclear radius and ΔR is the difference
between the major and minor semi-axes of the ellipse. Alternatively, the angular shape in a plane containing Oz' may be defined
by the expression

$$R(\theta) = R_0[1 + \beta Y_2^0(\theta, \phi)] \tag{8.8}$$

where θ is the angle with Oz', and $\beta = 1 \cdot 06\delta$, using the normalized
form of Y_2^0 (Appendix 1).

The deformation parameter δ (or β) is typically $0\cdot2$–$0\cdot3$ and is obtained from measurements of electromagnetic transition rates (Sect. 8.5) and from the pattern of rotational states (Sect. 8.2) as well as from isotope shifts. All evidence confirms that the jump found between $N = 88$ and $N = 90$ in Fig. 8.2 represents a transition between the vibrational and rotational type of collective motion.

8.2 Rotational states

8.2.1 Coupling of angular momenta

The coupling of angular momenta in a deformed nucleus resembles that in molecules and is shown in Fig. 8.5. It is assumed, in the circumstances outlined in Section 8.1.3, that the nuclear wave-function is a product of an intrinsic function and a rotational function, and it is the latter with which we are mainly concerned in this section. The intrinsic motion, with resultant angular momentum

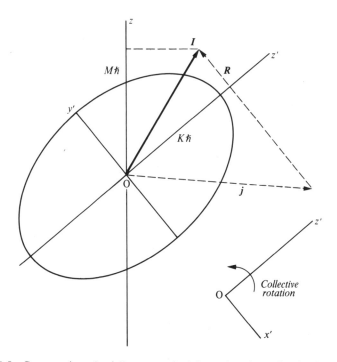

Fig. 8.5 Cross-section of axially symmetric deformed nucleus, showing body-fixed axis Oz' and space-fixed axis Oz. The total angular momentum vector **I** is indicated; $M\hbar$ is the resolved part of **I** along Oz and $K\hbar$ the resolved part along Oz'. **R** is the vector representing collective rotation, which adds vectorially to the intrinsic angular momentum **j** to form **I**.

j, is related to the body-fixed axis Oz', but in the non-spherical field j is not a constant of the motion. Rather, it couples with any collective rotational angular momentum R to give the total resultant angular momentum I of the nucleus, which *is* a constant of the motion, as is its component $M\hbar$ along the space-fixed axis Oz.

Because Oz' is an axis of symmetry the orientation of the body-fixed axes cannot be uniquely specified and this means (Ref. 8.1) that there is no collective rotation about Oz'. This is so for the case of diatomic molecules and also for spherical nuclei, which do not show rotational bands of the type discussed in this section. Consequently, R is perpendicular to the symmetry axis. Moreover, since the system is invariant to rotations about Oz', the component $K\hbar$ of the angular momentum I is also a constant of the motion. Figure 8.5 shows that $K\hbar$ is also the component of the intrinsic angular momentum along Oz'. Together the three observables I^2, $M\hbar$ and $K\hbar$ completely specify the state of motion.

From Fig. 8.5, the rotational angular momentum is given by $R = I - j$. If however averaging of the intrinsic motion is assumed it is possible to write

$$R^2 = I^2 + j^2 - 2I \cdot j = I^2 + j^2 - 2K^2\hbar^2 \qquad (8.9)$$

except in the case of $K = \frac{1}{2}$, and this expression will be used in the following discussion.

8.2.2 Energy levels

The energy of collective rotation may be written in the classical form

$$E_R = \tfrac{1}{2}\mathcal{I}\omega^2 \qquad (8.10)$$

where ω is the angular velocity about the axis of rotation, which is perpendicular to the symmetry axis, as shown in Fig. 8.5, and $\mathcal{I}$ is a moment of inertia. If the rotational angular momentum $\mathcal{I}\omega = |R|$ is inserted in (8.10) we find, except in the case of $K = \frac{1}{2}$,

$$E_R = (\hbar^2/2\mathcal{I})[I(I+1) + j(j+1) - 2K^2] \qquad (8.11)$$

using the eigenvalues for I^2 and j^2, and for the rotational energies referred to the ground state

$$E_{\text{rot}} = E_I - E_g = (\hbar^2/2\mathcal{I})[I(I+1) - I_g(I_g+1)] \qquad (8.12)$$

The value of K ($\leq I_g$) arising from the intrinsic motion thus determines a *rotational band* of levels, whose extra energies are added to that of the intrinsic motion, as in the case of molecular rotational states.

For *even–even nuclei in their ground state* the individual particles are paired and fall alternately into states of opposite K, so that the resultant K-value is zero with even parity. Because of the symmetry about a plane perpendicular to Oz', only even I-values occur (as in the case of the homonuclear diatomic molecule obeying Bose statistics), and the spin-parity values for the rotational band are

$$K = 0 \qquad I = 0^+, 2^+, 4^+, 6^+, \ldots \tag{8.13}$$

The total angular momentum arises solely from the collective rotation.

Clearly defined rotational bands are found in even–even nuclei with N and Z values lying within the ranges shown in Fig. 8.6a, and a well-known rotational band is shown in Fig. 8.6b. The levels of the band may be excited electromagnetically (Coulomb excitation, Sect. 9.4.2) and they decay mainly by the emission of electric quadrupole radiation (Sect. 9.2). The ratios of excitation energies in a rotational band are (from (8.12) with $K = I_g = 0$) $E_4/E_2 = 10/3$, $E_6/E_2 = 7$, and $E_8/E_2 = 12$, independently of the moment of inertia, providing that this is independent of rotational frequency.

The actual value of the effective moment of inertia $\mathscr{I}$ is model-dependent. If the nucleus behaved as a rigid object of mass AM where A is the number of component nucleons of mass M then the classical rigid-body value would be

$$\mathscr{I}_{\text{rig}} = (\tfrac{2}{5})AMR^2 \tag{8.14}$$

where R is a mean radius and would be found by analysis of the spacings in the rotational band. It can be shown that this rigid-body value is also obtained in the shell model with wholly independent particle motion. Pairing forces, however, inhibit this independence and the motion then exhibits a lower moment of inertia; observed values are, in fact, smaller than the rigid value by a factor of 2–3 for low I-values.

The moment of inertia is not always constant, as would be expected for a rigid body, throughout a rotational band. From an experimental spectrum such as that shown in Fig. 8.6b, but more particularly for the even–even rare-earth nuclei, the moment of inertia is given by

$$2\mathscr{I}/\hbar^2 = (4I - 2)/(E_I - E_{I-2}) \tag{8.15}$$

Also, the angular velocity of rotation for the state I is obtained with a small approximation for $I \gg 1$ as

$$(\hbar\omega)^2 = \tfrac{1}{4}(E_I - E_{I-2})^2 \tag{8.16}$$

where E_I and E_{I-2} are the energies E_R of equation (8.11) with $K = 0$. A plot of $2\mathscr{I}/\hbar^2$ against $(\hbar\omega)^2$ is given in Fig. 8.7. It shows firstly an increase of $\mathscr{I}$ with ω up to spin values ≈ 15 and this can be understood in terms of a centrifugal stretching as ω increases (variable moment of inertia). For some even–even nuclei there then

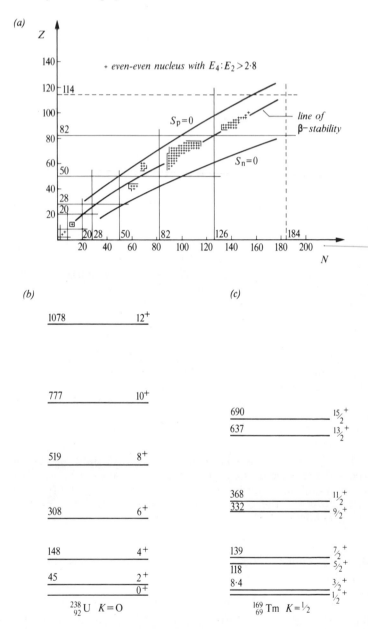

Fig. 8.6 (a) Regions of deformed even–even nuclei (+). The lines $S_p = 0$, $S_n = 0$ give the stability limits for proton and neutron emission (Ref. 8.1). (b) Ground-state rotational band in ^{238}U. (c) Ground-state rotational band in ^{169}Tm. Energies in keV and spin-parities I^π are shown.

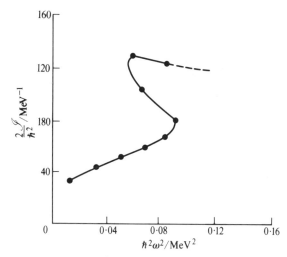

Fig. 8.7 Collective moment of inertia as a function of angular velocity squared for $^{158}_{68}$Er (Grosse, E. *et al.*, *Phys. Rev. Letters* **31**, 840, 1973).

ensues the striking phenomenon of *back-bending*, in which there is a sharp increase of $\mathscr{I}$ towards the rigid value. One explanation of this effect is that it is a disappearance of the pairing correlation due to the action of Coriolis forces; the nucleus then undergoes a phase transition from a superfluid state to a state of independent-particle motion. Alternatively, just one pair of neutrons, for instance, is broken, providing a large angular momentum which may couple with the collective rotation to produce a new band. Such effects will depend on the actual value of the pairing energy and will not be seen for all nuclei. It may be noted at this point that, in general, the rotational states of an even–even nucleus lie *within* the energy gap predicted in the spectrum of intrinsic states (Sect. 6.4.3).

In even–even nuclei with an excitation from the intrinsic motion, and in *odd-A nuclei*, K is finite; in the latter case the last odd particle at least contributes a component of angular momentum along the symmetry axis Oz'. The energy levels have spin values given by

$$K \neq 0 \qquad I = K, K+1, K+2, \ldots \qquad (8.17)$$

and these are integral for A even and half-integral for A odd. The parity is that of the intrinsic motion. The states with $I_{z'} = +K$ have equal energy, and for $K = \frac{1}{2}$ the Coriolis forces in the rotating system connect these states and give rise to an extra term in the rotational energy. For a state of spin I this is found to be, apart from a constant term,

$$E_R = (\hbar^2/2\mathscr{I})[I(I+1) + a(-1)^{I+1/2}(I+\tfrac{1}{2})] \qquad (8.18)$$

The decoupling term $a(I+\tfrac{1}{2})$ has opposite sign for the successive levels of a $K=\tfrac{1}{2}$ band and the level spacing is considerably modified from the $I(I+1)$ interval found for $K=0$ (Fig. 8.6c).

For $K>\tfrac{1}{2}$ the additional term is much smaller and the excitation energies with respect to the ground state $I_g=K$ are well approximated by equation (8.12).

Although the rotational bands so far discussed have a relatively simple structure, the nuclear spectrum contains many more states deriving from vibrational modes (Sect. 8.4) and from special couplings. The lowest level of given I, however, has a special significance because of its intimate connection with the joint effects of collective motion and pair correlation; it is referred to as the *yrast* level.

8.3 Single-particle motion in a deformed potential

For nuclei which have an ellipsoidal deformation, the nuclear radius is given by equation (8.8) and the nuclear potential would be expected to follow this form, e.g.

$$V(r, \theta) = V_0(r) + V_2(r) Y_2^0(\theta, \phi) \tag{8.19}$$

In this expression the first term may be identified with the customary Saxon–Woods form of a spherical potential (Sect. 7.2.4).

For *small deformations* it will be permissible to use the single-particle quantum numbers n, l, j appropriate to a spherical potential, remembering that for axial symmetry the projection Ω of the total angular momentum j along the symmetry axis is a constant of the motion. The energies may then be obtained by treating the second term of (8.19) as a perturbation, and are found to be (Ref. 8.1)

$$\varepsilon(nlj\Omega) = \varepsilon_0(nlj) - \frac{3\Omega^2 - j(j+1)}{4j(j+1)} \langle V_2(r)\rangle_{nlj} \tag{8.20}$$

with $\Omega = \pm\tfrac{1}{2}, \pm\tfrac{3}{2}, \ldots, \pm j$. There is degeneracy with respect to Ω since positive and negative values have the same energy. Each undeformed shell-model level then splits into $\tfrac{1}{2}(2j+1)$ distinct states labelled by Ω, which is also equal to K as used in Section 8.2.

For rather *large deformations*, more characteristic of the collective model, neither j nor l is a good quantum number. An approximate description of the single-particle motion is obtained by considering an anisotropic harmonic oscillator potential

$$V = \tfrac{1}{2}M[\omega_3^2 x_3^2 + \omega_\perp^2 (x_1^2 + x_2^2)] \tag{8.21}$$

where x_1, x_2 and x_3 are Cartesian coordinates and ω_3 and $\omega_\perp$ are the axial and transverse oscillation frequencies. The energies are then

$$\varepsilon(n_3 n_\perp) = (n_3 + \tfrac{1}{2})\hbar\omega_3 + (n_\perp + 1)\hbar\omega_\perp \tag{8.22}$$

where $n_\perp = n_1 + n_2$ is the oscillator number for the motion perpendicular to the symmetry axis. There is a degeneracy associated with $n_\perp$ but it may be labelled by the component Λ of the orbital momentum along the symmetry axis. Further degeneracy arises because of the possible positive and negative alignment of the intrinsic spin Σ $(=\frac{1}{2})$ of the nucleon along the symmetry axis. Together with Λ this produces the component $\Omega = \Lambda \pm \Sigma$ already discussed.

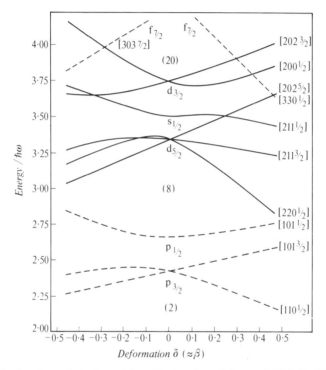

Fig. 8.8 Spectrum of single-particle orbits in spheroidal potential ($N, Z < 20$) (Ref. 8.1). The level energies are in units of the oscillator spacing (eqn (7.3)) and are plotted as a function of the deformation parameter δ ($\approx \beta$) (eqn (8.7)). The individual labels are the asymptotic quantum numbers (Sect. 8.3).

Couplings exist which remove these degeneracies and it is then useful to label the resulting states by the numbers N ($= n_3 + n_\perp$), n_3, Λ, Ω. They are known as *asymptotic quantum numbers* since they really apply in the limit of large deformations when the (n, l, j, Ω) scheme is invalid. The one-particle spectra obtained by use of an oscillator potential of the form (8.21) are shown for N and $Z < 20$ in Fig. 8.8; these states, which can each hold two neutrons or protons, replace those of the spherical shell model.

8.4 Vibrational states

8.4.1 Vibrational quanta (phonons)

In Section 8.1.3 it was noted that nuclei with a few nucleons outside closed shells, and the closed-shell nuclei themselves, have a spherical equilibrium shape. The simplest collective motion of such a system is a simple harmonic vibration of the surface about equilibrium. The corresponding energies give rise to levels in the nuclear spectrum which, as in the case of the corresponding levels in molecular physics, have a somewhat greater spacing than the levels of a rotational band.

Nuclear surface vibrations may be discussed classically using the model of an incompressible liquid drop. The instantaneous form of the surface may be described generally by an expansion in terms of spherical harmonics

$$R(\theta, \phi) = R_0 \left(1 + \sum_{\lambda, \mu} \alpha_{\lambda\mu}(t) Y_\lambda^\mu(\theta, \phi) \right) \qquad (8.23)$$

of which equation (8.8) is a particular case, with $\lambda = 2$, $\mu = 0$. The vibrations about equilibrium are governed by a potential-energy curve of the form shown in Fig. 8.3a and the total energy for a given λ may be written in terms of a force parameter C and a mass parameter D

$$E = \tfrac{1}{2} C \alpha^2 + \tfrac{1}{2} D \dot{\alpha}^2 \qquad (8.24)$$

with an oscillation frequency given by

$$\omega_\lambda = \sqrt{C/D} \qquad (8.25)$$

Solution of the equation of motion shows that a surface vibration may be described classically in terms of waves that travel round the z-axis $\theta = 0$. These waves carry angular momentum.

If the oscillations are quantized (compare the treatment of the single particle in an oscillator potential, Sect. 1.4.2) the energies have the values

$$E_n = \hbar\omega_\lambda \sum_\mu (\tfrac{1}{2} + n_\mu) = \hbar\omega_\lambda \left(\frac{2\lambda + 1}{2} + n_\lambda \right) \qquad (8.26)$$

where λ is, of course, integral, $|\mu| \leq \lambda$ and n_μ is the number of oscillator quanta in a state (λ, μ). The oscillator number $n_\lambda = \sum_\mu n_\mu = 0, 1, 2, \ldots$ is the total number of such quanta. The excitation energy is thus determined by the number of quanta, or *phonons* present in the state. Phonons may be regarded as particles each carrying an angular momentum $\lambda\hbar$, parity $(-1)^\lambda$ and energy $\hbar\omega_\lambda$; like photons they obey Bose–Einstein statistics. The spin values of vibrational states follow directly from their phonon content.

The inertial parameters C and D are functions of λ and may be estimated by classical calculations which show that ω_3 (octupole phonon) $\approx 2\omega_2$ (quadrupole phonon). The phonon energies indicated by experiment (cf. Fig. 8.11) are ≈ 0.5 MeV and must be distinguished from the very much larger energies (≈ 10 MeV) associated with the single-particle oscillator spacing of the shell model. In terms of that model, the surface waves correspond with excitations not requiring any change of the single-particle oscillator number N.

8.4.2 Energy levels (spherical equilibrium)

In the following sequence, $\lambda = 0$ and $\lambda = 1$ excitations are included for completeness, although they are high-energy modes and are not appropriately described as surface vibrations.
(i) $\lambda = 0$ (Fig. 8.9a). This is a wholly radial oscillation and is not possible for an incompressible fluid. Nuclear matter is not incompressible, but no monopole state of this sort ('breathing mode') has

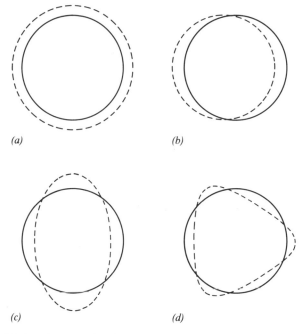

(a)   *(b)*

(c)   *(d)*

Fig. 8.9 Modes of nuclear vibration about a spherical equilibrium state (solid line). The dotted line shows one extreme excursion of the nuclear surface. (a) $\lambda = 0$. (b) $\lambda = 1$, relevant to opposite motions of neutron and proton fluids. (c) $\lambda = 2$. (d) $\lambda = 3$.

been identified with certainty and such 0^+ states if they exist probably lie high in the nuclear spectrum (Fig. 8.16*b*). The 0^+ states known at fairly low excitation in even–even nuclei, such as ^{16}O, ^{72}Ge and ^{90}Zr can be given a shell-model interpretation.

(ii) $\lambda = 1$ (Fig. 8.9*b*). This is a dipole deformation which corresponds with a displacement of the centre of mass and can only be produced by external forces. If such a displacement is imposed on a nucleus in which the neutrons and protons behave as a homogeneous fluid, and always move in phase, the centre of charge always coincides with the centre of mass and no electric dipole moment can arise. If, however, the neutrons and protons behave as separate fluids and move in antiphase, the centres of charge and mass separate and a dipole moment develops.

The corresponding collective state has a spin of 1^- in even–even nuclei and is at a high excitation, perhaps 10–25 MeV above the ground state. It is known as the *giant (electric) dipole resonance* and is a prominent feature in many nuclear reactions, especially in the photodisintegration of both even- and odd-mass nuclei (Fig. 8.10). The dipole resonance can also be given a shell-model interpretation (Sect. 8.6.1) and the parity change shows that transitions between two major shells must be involved. The excitation energy as a function of A is about twice the value predicted for the shell spacing (eqn (7.3)) but the introduction of the proper residual interactions in a system containing both an excited particle and a hole is found to predict a collective dipole state that is pushed up in energy. The effect is similar, but opposite in sign, to the lowering of ground states by the pairing interaction discussed in Section 6.4.3.

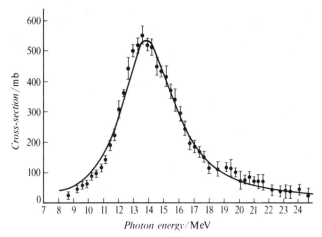

Fig. 8.10 Giant resonance of photodisintegration in ^{197}Au. The yield of neutrons is shown as a function of the energy of the monochromatic photons used to produce the reaction (Fultz, S. C. *et al.*, *Phys. Rev.*, **127**, 1273, 1963).

(iii) $\lambda = 2$ (Fig. 8.9c).   Among the lowest-lying states, quadrupole phonons describe the most important excitations since the $\lambda = 0, 1$ modes have the special features just discussed. In the shell-model picture quadrupole excitation does not require a parity change and may, therefore, take place *within* a major shell provided that it is not full, as in closed-shell nuclei. For even–even nuclei the excitation energies above the 0^+ ground state are

$$E(n_2) = n_2 \hbar \omega_2 \qquad (8.27)$$

where n_2 is the number of quadrupole phonons and the total angular momentum permitted for a given n_2 is determined by the occupation numbers allowed by the Bose–Einstein statistics. The spin values for a few values of n_2 are shown in Table 8.1.

TABLE 8.1 States deriving from quadrupole phonons (from Ref. 8.1, p. 347)

n_2 \ Spin I	0	1	2	3	4	5	6	7	8
0	1								
1			1						
2	1		1		1				
3	1		1	1	1		1		
4	1		2		2	1	1		1

The systematic appearance of the 2^+ single phonon vibrational state is an outstanding feature of the spectra of even–even nuclei, and the triplet of 2-phonon states 0^+, 2^+, 4^+ at approximately double the energy of the first excited states is also frequently seen, though other states are often present in the same region; Figure 8.11 shows the low levels of ^{114}Cd, which illustrate the point; they lie within the energy gap in the spectrum of intrinsic states.

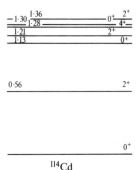

Fig. 8.11 Vibrational spectrum of ^{114}Cd. The level energies in MeV are shown together with spin-parity I^π.

Quadrupole modes of much higher energy, corresponding with particle excitations across two major shells, have also been seen and form a *giant quadrupole resonance* in the region of unbound levels near the giant dipole state. Evidence for this resonance is discussed in Reference 8.3.

(iv) $\lambda = 3$ (Fig. 8.9*d*). Octupole phonons have an energy approximately twice that of a quadrupole vibration, and the single-phonon octupole state (3^-) is found in the spectrum of even–even nuclei near the two-phonon quadrupole states. The collective 3^- excitation is a prominent feature among the low-lying states of these nuclei as revealed by inelastic scattering experiments.

(v) *Coupling of vibration with intrinsic motion.* For an intrinsic state with angular momentum $\boldsymbol{J}$ and a vibrational state with angular momentum $\boldsymbol{R}$, a multiplet of states arises with quantum number I given by

$$I = J + R, J + R - 1, \ldots, |J - R| \tag{8.28}$$

For weak coupling between the two motions, in which the energy differences produced are small compared with the separation of intrinsic states and the two motions are independent, collective effects are due to the vibrational transition. A well-known example of this coupling is found in the spectrum of ^{209}Bi in which a 3^- octupole phonon, effectively an excitation of the ^{208}Pb core, is coupled to a $h_{9/2}$ single-proton state, yielding resultant I-values from $\frac{3}{2}$ to $\frac{15}{2}$.

Often the coupling between the two motions is strong, the pattern of excited states does not show such clear regularities, and the nucleus may acquire a stable deformation.

8.4.3 Energy levels (deformed equilibrium)

When the deformed nuclear shape becomes stable, in accordance with the potential-energy curve shown in Fig. 8.3*b*, vibrational modes may be based upon this equilibrium shape. The main types of vibration for a quadrupole deformation, for instance, may also be discussed in terms of phonons, but the order λ is not a good quantum number in a non-spherical field. The projection ν of the phonon angular momentum along the symmetry axis may be specified and two simple types of vibration, with $\nu = 0$ (β-vibration) and $\nu = \pm 2$ (γ-vibration), may be identified. These are illustrated in Fig. 8.12, which shows that β-vibrations preserve the axis of symmetry, while γ-vibrations do not.

Phonon numbers n_β and n_γ may be ascribed to the two types of vibration and, together with the K-value from any intrinsic motion, define the spin of the vibrational state. Rotational bands may be built upon each such state; the direction of rotation is indicated in Fig. 8.12.

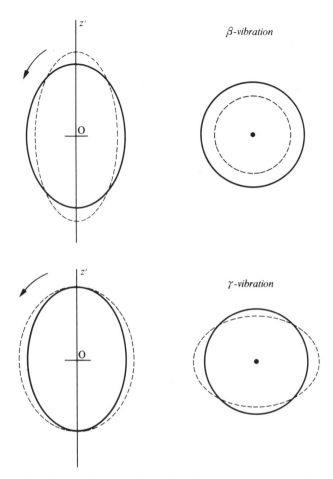

Fig. 8.12 β and γ vibrations of a distorted nucleus. The symmetry axis Oz' is shown and the diagrams at the right are sections in the equatorial plane. The full line is the equilibrium contour, the dotted line is one extreme excursion. The arrow shows the direction of rotation in a rotational band.

8.4.4 *Pairing vibrations; superfluidity*

The readiness with which like nucleons or nucleon holes form pairs with spin-parity 0^+ has frequently been referred to; it is the basis of the success of the single-particle shell model, and implies a strong correlation due to short-range attractive forces in the motion of the two particles. A pair may be regarded as a special form of collective excitation, resembling a vibrational quantum with $\lambda^\pi = 0^+$, $T = 1$ obeying Bose–Einstein statistics. A nucleus excited by one such quantum, e.g. a neutron pair, becomes the corresponding nucleus with two extra neutrons and with the 'pair-vibrational' quantum energy.

Starting with a closed shell of neutrons, successive addition of such quanta produces a sequence of nuclei (e.g. $^{204}\text{Pb} \rightarrow {}^{206}\text{Pb} \rightarrow {}^{208}\text{Pb}$) with neutron numbers differing by 2 and with uniformly spaced ground state energies, if perturbations are neglected. This is analogous to the spectrum of a harmonic oscillator; in the present case the states may be labelled by the total isospin whereas for the oscillator the corresponding label is l, the angular momentum (Sect. 1.4.2). These states may be studied in two-particle transfer reactions of the type (t, p) or (p, t) (Sect. 11.5).

The ground state of a nucleus with many particles outside a closed shell contains a number of 0^+ pairs and is potentially in a superfluid condition as discussed for nuclear matter (Sect. 6.4.3). In a macroscopic superfluid such as HeII, correlations arise because of coupling between electronic motions and lattice vibrations and there is a coherence length of the order of atomic dimensions. In nuclei the 0^+ pairs can only be located within the nuclear dimension and the relative coherence is not so marked. The familiar properties of superfluid systems are, therefore, not clearly apparent in nuclear behaviour, except for the existence of the energy gap in the spectrum of intrinsic states in an even–even nucleus (Sect. 6.4.3).

8.5 Static and transition moments

The collective model was developed to explain the large static electric quadrupole moments observed, for example, in the rare-earth nuclei (Sect. 7.2.3). The quadrupole moment Q_0 in the body-fixed system is given by equation (3.32) and for a simple axially symmetrical deformation (Fig. 8.4) specified by the parameter δ we have, if the charge density is uniform (Ex. 8.3),

$$Q_0 = \tfrac{4}{5} Z R^2 \delta \tag{8.29}$$

where R is the equivalent spherical radius. The spectroscopically observed quadrupole moment is related to Q_0 by the expression

$$Q_I = \frac{3K^2 - I(I+1)}{(I+1)(2I+3)} \cdot Q_0 \tag{8.30}$$

for the state $K = I$, i.e.

$$Q_I = \frac{I(2I-1)}{(I+1)(2I+3)} \cdot Q_0 \tag{8.31}$$

where the ratio between Q_I and Q_0 is required because of averaging of the direction of the nuclear axis due to the rotational motion. For $I = 0$ and $I = \tfrac{1}{2}$, Q_I vanishes, although Q_0 may still exist. For very large I, $Q_I \rightarrow Q_0$. The Q_0 values obtained in this way for the rare-earth nuclei are about 10 times the single particle value, which is of the order of R^2, and suggest δ-values of about 0·2, using (8.29). The states of a rotational band are connected by electric quadrupole radiation ($L = 2$) and the probability of this process depends on Q_0^2. The corresponding transition moments for deformed nuclei are very much larger than the single-particle values (Sect. 9.2) as illustrated in Fig. 8.13, and the values of δ obtained from transition rates agree with the evidence from static moments.

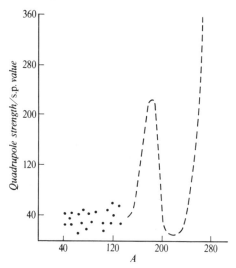

Fig. 8.13   Ground state electric quadrupole ($2^+ \leftrightarrow 0^+$) transition strength, in units of single-particle value, as a function of mass number A (adapted from Ref. 8.1).

Magnetic moments are also better predicted by the collective model than by the single-particle shell model. In a $K = 0$ rotational band there is no magnetic moment due to the intrinsic motion and

$$\mu_I = g_R I \mu_N \tag{8.32}$$

where g_R is the *rotational* g-factor. The magnetic moment is also a factor in the probability for magnetic dipole radiation but this cannot occur between the states of a rotational band with $K = 0$ because it requires $\Delta I = 1$. Measured g_R-factors are $\leqslant Z/A$, which is the classical value for a charged rigid body.

In rotational bands with $K \neq 0$ there is a contribution to the magnetic moment from the intrinsic motion, with a g-factor g_K. The static moment for $K > \frac{1}{2}$ is given in Reference 8.1 as

$$\mu_I = \left\{ g_R I + (g_K - g_R) \frac{K^2}{I+1} \right\} \mu_N \tag{8.33}$$

and the magnetic dipole transition probability includes a factor $(g_K - g_R)^2$. The factor g_K is smaller than the single particle or rigid body value, and the magnetic moments move off the Schmidt lines (Sect. 7.2.3) as required by observation.

8.6 The extended shell model

8.6.1 General approach

The energy of a nucleus containing A nucleons may be written as a sum of kinetic and potential energy terms, i.e.

$$\text{Total energy } H = \sum_{i=1}^{A} T_i + \sum_{i<j} V_{ij} \qquad (8.34)$$

where V_{ij} is the mutual potential energy for nucleons i, j. The SPSM, described in Chapter 7, approximates this to the form

$$H_{SP} = \sum_{i=1}^{A} (T_i + U_i) + \sum_{i<j} v_{ij} \qquad (8.35)$$

where $T_i + U_i$ is now the total energy of a single nucleon obtained by solving the wave equation for a nucleon moving independently in a central field in which the one-body potential energy is U_i. The residual interactions v_{ij} are neglected except for the assumption that nucleons outside closed shells pair off as far as possible with antiparallel spins and orbital momenta. The model has been surprisingly successful in predicting the level sequences for many odd-mass nuclei, and especially for those with nucleon numbers near major closed shells, e.g. $^{209}_{82}\text{Pb}$, but, as noted in Chapter 7, it fails to predict many observed nuclear moments. Moreover, it cannot account for the fragmentation of the single-particle levels observed in nuclei with several 'valence' particles outside the shells. Extension of the model involves consideration of both its major assumptions, namely the independence of motion in a central potential, and the incomplete treatment of the residual interactions.

It has been seen in Section 6.4 that an average field, which essentially creates the one-body shell model potential, is a property of nuclear matter; it will, therefore, plausibly be a property of finite nuclei. For a nucleus it might therefore be thought that the ground state wavefunction could be obtained by a calculation of the *Hartree–Fock* type that has been so successful for atoms. In such calculations the central field is made *self-consistent* by allowing the wavefunctions that it predicts to modify the field itself in which the particles move. A variational procedure then determines both the potential and the wavefunctions that give a minimum energy. Unfortunately, this procedure will not work so simply for nuclei, because of the existence of the repulsive core in the internucleon force (potential V_{ij}, eqn (8.34)). It is necessary, in fact, to use the methods developed for nuclear matter calculations and in particular the Brueckner effective potential or G-matrix (Sect. 6.4). The calculations are complex and difficult but they have been made for the double-closed-shell nuclei ^{16}O, ^{40}Ca and ^{208}Pb and may be used to check the validity of more approximate calculations in which simplifying assumptions are made.

In many such calculations the single-particle energies $(T_i + U_i)$ of equation (8.35), in which kinetic energies appear, are taken directly from experimental data. Attention is then concentrated on the potential energy shifts due to the residual interaction v_{ij}, and on the single-particle states that may be occupied by the particles between which the residual forces act; these states define the *model space* of the IPM (independent-particle model) introduced at the beginning of this chapter. Levels in this model are specified by the symbols (I^π, T) where π denotes the parity. The procedure for calculating a level spectrum is to specify the nucleon configuration that may contribute to the wavefunction of a given level and then to evaluate the matrix elements of the assumed residual interaction between the states of the model space. These matrix elements are to be understood in terms of collisions between two interacting particles (see Fig. 8.14*b*) in which there may be an exchange of orbital motion although angular momentum j is conserved overall. Finally, using these matrix elements augmented by single-particle terms and spin-orbit energies the full Hamiltonian matrix is set up. From this the eigenvalues giving the energies of the states and the eigenvectors giving their wavefunctions are obtained by standard methods. The eigenfunctions can then be used in the evaluation of nuclear moments and transition probabilities.

The calculation of the matrix elements should ideally employ self-consistent wavefunctions, but oscillator functions are generally used for convenience, with an extent chosen to match the nuclear radius. Their use has been justified in some cases by comparison with the results of a Hartree–Fock calculation. The residual interaction v_{ij} should be consistent with what is known of the two-body force (Sect. 5.8) in nuclear matter (Sect. 6.4), but the general trend of level energies can be shown by using rather simple forms of force. In the early work of Edmonds and Flowers and of Elliott and Flowers, particularly for p-shell nuclei, Yukawa or Gaussian forces were used with a specific set of exchange operators and the energy-level spectrum was studied as a function of the range of the force and of the angular momentum coupling scheme (LS or jj). An even simpler force is the *surface-delta* interaction, which exists only in the nuclear periphery and for identical coordinates of the interacting particles. An example of the use of this interaction will be given in the next section.

More realistic calculations must introduce the repulsive core of the internucleon force and a method of doing this was developed by Kuo and Brown (Ref. 8.2) using the techniques of the nuclear matter problem, but confining attention still to only a small number of configurations. When the residual interaction is set up in this way, so that the resulting Hamiltonian acting on a limited range of wavefunctions produces a result hoped to be equivalent to those of a full, unlimited calculation, it is said to be an *effective interaction*. Calculations using effective interactions and experimental single-particle energies can now be made on a very large scale, using several nucleons distributed between several single-particle levels. Such work, in fact, predicts some levels which would normally be thought to form a band, and this emphasizes the basic connection between shell-model states and nuclear deformations.

Finally, it may be noted that in the work of Kuo and Brown, allowance was made for levels resulting from the one-particle, one-hole (1p–1h) configuration in which a nucleon is elevated from the core to a valence state. In light nuclei at least such excitations may form both the 1^- ($T = 1$) giant dipole state and the 3^- ($T = 0$) octupole state, which were previously given a collective interpretation. The observed energies of these states can be understood in this shell-model picture by introducing the particle–hole interaction, which is attractive in $T = 0$ states and repulsive in states of $T = 1$. Intervention of such core excitations also seems necessary in order to remove remaining discrepancies in predicted electromagnetic transition probabilities.

8.6.2 *Typical calculation – the nucleus* ^{18}O

The low-lying energy levels of the nucleus ^{18}O are shown, together with their spins and parities, in Fig. 8.14a. From the neutron and proton numbers $N = 10$, $Z = 8$ the isobaric spin component is $T_z = -1$, so that the isobaric spin T itself must be at least 1 for all levels. Since the lowest levels have analogue states in just two other nuclei (^{18}F, ^{18}Ne) we conclude that these levels have $T = 1$.

The SPSM suggests that for an inert ^{16}O core ($Z = N = 8$, $I^\pi = 0^+$) the two extra neutrons may occupy the 2s or 1d levels or both. Possible configurations for a state $(I, T) = (0, 1)$ with jj-coupling would be

$$(1d_{5/2}^2; 0, 1), \qquad (2s_{1/2}^2; 0, 1), \qquad (1d_{3/2}^2; 0, 1) \qquad (8.36)$$

in an obvious notation and for the state $(2, 1)$ the first and last of these are also possible together with

$$(1d_{5/2}2s_{1/2}; 2, 1), \qquad (1d_{5/2}1d_{3/2}; 2, 1), \qquad (1d_{3/2}2s_{1/2}; 2, 1) \quad (8.37)$$

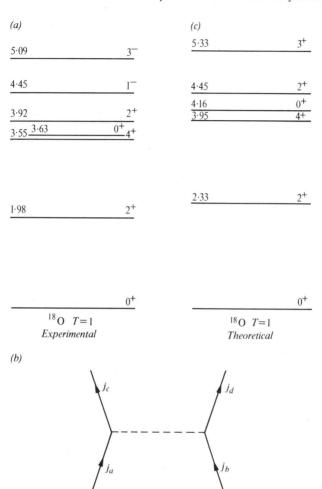

Fig. 8.14 (a) Low-lying levels of ^{18}O (experimental). (b) Interaction between two nucleons in a bound state, $j_a + j_b = j_c + j_d$. (c) Even-parity ^{18}O levels predicted by T. T. S. Kuo and G. E. Brown (Ref. 8.2) allowing for core excitation.

making five in all for this state. All the states (8.36) and (8.37) are symmetrical for exchange in isospin because they relate to two identical particles and they are of even parity because only the angular momenta $l = 0, 2$ are involved. Levels (3, 1) and (4, 1) can also be formed from some of these configurations.

The space wavefunctions ϕ of the two nucleons are usually taken to be eigenfunctions $R_{nl}Y_l^m$ of the oscillator potential. The two-nucleon wavefunction is then constructed, in analogy with similar constructions in the theory of the helium atom, using the standard

245

coupling procedure for angular momenta described in Appendix 4. It is an eigenfunction of the total angular momentum operator and has the form

$$\psi(j_1 j_2; IM) = \sum_{m_1} (j_1 j_2 m_1 m_2 \mid IM) \phi_1(j_1 m_1) \phi_2(j_2 m_2) \qquad (8.38)$$

where j_1, j_2 are the individual angular momenta and m_1, m_2 ($= M - m_1$) their resolved parts, coupling to the two-particle state of spin I and resolved part M. If $j_1 \neq j_2$ (8.38) should be anti-symmetrized, but for $j_1 = j_2$ it is, in fact antisymmetric for even values of I (see Ex. 8.10). Since only $T = 1$ states are involved the isobaric spin coupling coefficient $(\frac{1}{2}\frac{1}{2} - \frac{1}{2} - \frac{1}{2} \mid 1 - 1)$ is unity. Each individual j is composed of a spin part and an orbital part, and the component wavefunctions may be expressed

$$\phi_1(j_1 m_1) = \sum_{m_l m_s} (l_1 s_1 m_l m_s \mid j_1 m_1) \phi_L(lm_l) \phi_S(sm_s) \qquad (8.39)$$

The wavefunction for a given state of ^{18}O is a linear combination of the two-particle states (8.38) which form a model space in this example.

For the surface-delta interaction a specific formula for the matrix elements $\langle j_a j_b \mid v \mid j_c j_d \rangle$ of the residual interaction v is given by Glaudemans *et al.* (*Nuclear Physics* **A102**, 593 (1967)), together with an empirical value of the interaction strength. Matrices can, there-fore, be set up for the 0^+, 2^+, 3^+ and 4^+ states (see Ex. 8.11). Each matrix element is adjusted for the spin-orbit energy, which is known empirically from the $d_{5/2}$–$d_{3/2}$ doublet separation in ^{17}O. One of the states, e.g. 4^+, is then made to have its observed energy and this gives a further term to be added to the diagonal matrix elements; this is the same for all states if the 2s and 1d nucleons are described by a harmonic oscillator potential (in which these two states are degenerate) and if the effect of the ^{16}O core is the same in all states. The energies of the remaining states 0^+, 2^+, 3^+ then follow as solutions of the eigenvalue problem. Orthogonal states of the same spin-parity but at higher energy are also found in the calculation.

Example 8.11 gives numerical results for an even further simp-lified calculation for ^{18}O, excluding the $d_{3/2}$ states and the 3^+ level. The more realistic predictions given by Kuo and Brown are shown in Fig. 8.14c. These reproduce the lowering of the 0^+ ground state that is an example of the energy gap well known in fermion systems. The 2^+ state is also lowered for similar reasons.

8.7 Study of the nuclear level spectrum

Experimentally, level sequences and properties may indicate the nature of the nuclear Hamiltonian, revealing the presence or ab-sence of collective or single-particle effects. By far the greatest amount of information on these properties, for comparison with the theories outlined in this and the preceding chapter, is obtained from nuclear reaction experiments. Such dynamical processes have an interest of their own, particularly in the processes of fission and fusion, and they will be described in the following chapters with some emphasis on the information that they offer on nuclear levels. The processes chiefly used are summarized below for particular properties.

(i) *Wavefunction.* Static moments of levels (Sect. 3.4) and radiative transition probabilities between them (Sect. 9.2) are obtained directly from wavefunctions by the application of suitable operators. In nuclear reactions the spectroscopic factor (Sect. 11.5.2) measures the extent to which a wavefunction represents that of a single-particle level.

(ii) *Excitation energy.* Discrete levels are studied by nuclear reactions such as deuteron stripping or pickup. Such transfer reactions strongly excite single particle or hole levels, whereas inelastic scattering preferentially excites collective modes. Low-lying levels are excited by α- and β-decay of radioactive parents.

(iii) *Width or lifetime.* The (radiative) width of bound levels is generally so small that direct measurement is impossible except under the special circumstances offered by the Mössbauer effect (Sect. 9.4.4). The widths Γ are obtained from lifetime measurements (see eqn (9.11) and Sect. 9.4) or from the cross-section for Coulomb excitation (Sect. 9.4.2).

Virtual levels have particle widths as well as radiative widths and may be determined from the shape of resonance or transmission curves.

(iv) *Spin, parity and isospin.* Information on these quantum numbers comes from angular distribution experiments, both for production and decay of the states, and from the operation of selection rules in these processes. Identification of isobaric analogue states in a range of nuclei may fix the total isospin number T.

In discussing these processes in the following chapters it will be convenient to group them under the headings of the three main interactions, electromagnetic, weak and strong, though it will rapidly become clear that rigid boundaries cannot be drawn.

8.8 Summary and survey of nuclear spectra

Collective and independent-particle modes of motion in nuclei are superficially distinct and each describes a special group of experimental facts. More fundamentally, however, these two models must be unified, and there is evidence from Hartree–Fock calculations that this is possible. An intermediate step is the Nilsson model in which shell-model orbits are set up in a deformed potential that is chosen semi-empirically.

From a phenomenological point of view, it is apparent that the most important parameter for an understanding of the general nature of the low-lying (bound) levels is the number of particles outside a closed shell or the number needed to complete a shell. Figure 8.15 illustrates the change in the pattern of observed levels as the number of holes increases from the doubly-magic nucleus $^{208}_{82}$Pb. In ^{207}Pb the intrinsic single-particle (i.e. hole) levels of the spherical potential are found, in ^{206}Pb there is a two-hole spectrum showing the depression of the 0^+ ground state and 2^+ excited state with respect to the higher levels as a result of coherence associated with the pairing interaction.

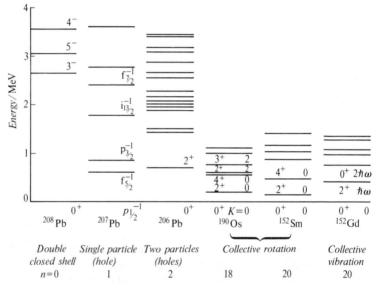

Fig. 8.15 Sketch of level spectra with increasing number (n) of particles outside, or holes in, a double-closed shell.

From ^{190}Os to ^{152}Sm, the even–even nuclei show the rotational spectra of the deformed shape, while at ^{152}Gd the deformation is small and the vibrational spectrum of the spherical shape is seen. When allowance is made for the $A^{-1/3}$ dependence of level spacings (eqn (7.3)), it is found that the lower rotational and vibrational states lie within the energy gap in the intrinsic states of an even–even nucleus.

At excitation energies well above the nucleon binding energy, the most prominent features of nuclear spectra are the giant states, which are envelopes of several MeV width corresponding with excitations that traverse one or two major shells. In the same region of virtual levels, the isobaric analogue states of ≈ 100 keV width (Sect. 5.6.1) are located. Figure 8.16a, taken from Reference 8.3a, relates the giant resonance states to the bound levels and to the so-called 'statistical' region of levels (Sect. 11.3) in which the concept of a level spectrum loses its meaning. Figure 8.16b is an energy-level schematic conveying similar information.

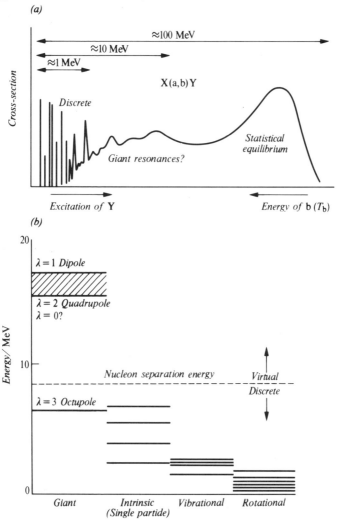

Fig. 8.16   (a) The nuclear level spectrum, as seen in the energies of particles b emerging from a nuclear reaction $X(a, b)Y^*$ with bombarding energy ≈ 100 MeV. The energy scale gives both the energy T_b and the resulting excitation of the nucleus Y (Ref. 8.3a). (b) Schematic of main features of a nuclear level spectrum, showing different types of state.

Elements of nuclear physics

Examples 8

8.1 Find the excitation of the compound nucleus ^{108}Ag in process (8.3) if it is produced by the absorption of a neutron of energy 1 eV. Assume the values of $M - A$ in keV for the nuclei concerned are n, 8072; ^{107}Ag, $-88\,408$; ^{108}Ag, $-87\,605$. What precision in the measurement of $(M - A)$ would justify the inclusion of a correction for the neutron energy? [7269 keV]

8.2 Consider two masses M bound together with an equilibrium separation R. Estimate the order of magnitude of the lowest rotational frequency of the system, and compare it with the vibrational frequency due to a vibrational amplitude βR. $[\omega_v/\omega_r \approx 1/\beta^2]$

8.3* Calculate the intrinsic quadrupole moment of a uniformly charged ellipsoidal nucleus with major and minor semi-axes a, b.

If the deformation parameter δ is defined by the equation

$$\delta = \tfrac{3}{2}(a^2 - b^2)/(a^2 + 2b^2)$$

show that the quadrupole moment of a nucleus with Z protons located at points r_k may be written

$$Q_0 = \frac{4}{3}\left\langle \sum_{k=1}^{Z} r_k^2 \right\rangle \delta$$

By finding $\langle \sum r_k^2 \rangle$ for a spherical distribution of sharp radius R show also that

$$Q_0 = \tfrac{4}{5}ZR^2\delta$$

8.4 A rotational band based on the state $K = \tfrac{7}{2}^-$ is known in $^{179}_{74}$W. The $\tfrac{9}{2}^-$ state is at an excitation of 120 keV; predict the excitations of the $\tfrac{11}{2}^-$, $\tfrac{13}{2}^-$ and $\tfrac{15}{2}^-$ levels. [267, 440, 640 keV]

8.5* The nucleus $^{234}_{92}$U has levels of spin-parity 0^+, 2^+, 4^+, 6^+, 8^+ at energies of 0, 44, 143, 296 and 500 keV. Show that these form a rotational band, predict the energy of the 10^+ state, and check the variation of the moment of inertia of the nucleus with rotational frequency. Compare the moment of inertia with the rigid-body value.

8.6* Prove equations (8.15) and (8.16).

8.7 Using equations (8.15) and (8.16) study the variation of moment of inertia with rotational frequency for the ground state band of ^{158}Er using the following energies for the successive electric quadrupole transitions upwards through the band from $I = 2 \to 0$ to $I = 16 \to 14$: 192·7, 335·7, 443·8, 523·8, 578·9, 608·1, 510·0, 473·2 keV. (Ward, D. et al. Phys. Rev. Lett. **30**, 493, 1973)

8.8 Using equation (8.18) and the information on ^{169}Tm in Fig. 8.6c, estimate the decoupling parameter a. [$-0\cdot78$]

8.9* Write down the μ-values for two quadrupole phonons and verify that the states which are symmetrical under exchange have the angular momenta shown in Table 8.1 for $n_2 = 2$.

8.10* Show that for two nucleons in the same j-shell, the $T = 0$ states have I odd and the $T = 1$ states have I even.

[Express the angular momentum wavefunction in terms of single-particle functions as in equation (8.38) and consider the effect of nucleon exchange. Use the Clebsch–Gordan property

$$(j_1 j_2 m_1 m_2 \mid IM) = (-1)^{j_1 + j_2 - I}(j_1 j_2 m_2 m_1 \mid IM)]$$

For the case of two $p_{3/2}$-protons, write out the m_l and m_s values explicitly and verify the grouping into states of $I = 0$ and $I = 2$.

8.11* Calculate the spacing of the 0^+ and 2^+ levels of the nucleus ^{18}O relative to the 4^+ level due to an effective interaction in the $d_{5/2}$ and $s_{1/2}$ states only (i.e. omitting $d_{3/2}$ and the 3^+ level). Take the energy of the 4^+ state to be -0.35 MeV, the $(d_{5/2}-d_{3/2})$ doublet splitting in ^{17}O to be 5.08 MeV, and use the following matrix elements of Kuo and Brown (in MeV) for the residual interaction:

	$\frac{5}{2}\frac{5}{2}\lvert v \rvert \frac{5}{2}\frac{5}{2}$	$\frac{5}{2}\frac{5}{2}\lvert v \rvert \frac{5}{2}\frac{1}{2}$	$\frac{5}{2}\frac{1}{2}\lvert v \rvert \frac{5}{2}\frac{1}{2}$
$I = 4^+$	0.14		
$I = 2^+$	-0.94	-0.81	-1.09
$I = 0$	-2.53	-1.09	-2.35

Find the corresponding eigenfunctions.

III Nuclear interactions

9 The electromagnetic interaction

The electromagnetic interaction, specified by the laws associated with the names of Coulomb, Ampère and Faraday, is well tested over the whole range of energies significant to atomic, nuclear and particle physics. We have already seen that electromagnetic forces are of long range, operating on a macroscopic scale, and that Maxwell equations, which describe interference and diffraction of light, are relativistically invariant. The quantization of the electromagnetic field must, therefore, be such as to leave these general classical properties unaffected, and this is achieved by the introduction of the massless *photon* as the field quantum.

The photon is a desirable concept, beyond its role in field theory, in order to simplify the description of phenomena such as the emission and absorption of radiation, the photoelectric effect and the Compton effect (Sect. 3.1.4). The field itself may be pictured as in Fig. 2.8, redrawn in Fig. 9.1a, in which the exchange of a virtual photon between two charges affects the energy of the system and leads to a force between the charges. In terms of the uncertainty principle argument given in Section 2.2.1, the zero mass of the (real) photon would correspond with a force of infinite range. The virtual photon may also transform itself into a positron–electron pair and these charges separate in the electric field of the original charge creating what is known as a *vacuum polarization*. The strength of the interaction mediated by these virtual processes is characterized by the dimensionless fine-structure constant

$$\alpha = (\mu_0 c^2/4\pi)e^2/\hbar c = \tfrac{1}{137} \qquad (9.1)$$

also known as the *electromagnetic coupling constant*.

The virtual radiations just discussed will be re-absorbed by the particle from which they emerge (Fig. 9.1b). The theoretical consequences of such processes are profound and have led to the *renormalization* techniques of field theory which are necessary to permit the normal particle-mass to be used meaningfully. Experimentally, these essentially quantum effects account for the Lamb shift of atomic energy levels and for the small anomalous part of the

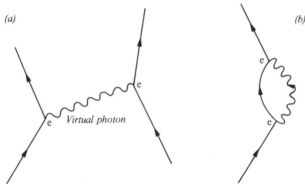

Fig. 9.1 The electromagnetic interaction. (*a*) Electron–electron force, or scattering, described by one-photon exchange. (*b*) Emission and re-absorption of a virtual photon.

magnetic moment of the electron and of the muon. The fact that these quantities are small is due to the actual value of α (eqn (9.1)) and is responsible for the validity of semi-classical calculations of electromagnetic processes in which only *single-photon exchange* between (for instance) a nucleus and a radiation field is considered and the perturbation methods of quantum mechanics are used. These methods will be applied in the present chapter. The smallness of α also makes it useful to classify radiative processes according to the power of α, or of e^2, that enters into the transition probability or interaction cross-section.

Quantum electrodynamics has been exhaustively tested in high-energy electron scattering and bremsstrahlung experiments and in the prediction of the anomalous magnetic moments of the electron and muon. There is no evidence to suggest that these particles behave differently from structureless point charges, and there is no breakdown of the electromagnetic interaction between such charges down to distances of considerably less than the 'classical' radius of the electron, which is $2 \cdot 8 \times 10^{-15}$ m. This interaction is, therefore, likely to be a reliable means of studying nuclear properties, especially the nuclear charge distribution (Sect. 6.2) and the quantum numbers and radiative properties of nuclear levels, with which we will be concerned in the following sections. The role of the static Coulomb interaction in nuclear reactions, in the phenomena of α- and β-decay, in fission and in heavy-ion reactions will be considered in Chapters 10 and 11.

9.1 General properties

9.1.1 Classification and selection rules

The invariance properties of the electromagnetic interaction are listed in Tables 1.1 and 2.3*b*. Classification of radiative processes

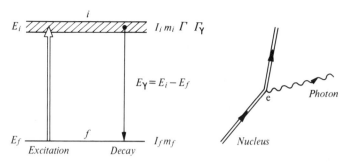

Fig. 9.2 Emission of radiation from a nuclear level of width Γ and radiative width Γ_γ. The sketch at the right is a graph for a first-order decay process. The converse excitation between the levels is also indicated.

such as the nuclear transition shown in Fig. 9.2 is based on the conservation of angular momentum and of parity between the radiating system and the radiation field.

Thus, if an emitted photon transports angular momentum L with respect to the origin, with resolved part $M\hbar$, we have for the level spins (I, m) shown in the figure,

$$\boldsymbol{I}_i = \boldsymbol{L} + \boldsymbol{I}_f \qquad (9.2)$$

where in accordance with equation (5.48)

$$|\boldsymbol{L}|^2 = L(L+1)\hbar^2, \qquad |M| \leqslant L$$

From equation (9.2) we then obtain

$$\left. \begin{array}{c} I_i + I_f \geqslant L \geqslant |I_i - I_f| \\ m_i = M + m_f \end{array} \right\} \qquad (9.3)$$

These are the *selection rules* for angular momentum, which limit the *multipolarity* L of the transition, i.e. the number of units of angular momentum $\hbar$ transferred in the radiative process. The meanings of the quantum numbers are shown in the vector diagram, Fig. 9.3a.

Despite a small effect due to weak currents (Ch. 10) parity (P) is conserved to a high degree of accuracy in electromagnetic processes and a *parity selection rule* must also be obeyed. The field, however, can be of either even or odd parity for a given multipolarity and the classification of radiative transitions is, therefore, as shown in Table 9.1. The symbols E, M denote fields of *electric* or *magnetic* type respectively and pay due regard to the corresponding classical oscillator.

As indicated in Table 2.3b the electromagnetic interaction conserves charge but not isobaric spin. The selection rules for this

TABLE 9.1 Classification of radiative transitions

Type of radiation*	E1	E2, M1	E3, M2
Name	Electric dipole	Electric quadrupole, magnetic dipole	Electric octupole, magnetic quadrupole
Multipolarity	1	2, 1	3, 2
Parity change in transition	Yes	No	Yes

* Monopole transitions, e.g. E0, do not exist in the radiation field (Sect. 9.1.2).

quantity are

$$\Delta T = 0, \pm 1 \quad \text{(isospin change)}$$

$$\Delta T_z = 0 \quad \text{(charge conservation)}$$

The interaction is invariant under time reversal (T) as is known from experiments on radiative transitions from oriented nuclei, and also under charge conjugation (C) (Table 1.1) since positrons behave electromagnetically in complete analogy to electrons.

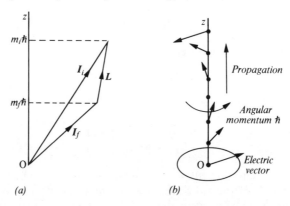

Fig. 9.3 (*a*) Vector diagram for emission of radiation of multipolarity L. (*b*) For propagation along Oz, $M = m_i - m_f = \pm 1$ and radiation is circularly polarized.

9.1.2 The multipole fields (Ref. 9.1)

Because the electromagnetic field of force is a vector field, and the vector potential $\mathbf{A}$ is odd under the parity operation, the photon is assumed to have odd *intrinsic* parity. Moreover, a vector field transforms under rotations like a particle of spin 1 and the spin-parity character of the photon is, therefore, 1^-.

Field vectors are solutions of a wave equation and any arbitrary field specified, for instance, by a vector potential $\mathbf{A}$ can be expanded in terms of a set of eigenfunctions. For nuclear problems it is convenient (as in the case of the description of a beam of particles, Sect. 1.2.7) to choose eigenfunctions of the angular momentum operators, which will be written $\mathbf{L}^2$ and L_z, where for an eigenfunction $\mathbf{A}_{LM}$

$$\mathbf{L}^2\mathbf{A}_{LM} = L(L+1)\hbar^2\mathbf{A}_{LM} \qquad (9.4a)$$

$$L_z\mathbf{A}_{LM} = M\hbar\mathbf{A}_{LM} \qquad (9.4b)$$

The angular part of such eigenfunctions can be constructed by coupling together the intrinsic spin of the field s with an orbital angular momentum l using the procedures of Appendix 4. For $s = 1$ and a given L, M there are three angular eigenfunctions, known as *vector spherical harmonics*, corresponding with $l = L + 1$, $l = L$, and $l = L - 1$, and these multiplied by radial functions and added together form a *pure multipole field*, labelled by L, M, for which equations (9.4) hold.

For each multipole field the terms with $l = L \pm 1$ have an orbital parity opposite to that of the term with $l = L$ since the orbital angular functions are the ordinary spherical harmonics Y_l^m. These fields are those of electric and magnetic radiation respectively, and after multiplication by the parity of the photon, give the overall parities shown in Table 9.1. Because of the transverse nature of the electromagnetic field it may be shown (Ref. 9.1) that there is no multipole field in free space for $L = 0$. For nuclear spins $I_i = I_f = 0$, radiation as such is *strictly forbidden*; alternative electromagnetic processes for the $0 \to 0$ transitions are discussed in Section 9.3.3.

In the present formalism, a photon propagating along the axis Oz may be represented as in Fig. 9.3b. There is no component l_z of orbital momentum along the axis and the spin can only have components $\pm\hbar$ if the wave is transverse; these correspond with the two possible signs of circular polarization.

The multipole fields can be used to describe the radiation pattern observed as plane waves at a distance of many wavelengths from the source ($kr \gg 1$, where k is the wavenumber). The angular distribution for a particular multipole indicated by the selection rules is proportional to the square of a vector spherical harmonic. The fields can also be connected with the source densities that give rise to the radiation in the other limiting approximation ($kr \ll 1$). This leads directly to the nuclear matrix element which determines the multipole transition probability, apart from energy-dependent factors.

9.1.3 *The electromagnetic matrix element*

If the coupling between the nuclear current and magnetization distributions is sufficiently weak that second-order effects may be neglected, the probability per unit time of an electromagnetic transition between the levels i and f shown in Fig. 9.2 is given by time-dependent perturbation theory as

$$T_{if} = (2\pi/\hbar) |\langle f| H_{em} |i\rangle|^2 \rho(E) \tag{9.5}$$

where $\langle f| H_{em} |i\rangle = H_{if}$ is the matrix element of the electromagnetic interaction between the initial and final states, and $\rho(E)$ is the phase-space factor giving the number of momentum states per unit energy range at the transition energy in the volume V in which the process is considered to take place. The quantity $\rho(E)$ has the dimensions of (energy)$^{-1}$ and the matrix element itself is an energy. If the wavefunctions occurring in H_{if} are suitably normalized the system volume V will not appear in the final expression for T_{if}. The brackets $\langle \ \rangle$ indicate an integral of the interaction operator between two states and over the volume V in which their wavefunctions exist. Both the specification of the initial and final states and of the interaction Hamiltonian H_{em} depend on the process considered, e.g. electron scattering, Coulomb excitation, internal conversion or absorption and emission of radiation; in the case of elastic electron scattering the transition probability leads to a cross-section which is exactly that predicted by the plane-wave Born approximation (Sect. 6.2).

For purposes of illustration we confine attention to the radiation problem. Although formula (9.5) may be applied directly to an absorption process we shall assume that it is also valid for emission because of time-reversal invariance. The matrix elements are actually complex conjugates in the two cases, and the assumption is justified by a more rigorous calculation using quantum field theory. If the radiation is produced by an oscillating current distribution of density j (A m^{-2}) in the nucleus then the classical interaction energy is

$$H_{em} = \tfrac{1}{2} \int j \cdot A \, dV \tag{9.6}$$

where A is the vector potential of the radiation field, which may be written as an outgoing wave

$$A = A_0 \exp i(k \cdot r - \omega t) \tag{9.7}$$

where k is the wavevector and ω is the frequency. This form of the interaction energy has an immediate application, by analogy, in the theory of weak interactions (Ch. 10) where the energy arises by the interaction of one current with another. Until recently it was thought that only charged currents were effective in such a scheme but, as mentioned in Section 2.2.3, neutral currents are now known also to exist.

In evaluating the matrix element the current density is replaced by the proper quantum operator, which is shown, apart from the ratio e/m, in equation (1.90). This is in the first place applied to a single proton but a sum over all protons in the nucleus may also be taken if necessary. The time factor $e^{-i\omega t}$ disappears, essentially because energy is conserved between the nucleus and the radiation field and we are left with the factor $e^{i\mathbf{k}\cdot\mathbf{r}}$. For radiation of energy ≈ 500 keV the wavelength $\lambda = 2\pi/k$ is 2×10^{-12} m so that kr is a small quantity for an average nucleus and in this *long-wavelength approximation* the exponential may be expanded directly, or the interaction may be expressed as a series of multipole fields, beginning with $L = 1$. In the latter case the radial part of the wavefunction for the multipole L is a spherical Bessel function, which for $kr \ll 1$ has the form

$$j_L(kr) \approx (kr)^L/(2L+1)!! \qquad (9.8)$$

where $(2L+1)!! = 1 \cdot 3 \cdot 5 \cdot \cdots \cdot (2L+1)$.

In either case the interaction finally appears as a set of multipole operators, which for electric radiation have the structure (for a point charge e)

$$er^L Y_L^m(\theta, \phi) \qquad (L \geqslant 1) \qquad (9.9)$$

but for magnetic radiation due to the currents are reduced in value by a factor $\approx v/c$ where v is a typical nucleon velocity in the nucleus, say $v \approx 0 \cdot 1c$. The matrix elements of these operators between the nuclear states $i(I_i, m_i)$ and $f(I_f, m_f)$ are the basic quantities determined by electromagnetic experiments; they are non-zero only for the transitions specified by the selection rules. From (9.8) it may be seen that the intensity of successive multipoles, if allowed by the selection rules, decreases by a factor $\approx (kR)^2$ where R is the nuclear radius, for each increase of 1 in multipolarity. Low multipolarities are, therefore, strongly favoured in nuclear radiation unless the corresponding matrix element is abnormally small.

The treatment so far sketched must be extended to include radiation due to the intrinsic magnetic moments of the nucleons whose state of motion in a nucleus is changing.

9.2 Lifetime-energy relations

9.2.1 General: the Weisskopf formula

For the level i shown in Fig. 9.2 the radiative transition probability T_{if} is the reciprocal of the mean life τ_γ for the emission of radiation and is connected with the radiative width Γ_γ by the equation

$$T_{if} = 1/\tau_\gamma = \Gamma_\gamma/\hbar \qquad (9.10)$$

so that

$$\Gamma_\gamma \tau_\gamma = \hbar = 6\cdot 6 \times 10^{-16} \text{ eV s} \tag{9.11}$$

It is, of course, just equal to the 'radioactive' decay constant λ of the exponential decay of the state or to $0\cdot 693/t_{1/2}$ where $t_{1/2}$ is the halflife.

If neither I_i nor I_f is zero, transitions of mixed multipolarity, e.g. M1 + E2, may occur between the states. The mixing ratio δ is defined as a ratio of matrix elements but for most purposes it is sufficient to note that

$$\delta^2 = \Gamma_\gamma(L+1)/\Gamma_\gamma(L) \quad \text{e.g.} \quad \Gamma_\gamma(\text{E2})/\Gamma_\gamma(\text{M1}) \tag{9.12}$$

This ratio of partial widths must be obtained using conversion coefficients (Sect. 9.3) or angular correlation experiments (Sect. 9.5) before the individual radiative widths can be extracted from the total radiative width Γ_γ.

A crude estimate of the electric dipole (E1) transition probability may be obtained from the classical formula for the averaged rate of radiation of energy from a simple harmonic oscillator of dipole moment p and angular frequency ω. This is

$$\frac{\overline{dW}}{dt} = \left(\frac{\mu_0 c^2}{4\pi}\right)\left(\frac{p^2 \omega^4}{3c^3}\right) \tag{9.13}$$

If this radiation is emitted as photons of energy $E_\gamma = \hbar\omega = E_i - E_f$ and if p is replaced by a suitable quantum-mechanical matrix element er_{if}, for the intial and final states shown in Fig. 9.2, then the transition probability for level i is given by

$$T_{if} = \frac{1}{\tau_\gamma} = \frac{1}{\hbar\omega}\left(\frac{\overline{dW}}{dt}\right) \propto \left(\frac{E_\gamma}{\hbar c}\right)^3 |\mathbf{r}_{if}|^2 \tag{9.14}$$

The transition probability is thus proportional to the transition energy cubed.

Alternatively, and more generally, use may be made of the formula (9.5) for the transition probability, and the multipole expansion of the interaction. The result for multipolarity L is

$$T_{if}(L) = \frac{1}{\tau_\gamma} = \left(\frac{\mu_0 c^2}{4\pi}\right)\frac{8\pi(L+1)}{L[(2L+1)!!]^2} \cdot \frac{1}{\hbar}\left(\frac{E_\gamma}{\hbar c}\right)^{2L+1} B_{if}(L) \tag{9.15}$$

where $B_{if}(L)$, sometimes written $B(L)\downarrow$ to signify an emission process, is the *reduced transition probability* and is the 'internal' factor containing nuclear information. It is essentially the square of the matrix element of the appropriate multipole operator averaged over the $(2I_i + 1)$ substates m_i and summed over the accessible substates m_f. For electric transitions $B(L)$ is measured in the units

e^2 fm^{2L} and for magnetic transitions in $(\mu_N/c)^2$ fm^{2L-2}, where μ_N is the nuclear magneton.

In some experiments such as Coulomb excitation (Sect. 9.4.2) the upward probability $B_{fi}(L)$ or $B(L)\uparrow$ is measured. This has the same matrix element and is thus related to $B(L)\downarrow$ according to the definition by the formula

$$(2I_i+1)B_{if}(L)=(2I_f+1)B_{fi}(L) \tag{9.16}$$

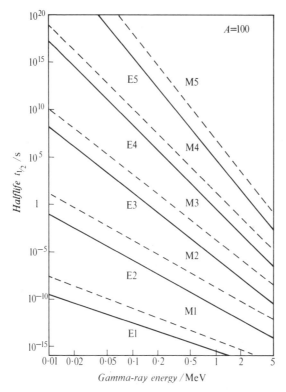

Fig. 9.4 Lifetime–energy relations for γ-radiation according to the single-particle formula of Weisskopf, for a nucleus of mass number $A = 100$.

For practical purposes it is convenient to introduce an estimate of the quantity B_{if}, so that a lifetime–energy relation may be established. The *Weisskopf* formula is based on the *single-particle shell model* and assumes radiation to result from the transition of a single proton from an initial orbital state to a final state of zero orbital angular momentum. The resulting reduced transition probabilities are

$$B(EL)=(e^2/4\pi)(3R^L/L+3)^2 \quad \text{for electric radiation} \tag{9.17a}$$

and

$$B(ML) = 10(\hbar/m_p cR)^2 B(EL) \quad \text{for magnetic radiation} \quad (9.17b)$$

where R is of the order of magnitude of the nuclear radius. The factor 10 in $B(ML)$ is introduced to allow for magnetic radiation originating from reorientation of intrinsic spins. The lifetime–energy relations based on these estimates have been very widely used; they are shown in Fig. 9.4 for the particular case of $A = 100$ and it is clearly seen that lifetimes become long as energy decreases and as multipolarity increases. The formulae also give *single-particle radiative widths* directly, e.g.

$$\left. \begin{array}{l} \Gamma_\gamma(E1) = 0 \cdot 07 \, E_\gamma^3 A^{2/3} \\[4pt] \Gamma_\gamma(M1) = 0 \cdot 021 \, E_\gamma^3 \\[4pt] \Gamma_\gamma(E2) = 4 \cdot 9 \times 10^{-8} \, A^{4/3} E_\gamma^5 \end{array} \right\} \qquad (9.18)$$

where Γ_γ is the radiative width in eV, E_γ the transition energy in MeV and A the mass number of the nucleus. In the individual-particle model radiative widths would be expected to be smaller, because of the sharing of the radiative moment among the several particles of the configuration. The M1 transition probability would vanish in the absence of spin-orbit coupling.

If the nuclear core participates in the radiative process, in-phase motion of particles gives radiative widths which are much larger than the single-particle value (Sect. 8.5).

9.2.2 Nuclear isomerism

Figure 9.4 shows that electromagnetic transitions of high multipolarity and low energy are relatively slow processes. Excited states of nuclei which can only decay by such transitions may, therefore, have a long life. Nuclei excited to these states will differ from unexcited nuclei in their radioactive properties and in their static moments, but not in charge or mass number. Such nuclei are said to be *isomeric* with respect to their ground state.

In nuclei stable against beta decay an isomeric transition is usually accompanied by internal conversion electrons (Sect. 9.3) since the conditions for isomerism are just those for high internal conversion coefficients. The experimentally observed radiation is then often an electron line of low energy decaying with a halflife short compared with that expected for a nuclear beta decay of corresponding energy. The X-rays following the electron emission are characteristic of the radiating atom itself and not of the daughter as in the case of beta decay or electron capture (Ch. 10). There is no clear delimitation of the range of isomeric lifetimes but it is customary to regard them as measurable without special methods, i.e. perhaps $t_{1/2} > 10^{-6}$ s.

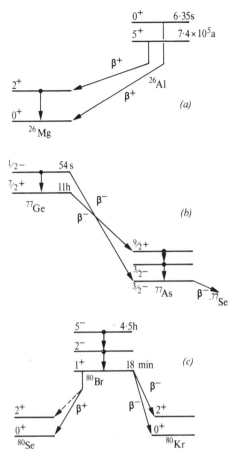

Fig. 9.5   Examples of nuclear isomerism. (*a*) ^{26}Al, in which the isomeric state decays by positron emission and no radiative transition is seen. (*b*) ^{77}Ge in which β- and γ-emission compete in the decay of the $\frac{1}{2}^-$ state. (*c*) ^{80}Br in which the 5$^-$ state decays wholly by radiative transition.

Typical examples of isomeric transitions are shown in Fig. 9.5. Isomers may be produced in all types of nuclear reaction including beta decay, but the yield may be low if the transfer of a large amount of angular momentum is necessary. In such cases it is possible to reach the isomeric level by exciting a higher level of lower spin, from which cascade transitions can take place. The frequency of occurrence of nuclear isomers is plotted as a function of the odd-nucleon number for nuclei of odd *A* in Fig. 9.6. There is a characteristic grouping just below the major closed shells at *Z* or *N* = 50 and 82 and a less marked distribution near *N* = 126. These so called 'islands of isomerism' receive an immediate interpretation

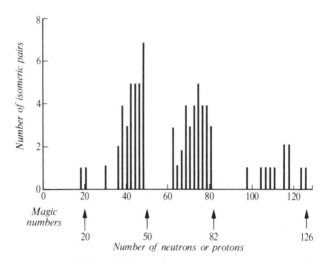

Fig. 9.6   Frequency distribution of odd-*A* isomeric nuclei (Ref. 2.1).

in terms of the single-particle shell model (Sect. 7.2.3) and form an important part of the evidence for the validity of this model.

An isomeric state is indicated by a superscript m, e.g. in Fig. 9.5 the states are ^{26}Al^{m}, ^{77}Ge^{m} and ^{80}Br^{m}.

9.3 Internal conversion

9.3.1 General

The lifetime formula of Weisskopf for radiative transitions gives the probability of decay of a bare point nucleus, completely stripped of the atomic electrons. Usually this is not the case, and since many electronic wavefunctions (in particular all wavefunctions of s-states) have finite amplitudes at or near the nucleus it is possible for nuclear excitation energy to be removed by the ejection of an atomic electron. The total probability per unit time of decay of the excited nucleus is then given by $\Gamma/\hbar$ where we write for a bound state

$$\Gamma = \Gamma_\gamma + \Gamma_e \qquad (9.19)$$

in which Γ_e is the width for emission of electrons. In both cases the total energy of nuclear excitation is removed, but the electron emission process is generally described for historical reasons as internal conversion. This does not imply that the process follows the emission of radiation and the two processes must be regarded as competing alternatives. That this is so has been demonstrated by experiments in which the lifetime of a nucleus for isomeric decay

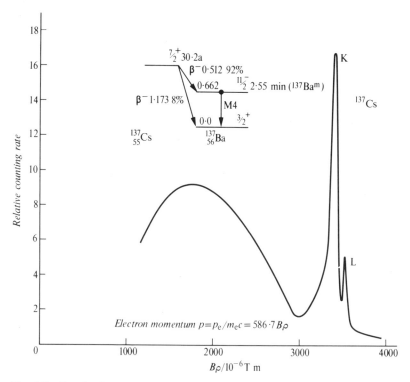

Fig. 9.7   Sketch of electron spectrum seen in a simple 180° spectrometer of the beta-decay electrons from ^{137}Cs, together with K and L internal conversion electrons from the 0.662 MeV transition in ^{137}Ba^{m}. The decay scheme is shown, with energies in MeV. The K- and L-vacancies lead to X-ray emission from the Ba atom.

has been significantly altered by chemical control of its electronic environment.

Figure 9.7 shows the electron spectrum obtained in a magnetic spectrometer for the decay of ^{137}Cs, whose long life of 30a is due to the forbidden nature of its beta-particle emission. Superimposed upon the continuous distribution of electrons arising from the nuclear beta decay are sharp internal conversion lines resulting from the subsequent electromagnetic decay of the isomeric state ^{137}Ba^{m} with a halflife of 2·6 min. Gamma radiation corresponding with this transition is also observed and the energy E_γ of the transition may be obtained very accurately from the conversion spectra because the kinetic energy of an electron ejected from the K-shell is

$$E_\gamma - E_K \qquad\qquad (9.20)$$

where E_K is the energy of the K-edge, which is known precisely from X-ray data. In experimental observations the internal-conversion electrons must be distinguished carefully from photo-electrons produced externally by the gamma radiation. If the number of electrons observed per excited nucleus is N_e and the number of γ-rays is N_γ we define the *internal conversion coefficient* α as

$$\alpha = N_e/N_\gamma = \Gamma_e/\Gamma_\gamma \tag{9.21}$$

and α may have any value between 0 and ∞. From (9.19) it follows that if the total width Γ is determined then the radiative width Γ_γ is given by

$$\Gamma_\gamma = \Gamma/(1+\alpha) \tag{9.22}$$

or, in terms of mean lives,

$$\tau = \tau_\gamma/(1+\alpha) \tag{9.23}$$

High-resolution magnetic spectra show that for each γ-ray there are in fact several conversion lines corresponding to the ejection of electrons from different atomic shells, e.g. K, L_I, L_{II}, L_{III}, M_I, M_{II}, ... and the total conversion coefficient is, therefore

$$\left. \begin{array}{l} \alpha = (N_K + N_L + N_M + \cdots)/N_\gamma \\ = \alpha_K + \alpha_L + \alpha_M + \cdots \end{array} \right\} \tag{9.24}$$

The differences in energy between the various groups of internal-conversion electrons corresponding with a single nuclear transition agree exactly with the energies of the lines of the X-ray spectrum of the atom containing the excited nucleus, e.g. Ba in the case shown in Fig. 9.7. The emission from an atom of an internal-conversion electron leaves a vacancy in one of the atomic shells. As in the case of the photoelectric effect with γ-radiation (Sect. 3.1.4), either X-rays or Auger electrons or both may then be emitted.

The importance of the experimental study of internal conversion lies in the information that it may give about the multipolarity of the nuclear transition. The total conversion coefficient α is a ratio of intensities and if a nucleus of infinitesimal size is assumed it does not depend upon the nuclear matrix element. This matrix element arises naturally when the scalar Coulomb interaction between the atomic electron and the nucleus is expanded in multipoles; the conversion coefficients are found to be relatively higher for the higher multipoles. The interaction may be represented as due to an exchange of virtual photons and a factor α enters the conversion probability in this way, making it a second-order process overall.

Extensive tables of K- and L-conversion coefficients based on relativistic wavefunctions now exist. Figure 9.8 shows how the coefficient α_K varies with transition energy for a number of multipolarities in a typical nucleus. The rapid decrease of α with energy

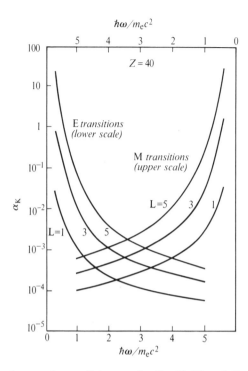

Fig. 9.8 Internal conversion coefficient α_K for $Z = 40$ (Blatt, J. B. and Weisskopf, V. F., *Theoretical Nuclear Physics*, Wiley, New York, 1952).

and the relatively poor discrimination between different multipolarities for high transition energies are made clear in this figure. Internal conversion coefficients are small for light atoms and their measurement is mainly useful for heavy elements and low transition energies.

Internal-conversion electrons from the L- and M-shells are also observable and the ratios of K to L and L to M intensities also depend on multipolarity. So do the relative intensities of internal conversion lines corresponding to the L- and M-subshells.

Corrections to internal conversion coefficients that are structure-dependent arise for a finite nucleus because the atomic electrons penetrate within the nuclear charge and current distribution. They are important for the E0 monopole transitions between states of zero spin (Sect. 9.3.3) for which internal conversion is the main mode of decay. These transitions may also compete with alternative transitions such as M1 or E2 when the spins of the initial and final states are the same and may become significant if the alternative transitions are retarded.

269

9.3.2 Pair internal conversion

If the transition energy E_γ exceeds $2m_ec^2$, i.e. if $E_\gamma > 1.02$ MeV, an excited nucleus may emit a positron–electron pair as an alternative to γ-ray emission and electron internal conversion. The theory of this process, which is electromagnetic in that it takes place in the Coulomb field of the excited nucleus, shows that the probability of pair internal conversion increases with transition energy, is greatest for small multipolarities and is almost independent of Z. These differences from ordinary internal conversion arise because the pair creation process according to hole theory requires only the elevation of an electron in a negative energy state to a positive energy and the supply of such electrons is unlimited, in contrast with the situation for ordinary electrons in an atom. Pair conversion thus becomes important under exactly the conditions under which ordinary internal conversion is becoming small and the phenomenon provides a powerful method of studying energetic radiative transitions in light nuclei. In such experiments care must be taken to distinguish between internal pairs and ordinary pair production in matter by the competing γ-ray.

Unfortunately, the theory of pair internal conversion shows that the discrimination between different multipoles is only good when the yield is low. A more sensitive method is to measure the angular correlation between directions of emission of the positron and the electron.

9.3.3 Zero–zero transitions

In the particular case of $I = 0$ to $I = 0$ transitions, single quantum radiation is strictly forbidden (Sect. 9.1) and the multipole fields do not exist outside the nucleus. The only contribution to internal conversion effects then involves electrons (such as the K-electrons) whose wavefunctions penetrate the nucleus, within which electromagnetic fields may still exist. Total internal conversion ($\alpha = \infty$) arising in this way is observed as a single homogeneous electron line, without associated γ-radiation, in the spectrum of RaC′ (1·414 MeV level) and of ^{72}Ge (0·7 MeV level). The lifetimes of these levels are longer than would be expected for radiative transitions of the particular energy and both transitions are assumed to be of the monopole or radial type $0^+ \rightarrow 0^+$. Pair internal conversion can also occur in $0^+ \rightarrow 0^+$ transitions and it is known for the first excited states of ^{16}O and ^{40}Ca.

Transitions of the form $0^+ \rightarrow 0^-$ cannot occur by any of the internal conversion processes just discussed. Two radiations are required, passing through a suitable virtual intermediate state, e.g. $0^+(\text{E1})1^-(\text{M1})0^-$ with the emission of two quanta would be a possible mode of de-excitation. Alternatively, two conversion electrons or one conversion electron and one photon might be emitted.

9.4 Determination of transition probabilities

9.4.1 Mean lifetimes of bound states ($\tau_\gamma < 10^{-6}$ s)

(i) *Delayed coincidence method* ($\tau_\gamma \approx 10^{-6}$–$10^{-11}$ s). If the time of formation of a nuclear excited state by a nuclear reaction or decay process can be defined by an appropriate signal, and a further signal is obtained from the radiative decay itself, the probability of an interval t between the two signals is given by

$$P(t) = e^{-t/\tau_\gamma} \tag{9.25}$$

Experimentally it is convenient to delay the first signal by a time Δt and to find the time distribution of coincidences between the delayed signal and the decay pulse. This is given by

$$N_c(\Delta t) = N_0 e^{-\Delta t/\tau_\gamma} \tag{9.26}$$

and yields τ_γ directly.

This method can be applied to cascade radiative transitions in nuclear decay and to γ-radiation in a reaction such as X (p, p′γ)X in which the inelastically scattered proton provides the first signal. In the latter case, and for many similar nuclear reactions, the time delay in the particle emission is negligible in comparison with the lifetime under study and a pulsed accelerator beam may then be used to define the time of production of the excited state. Figure 9.9 shows the results of such an experiment on the reaction $^{16}O(d, p)^{17}O^*$ in which a special time-compression system was used to produce pulses of 10^{-10} s duration on the target.

Both Ge(Li) and scintillation detectors may be used in delayed coincidence circuits, the best timing being afforded by plastic scintillators. Wideband amplifiers are required in the timing circuits.

(ii) *Recoil distance method (RDM)* ($\tau_\gamma \approx 10^{-7}$–$10^{-12}$ s). If excited nuclei are produced in a nuclear reaction with sufficient (and known) velocity, the radiative lifetime may be found by studying the distribution of decay points along the path of a beam of such recoil particles. For $\beta = 0.005$ and a decay time of 10^{-10} s the path length is 150 μm, a distance which may be determined by a mechanical assembly (micrometer head or *plunger*) or electrically by a capacitance measurement. The decay point is defined by a narrow collimator placed in front of the radiation detector.

The lifetimes of *atomic states* can be determined in a similar way using the technique of *beam foil spectroscopy*. Excited ions are produced by passing a beam of heavy ions through a thin foil and the intensity of radiation at right angles to the beam is determined as a function of distance from the foil.

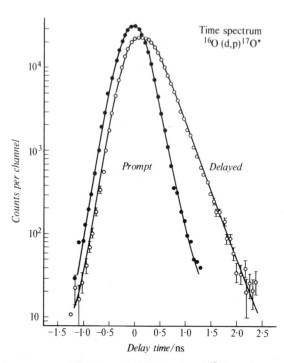

Fig. 9.9 Measurement of lifetime of 871 keV state of ^{17}O by delayed coincidence method, using a pulsed accelerator. The curve 'prompt' is for radiation from a state of effectively zero lifetime. A value $\tau_\gamma = (2 \cdot 63 \pm 0 \cdot 08) \times 10^{-10}$ s was obtained (Lowe, J. and McClelland, C. L., *Phys. Rev.*, **132**, 367, 1963).

The availability of Ge(Li) gamma-radiation detectors of high resolution has permitted an elegant development of the recoil distance method. If an excited nucleus emits a gamma ray while it is moving, the observed energy at angle θ with the direction of motion shows a Doppler shift and is $E_\gamma = E_0(1 \pm \beta \cos \theta)$; if the nucleus emits the radiation after it is brought to rest, the energy is unshifted, $E_\gamma = E_0$. If, therefore, the target is fixed and a stopping plate or plunger is used to receive the recoil ions, the spectrum observed at a certain angle with the beam direction and at a suitable distance shows two peaks whose relative height gives the fraction of nuclei decaying before and after reaching the plunger. This varies with plunger displacement and permits a direct calculation of lifetime; the recoil velocity is calculated from the reaction kinematics or obtained from the maximum observed Doppler shift, which can be an order of magnitude greater than the resolution width of the detector. Figure 9.10 shows results obtained by this method for the 871 keV state of ^{17}O formed in the reaction ^{2}H(^{16}O, p)^{17}O*. In all Doppler shift measurements it is advantageous to increase recoil

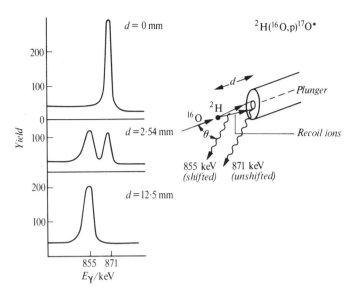

Fig. 9.10 Measurement of lifetime of 871 keV state of ^{17}O by RDM. Spectra are shown for three different plunger displacements and for an angle $\theta = 122°$. The lifetime deduced from the ratio of shifted to unshifted intensity was $(2\cdot33 \pm 0\cdot26) \times 10^{-10}$ s (adapted from Allen, K. W., *Electromagnetic Interaction in Nuclear Spectroscopy*, Ch. 9, North Holland, 1975).

velocities by the use of heavy rather than light projectiles whenever possible; in the case of ^{17}O, a deuterium target and ^{16}O beam were used.

(iii) *Doppler shift attenuation method* (DSAM) $(\tau_\gamma \approx 10^{-11}$–$10^{-14}$ s). For lifetimes less than 10^{-11}–10^{-12} s the recoil distance method becomes difficult because plunger displacements of $< 10^{-5}$ m have to be measured accurately. It is, however, still possible to use the Doppler shift to measure these lifetimes. The principle of the method is to relate the energy *spectrum* of the radiation from a beam of recoiling nuclei to the ratio between the radiative lifetime τ_γ of the excited state and the time (α) that the recoil nuclei take to slow down in an absorber. If $\tau_\gamma/\alpha \ll 1$ then nearly all decays will show the full Doppler shift $\Delta E_{\gamma 0} = E_0\beta \cos \theta$, while if $\tau_\gamma/\alpha \gg 1$ most of the decays are unshifted. If an attenuation factor $F(\tau_\gamma)$ is defined to be the percentage of the full Doppler shift $E_{\gamma 0}$ observed in a given spectrum then in these two extreme cases $F = 1$ and 0 respectively and only upper and lower limits can be established. In intermediate cases $0 < F(\tau_\gamma) < 1$, a value for τ_γ can be obtained.

In practical applications of the DSAM the recoil direction is often defined to a few degrees by requiring observations to be coincident with the detection of an associated reaction particle (usually at 180° with the beam direction to give maximum recoil velocity). The excited nuclei recoil into the target backing and slow down to rest. Their radiation, Doppler shifted according to their velocity, is detected over a certain range of angles by a Ge(Li) detector located at a known mean angle with the recoil direction.

The simplest analysis is to determine the centroid of the shifted line over a range of angles. From the indicated $F(\tau_\gamma)$ the lifetime τ_γ may be inferred. If good resolution is available and the Doppler shift is large enough, more information may be obtained from a detailed analysis of the line shape using τ_γ as a parameter. In both cases electronic and nuclear stopping-power data must be used and account taken of the scattering of recoils by nuclear collisions; the full Doppler shift $\Delta E_{\gamma 0}$ may be obtained by direct observation or by calculation using reaction kinematics.

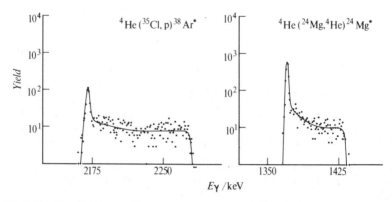

Fig. 9.11 Measurement of lifetime of 2168 keV state in ^{38}Ar and of 1368 keV state in ^{24}Mg by DSAM, line-shape analysis. Lifetimes of 0·72 and $1·82 \times 10^{-12}$ s ($\pm 7\%$) were deduced (adapted from Forster, J. S. *et al.*, *Phys. Letters*, **51B**, 133, 1974).

Figure 9.11 shows results obtained for ^{38}Ar* and ^{24}Mg* by heavy-ion bombardment of a ^{4}He target leading to lifetimes of about 10^{-12} s. The accuracy of DSAM results may be impaired by systematic errors due to lack of knowledge of stopping powers for slow heavy ions in matter. These errors are reduced if high recoil velocities can be used and an accuracy of about 5 per cent is attainable.

9.4.2 Coulomb excitation

The radiative lifetime measurements described in the preceding sections cover the region from the 'isomeric' lives of 10^{-6} s down to about 10^{-15} s, which corresponds with a width of 0·6 eV. It is, however, not always convenient to produce the recoil velocities necessary for measurement of the shorter lives. Fortunately, phenomena that are determined by the radiative *width* become easier to observe as the mean life diminishes and may be used to measure transition probabilities when direct lifetime measurements are impracticable. A generally applicable method is that of Coulomb excitation.

A charged particle passing near to a nucleus with an energy considerably less than the height of the mutual potential barrier so that nuclear effects may be disregarded, follows a classical trajectory prescribed by the Coulomb force $Z_1 Z_2 e^2 / 4\pi\varepsilon_0 r^2$ acting between mass centres. The total interaction however, for a finite target nucleus, must recognize the effect of the charge and current distributions. This may be described by making a multipole expansion of the scalar interaction and this in turn introduces the multipole matrix elements and reduced transition probabilities $B(L) \uparrow$ that apply in photon absorption (e.g. in the transition $f \to i$ in Fig. 9.2). The process is detected by observation of the de-excitation radiation ($i \to f$ in Fig. 9.2).

The cross-section for Coulomb excitation depends on the particle trajectory, which may be calculated, as well as on $B(L)$. The effect of multipolarity on transition probability indicated in Section 9.1.3 is less severe for Coulomb excitation than for spontaneous emission, because the effect is extra-nuclear. There is, however, an inhibition of magnetic transitions with respect to electric transitions of the same multipolarity by a factor $\approx (v/c)^2$ where v is the projectile velocity. Yield curves, such as that shown for ^{181}Ta in Fig. 9.12, may be used to determine the probability and multipolarity of a transition.

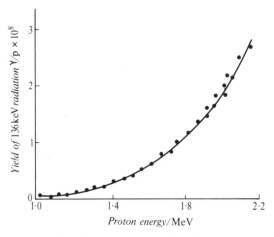

Fig. 9.12   Yield of 136-keV quanta from ^{181}Ta bombarded by protons as a function of proton energy, compared with theory for E2 excitation (solid curve) (Huus, T. and Zupančič, C., *Det. Kgl. Dansk. Vid. Selskab.*, **28**, 1, 1953).

Coulomb excitation has been observed throughout the periodic system and with many types of bombarding particle, including heavy ions such as ^{14}N and ^{40}A, which are particularly effective because of their high charge. An important application is in the study of rotational levels excited by E2 transitions and in fact a great many Coulomb excitation processes are of E2 type. In cases when $|\Delta I| = 1$ the subsequent radiative transition may be a mixture of M1 and E2 components and analysis of the mixture may be made by the methods normally applied to radioactive decay schemes.

The use of heavy ions has permitted the observation of higher-order processes, e.g. the excitation of a 4^+ level in an even–even nucleus by the double E2 sequence $0^+ \to 2^+ \to 4^+$. Another double E2 process of some importance is the *reorientation* effect in which a state of spin I which has been Coulomb-excited makes an E2 transition from one magnetic substate (I, m') to a second substate (I, m'') as a result of interaction of the electric field of the bombarding particle with the electric quadrupole moment Q of the excited state. The effect is a useful method for determination of this moment.

9.4.3 Inelastic scattering of electrons

Coulomb excitation by electrons is in principle possible, but in practice it is difficult to observe as an emission process because of the presence of bremsstrahlung quanta. The inelastically scattered electrons, however, may be observed with good discrimination against background at large angles of scattering, and the spectrum then shows peaks corresponding with the excitation of the lower excited levels of the target nucleus, in addition to the ground-state (elastic) peak. As with heavy-particle Coulomb excitation, the interaction between an electron and the nucleus may be expanded in multipoles, but no restriction on particle velocity is necessary in order to avoid nuclear effects. Magnetic transitions may, therefore, be more easily observed. Moreover, in ordinary Coulomb excitation only the normal (transverse) radiative matrix elements occur but in electron excitations the (longitudinal) static Coulomb interaction also contributes. This contains a monopole component so that $0^+ \to 0^+$ transitions may be excited. The most important difference between the two techniques, however, is the fact that in inelastic scattering the momentum transfer q to the nucleus may be varied independently of energy transfer, and the momentum dependence of matrix elements may be explored.

As in the case of elastic scattering (eqn (6.7)) the observed cross-section as a ratio to that for Mott scattering may be described by the quantity $|F(q)|^2$ where $F(q)$ is a form factor determined by matrix elements of the multipole operators. From the variation of scattering cross-section with q the ground-state charge density may be obtained for elastic scattering, and a 'transition' charge density in the inelastic case. This can be compared with the prediction of nuclear models.

In general, there are three form factors to be considered in an electron inelastic scattering experiment, namely $F_C(q)$, $F_e(q)$ and $F_m(q)$ corresponding with the longitudinal (or Coulomb) and transverse excitations, magnetic excitations being wholly transverse. The longitudinal form factor may be separated experimentally by observations at different angles for a given q. For low-momentum transfers the Coulomb transition probability reduces essentially to the electric multipole probability for the same L-value. Extrapolation of $F_C^2(q)$ to the 'photon' point corresponding to the actual transition energy, which is essentially $q \to 0$, then determines the upward transition probability $B(L) \uparrow$ to the excited state, while the spin-parity of the state and its transition charge density follow from the q-dependence of F_C^2.

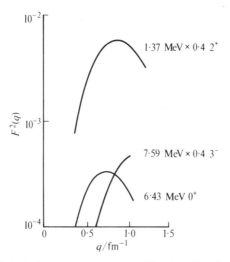

Fig. 9.13 Inelastic form factor for the scattering ^{24}Mg(e, e')^{24}Mg*, showing variation with momentum transfer q for a 0^+, 2^+ and 3^- transition (derived from Johnston, A. and Drake, T. E., *J. Phys.*, **A7**, 898, 1974).

Figure 9.13 shows this dependence for three transitions in ^{24}Mg, using electron energies of about 100 MeV and scattering angles of about 120°; one of the transitions is a monopole. For the 1·37 MeV level the $B(L) \uparrow$ value leads to a lifetime of $2 \cdot 0 \pm 0 \cdot 2$ ps in agreement with other methods (Sect. 9.4.1).

In inelastic electron scattering the kinematic factor $(kR)^2$ that reduces the intensity of high multipoles in spontaneous emission (eqn (9.8)) is replaced by $(qR)^2$ where q is normally very much greater than k (e.g. $q \approx 1 \, \text{fm}^{-1}$, $k \approx 10^{-3} \, \text{fm}^{-1}$). It follows that high-spin states can often be excited.

9.4.4 Nuclear resonance reactions

Suppose that the level i shown in Fig. 9.2 exists in a nucleus C, which may be represented as C* when excited. It may then be possible to produce the excitation by capture of a bombarding particle (a) in a suitable initial nucleus X, i.e.

$$X + a \rightarrow C^* \tag{9.27}$$

These processes are of great importance for the theory of nuclear reactions and were discussed in Section 8.1. For the present purpose we note only that the cross-section for the emission of radiation from C* by the process

$$C^* \rightarrow C + \gamma \tag{9.28}$$

assuming just one possible transition of energy $E_\gamma = E_i - E_f$ to the ground state f is

$$\sigma_{a\gamma} = \pi \lambdabar^2 g \Gamma_a \Gamma_\gamma / [(T - T_0)^2 + \Gamma^2/4] \tag{9.29}$$

where $\lambdabar$ is the c.m. wavelength of the bombarding particle a, T is the c.m. energy, T_0 is the c.m. energy at which there is exact resonance with the level E_i, and g is a statistical factor of the order unity depending on intrinsic spins. This is an example of the Breit–Wigner formula (Sect. 11.2) which has the present simple form when only one l-value contributes to the reaction; the value of g is given by equation (11.16).

In the present case the level i is virtual and is at an energy above the ground state f equal to T_0 plus the separation energy of particle a in nucleus C (cf. Fig. 11.6). The width Γ of the level measures the total probability of decay and in this case includes the width Γ_a for re-emission of the incident particle, together with the radiative width Γ_γ and the internal conversion width $\alpha \Gamma_\gamma$. If there are no other decay modes then

$$\Gamma = \Gamma_a + \Gamma_\gamma (1 + \alpha) \tag{9.30}$$

The internal conversion coefficient (Fig. 9.8) is normally negligible for $E_\gamma > 1 \, \text{MeV}$ and need not be included for high-energy capture transitions, $E_\gamma \approx 8$–$12 \, \text{MeV}$.

At resonance, $T = T_0$, the total cross-section is

$$\sigma_a = 4\pi \lambdabar^2 g \Gamma_a \Gamma / \Gamma^2 = 4\pi \lambdabar^2 g \Gamma_a / \Gamma \tag{9.31}$$

whereas the cross-section for production of radiation is

$$\sigma_{a\gamma} = 4\pi \lambdabar^2 g \Gamma_a \Gamma_\gamma / \Gamma^2 \tag{9.32}$$

If $\Gamma_a \gg \Gamma_\gamma$, the radiative cross-section yields $g\Gamma_\gamma/\Gamma$ and if Γ is known from the variation of yield with energy then $g\Gamma_\gamma$ for the level i can be obtained. This case arises for *proton and α-particle capture* in light elements at energies somewhat below the barrier height, for which $\Gamma_a \approx 10$ keV and $\Gamma_\gamma \approx 1$ eV.

If $\Gamma_a \ll \Gamma_\gamma$, then the resonant cross-section $\sigma_{a\gamma}$ yields $g\Gamma_a/\Gamma$ and $\Gamma_\gamma (\approx \Gamma)$ may be obtained from the variation of yield with energy. This case arises for *slow-neutron capture*, when $\Gamma \approx 0.1$ eV, $\Gamma_n \approx 10^{-3}$ eV and the virtual levels concerned are within a few electron volts of the neutron binding energy.

A special case of a resonance reaction arises when a is a photon and the only process possible is photon scattering. The Breit–Wigner formula is not strictly applicable in this case and should be replaced by the *Lorentzian* expression for the scattering of radiation by an oscillator. The difference, however, is not usually significant at energies near resonance and the Breit–Wigner form will be used. It is now possible for the level i to be bound, and of low excitation, so that it is necessary to take account of internal conversion. The absorption cross-section for monochromatic radiation at resonance then becomes

$$\sigma_\gamma = 4\pi \lambda^2 g/(1+\alpha) \qquad (9.33)$$

and for radiation of energy 500 keV this is of the order of 10^4 b.

The total electronic cross-section for attenuation of radiation of an energy of 500 keV in a light element such as oxygen or carbon is of the order of 10^2 b. It might therefore be expected that nuclear attenuation could be easily observed by increasing the energy of the incident radiation until resonance absorption appeared. However, if a bremsstrahlung spectrum is used for the experiment it is necessary to average the nuclear cross-section over the energy interval ΔE accepted at one setting of the detector. When equation (9.29) is averaged in this way (essentially by integration between energies plus and minus infinity) the mean absorption cross-section becomes

$$\bar{\sigma}_\gamma = 2\pi^2 \lambda^2 g\Gamma_\gamma/\Delta E \qquad (9.34)$$

and since the factor $\Gamma_\gamma/\Delta E$ may be 10^{-4} or less, the nuclear effect becomes small compared with electronic absorption. Discrimination against the latter is achieved by observing scattered intensities at backward angles, since the non-resonant electronic scattering is peaked in the forward direction. In this way radiative widths Γ_γ of levels of width ≈ 0.5 eV have been observed for a number of light elements.

For the narrowest states, say $\ll 0.5$ eV, it is usually necessary to use resonance radiation, as in optics, to excite a level by means of the radiation emitted in the decay of the same level in an active source or target. The absorption cross-section at exact resonance given by (9.33) does not determine Γ_γ, since this quantity has cancelled out, but in practice this particular situation is drastically altered in a useful way by the existence of Doppler shifts and recoil effects.

In the emission of the resonance radiation of energy E_γ from a stationary source atom of mass M, a momentum $p = E_\gamma/c$ must be imparted to the emitting body. In the absorption process the same momentum must be given to the absorber atom. The emission and absorption lines are therefore separated, because of recoil, by an energy

$$2 \times p^2/2M = E_\gamma^2/Mc^2 \qquad (9.35)$$

In addition, both source and absorber atoms are normally affected by thermal motion and the lines are broadened by the Doppler effect in liquids and gases and by the width of associated phonon transitions in solids. The recoil effect is negligible in atomic transitions and resonance absorption of broadened lines takes place, as shown in Fig. 9.14a. For a *nuclear* transition with free recoil, on the other hand, the recoil shift is always $\gg \Gamma_\gamma$ for low-lying bound levels, and is often greater than the Doppler width. Nuclear resonance is, therefore, a weak effect unless recoil losses are restored or eliminated; the situation is as shown in Fig. 9.14b. The introduction of the recoil and thermal terms into (9.33) creates a dependence of the resonant cross-sections on Γ_γ itself and permits measurement of this quantity if a sufficient effect can be achieved.

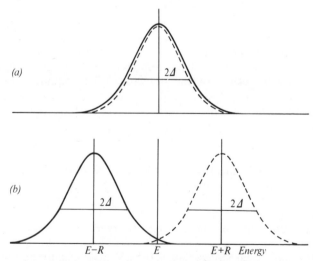

Fig. 9.14   Effect of recoil on atomic and nuclear resonance. The solid line represents the intensity distribution of the incident radiation and the dotted line the absorption cross-section of the absorber as a function of (homogeneous) incident energy. (*a*) Atomic resonance, recoil shift $R \ll$ Doppler width Δ. (*b*) Nuclear resonance, $R > \Delta$.

Recoil energy loss can be restored, and resonance between Doppler-broadened lines obtained, by the following means:

(a) By mounting a radioactive source on the tip of a high-speed rotor and using the Doppler effect to increase the frequency of the emitted radiation. From (9.35) the speed required is given by

$$E_\gamma(v/c) = E_\gamma^2/Mc^2 \qquad (9.36)$$

and for a nucleus of mass 200 and $E_\gamma = 500$ keV the value $v = 820$ m s^{-1} is found.

(b) By heating the radioactive source so that the Doppler or phonon broadening is largely increased. Temperatures of up to 1500 K have been used.

(c) By utilizing a preceding transition, such as β-decay, γ-decay or production in a nuclear reaction to provide the necessary velocity for Doppler shift of frequency.

Like Coulomb excitation, nuclear resonance shows largest effects in cases of large width, i.e. of small lifetime. The method is rather more limited than Coulomb excitation since gram quantities of the scatterer are required and it may be difficult to supply the velocity necessary to restore the recoil effect. Despite this, radiative widths have been obtained for many transitions by the resonance method and cover a range of lifetimes from 10^{-9} to 10^{-13} s.

Recoil energy loss in a gamma-ray resonant scattering or absorption experiment can be eliminated without associated broadening due to phonon transitions, in certain special cases. It was discovered by Mössbauer in 1958 that the radiation from a source of ^{191}Ir contained a narrow line of approximately natural width Γ_γ ($\approx 10^{-5}$ eV) as well as the expected thermal distribution. Nuclear resonance in an iridium absorber could be destroyed by motion of the source, with a velocity of less than 40 mm s^{-1}. Such narrow lines correspond to processes in which *no* phonon transitions take place, and they are unshifted because for photon energies less than about 200 keV the recoil energy is insufficient to remove the atom from its lattice site and the effective mass in equation (9.35) is that of the crystal as a whole, which absorbs momentum, but not energy, in the transition. Similar considerations were shown by Lamb, as early as 1939, to apply to the scattering of slow neutrons by crystals.

Although Mössbauer transitions have been observed in many nuclei, the limitation to low transition energies is rather restrictive. Moreover, the pure natural line width may not be observed, because of broadening by hyperfine interactions in the source and absorber. As far as nuclear physics is concerned, the effect has mainly exploited the high resolution available ($\Gamma_\gamma/E_\gamma \approx 10^{-13}$) and has been used, sometimes in conjunction with other techniques such as nuclear magnetic resonance (NMR), to measure nuclear magnetic moments and quadrupole moments of excited states and the isomer shifts which arise from the difference between the mean-square radius of nuclear ground and excited states. The applications of the Mössbauer effect in solid state physics, metallurgy and chemistry are extensive (Ref. 9.2).

9.5 Angular correlation experiments

9.5.1 Angular distribution of radiation

The most direct indication of the multipolarity L of a radiative transition is the angular distribution of the intensity of the radiation corresponding to a transition between initial and final substates $(I_2 m_2)$ and $(I_1 m_1)$ of two levels (Fig. 9.15). The axis of quantization is fixed experimentally, e.g. by the direction of an external magnetic field.

For atomic transitions it is well known from the Zeeman effect that the magnetic substates may easily be separated magnetically and the individual components of a line, two of which are shown in Fig. 9.15, may be studied. In the nuclear case, because of the

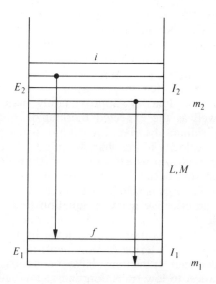

Fig. 9.15   Radiative line of multipolarity L between two nuclear levels of spin I_2 and I_1, each of which has $2I + 1$ substates characterized by the magnetic quantum number m. Two of the possible transitions (L, M) are indicated. If no perturbing fields act on the radiating nucleus the substates for a given I all have the same energy although they are shown distinct.

smallness of nuclear moments, a separation comparable with the experimental resolution is only possible under special conditions, such as those offered by the Mössbauer effect (Sect. 9.4.4). The total line, rather than its individual incoherent components, is then observed from an assembly of excited nuclei.

If in such an assembly the substates m of the initial level are uniformly populated the angular distribution of the line intensity is isotropic. If, however, the substate population $P(m)$ depends on m, the assembly is said to be *spin-oriented*. The case $P(m) = P(-m)$ defines an *aligned* system whereas if $P(m) \neq P(-m)$ the system is said to be *polarized*. The line intensity for an oriented system will, in general, be anisotropic.

The probability of the vectors $\boldsymbol{I}_1$, $\boldsymbol{L}$ combining to give a vector $\boldsymbol{I}_2$ with particular magnetic quantum numbers, is given by the Clebsch–Gordan coefficients, discussed in Appendix 4. The angular distribution arising from a given substate m_2 of the initial level can then be written

$$W(\theta) = \sum_{m_1} (I_1 L m_1 M \mid I_2 m_2)^2 F_L^M(\theta) \qquad (9.37)$$

where $F_L^M(\theta)$ is the angular distribution for the appropriate multipole field (Sect. 9.1), and is in fact the modulus squared of a vector spherical harmonic (Sect. 9.1.2). The quantum numbers L, M for the emitted radiation are limited by the selection rules (9.3), i.e.

$$I_2 + I_1 \geqslant L \geqslant |I_2 - I_1|$$
$$m_2 = M + m_1$$

If the initial substates have a weight $P(m_2)$ the total angular distribution is

$$W(\theta) = \sum_{m_1 m_2} P(m_2)(I_1 L m_1 M \mid I_2 m_2)^2 F_L^M(\theta) \qquad (9.38)$$

This must reduce to a constant if $P(m_2)$ is independent of m_2, in which case the initial levels are equally populated (random orientation). Thus, for example, if $I_2 = 1$, $I_1 = 0$, $L = 1$ (dipole) radiation is emitted; the specific angular distributions are $F_1^0 \propto \sin^2 \theta$, $F_1^{\pm 1} \propto (1 + \cos^2 \theta)/2$ and the Clebsch–Gordan coefficients are

$$(010 \pm 1 \mid 1 \pm 1) = 1, \qquad (0100 \mid 10) = 1$$

so that

$$W(\theta) \propto \sin^2 \theta + 2 \times \tfrac{1}{2}(1 + \cos^2 \theta)$$

= constant, as expected.

The functions $F_L^M(\theta)$ contain only Legendre polynomials of even order for pure multipole radiation so that we may also write

$$W(\theta) = a_0 + a_2 \cos^2 \theta + a_4 \cos^4 \theta + \cdots + a_{2L} \cos^{2L} \theta \quad (9.39)$$

which expresses the result that no term in $\cos \theta$ of power higher than $2L$ appears. The distribution is symmetric with respect to the plane $z = 0$, a consequence of the conservation of parity, i.e. invariance under the inversion operation, for the electromagnetic interaction.

The observation of a multipole angular distribution does not determine the electric or magnetic character of the transition, i.e. the parity change, since E1 and M1 radiation, for instance, have the same angular distribution patterns. A distinction may be made if the polarization of the radiation, e.g. the direction of the electric vector, can be observed.

Generally, the selection rules for radiation allow more than one multipolarity in the transitions $I_2 \rightarrow I_1$. The corresponding angular distribution with respect to the quantization axis then involves the mixing ratio δ, which is a model-dependent quantity of considerable interest.

9.5.2 Experimental methods

Nuclei may be oriented with respect to an axis of quantization defined by an external field by the application of a high magnetic field B_0 at low temperatures. If the normal Boltzmann distribution of nuclei between the magnetic substates is to be appreciably disturbed it is necessary to use fields and temperatures related by

$$\mu_I B_0 \approx kT \quad (9.40)$$

where μ_I is the nuclear magnetic moment. For $\mu_I = 1$ nuclear magneton this gives $B_0/T = 2 \cdot 8 \times 10^3$ tesla K^{-1}. Radioactive nuclei such as ^{60}Co, leading to daughter nuclei in which radiative transitions take place, have indeed been studied by this method.

The extreme conditions required by the direct orientation method soon led to the development of methods in which internal atomic fields, guided by a much smaller external field, produce the orientation. Dynamical methods in which the Boltzmann distribution is disturbed by microwave transitions or optical pumping techniques are also used, especially for the production of polarized beams and targets for accelerator experiments. For the study of radiative transitions, however, the most widely used technique has been that of angular correlation of successive radiations.

Figure 9.16a shows states of spin I_2, I_1, I_0 of a nucleus, linked by pure multipole transitions L_a, L_b. The first decay $I_2 \rightarrow I_1$ produces a non-uniform population of the substates of state I_1, with respect to

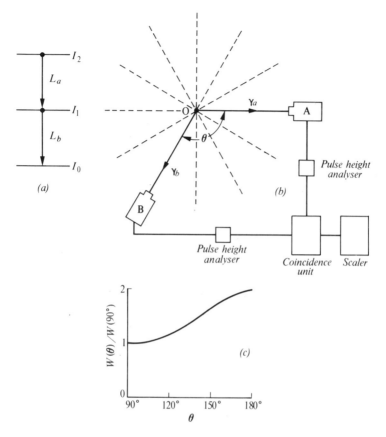

Fig. 9.16 Angular correlation of successive radiations from a radioactive nucleus at O. (*a*) Decay scheme. (*b*) Counters. (*c*) Correlation pattern.

an axis of quantization defined by the direction of emission of the photon γ_a (Fig. 9.16*b*). The nucleus I_1 is *aligned* with respect to this axis, since the first transition produces no difference between the substates of I_1 with equal and opposite *m*-values. The relative population is given explicitly by the expression

$$P(m_1) = \sum_{m_2} (I_1 L_a m_1 M_a \mid I_2 m_2)^2 F_{L_a}^{M_a}(0) \qquad (9.41)$$

where

$$m_2 = M_a + m_1 \qquad (9.42)$$

and the choice of axis restricts M_a to the values ∓ 1 corresponding to left or right circular polarization of the photon γ_a observed at $\theta = 0°$. The directional correlation γ_a, γ_b (Fig. 9.16*c*) is then given

285

by

$$W(\theta) = \sum_{m_0 m_1 m_2} (I_1 L_a m_1 \pm 1 \mid I_2 m_2)^2 F_{L_a}^{\pm 1}(0)$$
$$\times (I_0 L_b m_0 M_b \mid I_1 m_1)^2 F_{L_b}^{M_b}(\theta) \qquad (9.43)$$

where

$$m_1 = M_b + m_0 \qquad (9.44)$$

assuming that the state decays before its alignment is disturbed by perturbing fields.

In the particular case $I_2 = I_0 = 0$, $I_1 = 1$ the transitions permitted when γ_a is observed at $0°$ are $(I, m) = (0, 0) \rightleftarrows (1, \pm 1)$. If the appropriate Clebsch–Gordan coefficients are inserted in (9.43) the angular correlation is given by

$$W(\theta) \propto F_1^{\pm 1}(0)[F_1^1(\theta) + F_1^{-1}(\theta)]$$
$$\propto 1 + \cos^2 \theta$$

Similar calculations may be made for more complicated cascades but the numerical evaluation of the correlation function becomes very tedious because of the large number of summations over magnetic quantum numbers which are necessary. Fortunately, it is possible to avoid this through algebraic methods developed by Racah and it is now customary to write the general angular correlation function for transitions of pure multipolarity in the form

$$W(\theta) = \sum A_\nu P_\nu(\cos \theta) \qquad (9.45)$$

where

$$\nu = 0, 2, 4, \ldots$$

and

$$A_\nu = F_\nu(L_a I_2 I_1) F_\nu(L_b I_0 I_1) \qquad (9.46)$$

The functions F_ν are tabulated in Reference 9.3.

A typical experimental arrangement for the observation of angular correlations in cascade transitions is shown in Fig. 9.16b. The detector pulses are combined in the coincidence unit and the number of coincidences $W(\theta)$ is determined as a function of the angle between the counter axes. The anisotropy

$$\varepsilon = \frac{W(\frac{1}{2}\pi) - W(0)}{W(\frac{1}{2}\pi)} = \frac{W(90°) - W(180°)}{W(90°)}$$

can be compared with that predicted by (9.43) or a complete curve for $W(\theta)$ (Fig. 9.16c) may be obtained.

The formulae can be extended to include multipole mixing, and to permit evaluation of the mixing ratios for the pure multipoles permitted by the spin change. Polarization-sensitive detectors enable the parity change to be found.

9.6 Comparison between experiment and theory

In order to compare the matrix elements of multipole operators with the predictions of nuclear models, it is first of all necessary to establish the multipolarity of a transition, or the contributing multipolarities if it is mixed. Usually, one particular experiment does not provide the required information, but a combination of angular correlation data with internal conversion measurements may do so. At the same time the electric or magnetic multipole character of the radiation must be obtained, often from the energy–lifetime relations themselves, but sometimes from polarization measurements. When the appropriate L^π is known it is possible to extract the reduced transition probability $B(L)$ for a pure multipole from measurements of $T(L)$.

For a given multipole, the *transition strength* $|M|^2$ is conventionally measured in Weisskopf units (W.u.) using equation (9.15), i.e.

$$|M|^2 = \frac{T(L) \text{ observed}}{T(L) \text{ Weisskopf}} = \frac{B(L) \text{ observed}}{B(L) \text{ s.p.}} \quad (9.47)$$

where $B(L)$ s.p. indicates that the single-particle model is being used. This reduction of data has been extensively used for nuclei of mass number up to 40 though it is not confined to this range. For heavy deformed nuclei comparison may alternatively be made with the Nilsson model.

In terms of $|M|^2$, the following conclusions emerge:

(*a*) E1 transitions in the light nuclei have $|M|^2 \approx 0.03$. If the transition is isospin forbidden, as in the case of $\Delta T = 0$ transitions in self-conjugate nuclei, $|M|^2 \approx 0\cdot003$. The transition strength is then due to isospin admixtures in the nuclear states as a result of Coulomb forces, and the matrix element of the Coulomb interaction between the two states may be obtained from the observed strength.

(*b*) M1 transitions in the light nuclei, if isospin-favoured, have about the Weisskopf strength. For $\Delta T = 0$ transitions in self-conjugate $(N = Z)$ nuclei there is retardation by a factor of about 10 because of some cancellation between the neutron and proton magnetic moments.

(*c*) E1 and M1 transition strengths in medium-weight and heavy nuclei have been surveyed by the (n, γ) capture reaction and are generally less than 1.

(*d*) E2 strengths in both light and heavy nuclei also show a dependence on isospin, but in this case are retarded in self-conjugate nuclei for $\Delta T = \pm 1$. The T-allowed transitions may have strengths up to 100 for deformed nuclei.

(e) E3 and M4 transitions from isomeric states show relatively little spread of transition strength.

(f) There is no apparent difference in strength between transitions associated with a single odd neutron or a single odd proton.

The interpretation of the observed strengths follows the procedures already outlined for static moments (Ch. 8). For all but pure single-particle levels, the single-particle strength must be distributed among all transitions between states that derive from the basic levels, so that any specific transition has a reduced strength.

Calculations have been based on the individual-particle model for light nuclei near closed shells. As the number of loose particles increases, this type of calculation becomes difficult in detail, but still suggests that the single-particle transition probability should be multiplied by a factor (<1) depending on the number of available particles. Further away from closed shells the individual particle merges into the collective model as has already been discussed, and an explanation of the large observed E2 strengths is provided in terms of coherent motion of nucleons.

9.7 Summary

The electromagnetic interaction is well-described by Maxwell's equations and a complete classical theory exists. The electromagnetic field may be quantized by the introduction of the massless photon and many electromagnetic processes may be described in terms of single photon exchange. No breakdown of the laws of the interaction has been found even for interparticle distances of the order of 10^{-15} m; the electron and the muon behave essentially as point charges. These particles are therefore highly suitable for exploring nuclear charge distributions and this application is described.

Because the interaction is known, a precise formulation of the electromagnetic matrix element for a transition between two states may be given. Its properties determine selection rules for the transition and its squared magnitude gives the transition probability. Methods of measuring multipolarities and transition probabilities are reviewed and the information thus obtainable on the wavefunctions of the states involved is outlined.

Examples 9

9.1 Write down possible intrinsic and orbital angular momenta for the following radiations:

electric dipole (E1)   electric quadrupole (E2)

magnetic dipole (M1)   magnetic quadrupole (M2)

given that the parity of the multipole field EL is $(-1)^L$ and of ML is $-(-1)^L$. Assume that L^π for the photon is 1^-.

9.2* In the ^{207}Pb nucleus the magnetic moment of the ground state is 0·59 nuclear magnetons and the halflife of the first excited state at 570 keV is 130 ps. Use this information, together with equations (7.10), (7.11) and (9.18), to deduce the single-particle character of the two states.

9.3* In the decay of $^{137}_{55}$Cs (Fig. 9.7) the 662-keV gamma radiation arises from an isomeric transition of halflife 2·6 min in ^{137}Ba^m to the $\frac{3}{2}^+$ ground state. Use Fig. 9.4, correcting for the change in A-value, to confirm the spin of the isomeric state. Check from Fig. 9.8 that it was correct to use the observed halflife in making this deduction.

9.4* The isomeric state of ^{134}Cs^m (spin/parity 8^-) decays to the ground state (4^+) and to an excited state (5^+) by transitions of energy 137 and 127 keV. State the nature of the three transitions, and estimate the relative intensity of the 137- and 127-keV radiations.

9.5 Using the Weisskopf formulae (9.15) and (9.17) for a single-proton transition deduce an expression for the E3 radiative width.

9.6 The (upward) B(E2) value for the Coulomb excitation of the 2·938 MeV state (2^+) of ^{26}Mg is 40 e^2 fm^4. Calculate the radiative lifetime, if the branching ratio to the ground state is 10 per cent. $[5 \times 10^{-14}\,\text{s}]$

9.7 In an experiment to determine the lifetime of an excited nucleus by the recoil-distance plunger method, the counting rate was found to decrease by a factor of 2 for a source displacement of 0·07 mm. If the mean lifetime of the decaying state is 7×10^{-11} s find the velocity of recoil. $[1\cdot4 \times 10^6\,\text{m s}^{-1}]$

9.8 Ions of ^{32}S of energy 150 MeV were used to excite ^{64}Zn nuclei in order to measure an excited state lifetime by the plunger technique. When the plunger was 20 mm from the target 30 per cent of detected γ-rays were seen in the Doppler-shifted peak. Assuming that detection was in coincidence with back-scattered ^{32}S-ions, calculate the lifetime of the state. $[1\cdot9 \times 10^{-9}\,\text{s}]$

9.9 Calculate the recoil velocity of the ^{59}Cu* nucleus formed in the ^{58}Ni (pγ) reaction at a proton energy of 3547 keV and the full Doppler shift expected for a gamma-ray of energy 3820 keV radiated by this nucleus. The energy of the gamma radiation actually observed is 3824 keV in the forward direction; what is the $F(\tau)$ value for use in a DSAM calculation? $[v = 0\cdot0015c, 5\cdot7\,\text{keV},$ 0.7]

9.10 Calculate the full recoil shift between the emission and absorption lines for the 411 keV level in ^{198}Hg, and the source velocity necessary to give complete overlap. $[0\cdot92\,\text{eV}, 670\,\text{m s}^{-1}]$

9.11* In the experiment of Example 9·9, a resonance was also found at $T_p = $ 1833 keV, and the DSAM lifetime was determined to be $1\cdot7 \times 10^{-14}$ s. If the proton width is 0·005 eV, and the spin of the resonant state is $\frac{5}{2}$, find the cross-section at resonance.

9.12 The 14·4-keV γ-ray transition in ^{57}Fe has been extensively studied using the Mössbauer effect, (e.g. Hanna *et al.*, *Phys. Rev. Letters* **4**, 177, 1960) and it is found that the $I = \frac{1}{2}$ ground state is split with an energy correspond-ing with a source velocity of 3·96 mm s^{-1}. If the ground-state magnetic moment is 0·0955 μ_N calculate the internal magnetic field at the ^{57}Fe nucleus. [31·5 T]

9.13 A nucleus X emits a beta particle forming a residual nucleus Y in an excited state of energy 250 keV. What is the minimum energy of the beta transition necessary to permit the gamma radiation from Y to be absorbed resonantly in an external nucleus Y at rest? [0.56 MeV]

10 The weak interaction

10.1 General properties

In contrast with electromagnetic forces, the weak forces of nuclear physics are of short range, certainly less than nuclear dimensions, and are therefore not apparent as a property of matter in bulk. Recognition of the weak interaction as a fundamental natural process, therefore, had to await the advent of nuclear physics, but could, in principle, have been made at the very beginning, since the radiations observed in 1896 by Becquerel are now known to be essentially due to the weak decay of daughter products of the uranium nucleus.

The phenomena now classed as weak are the non-electromagnetic decay processes and interactions that proceed on a time scale slow compared with a typical 'nuclear' time, say $R/c \approx 10^{-21}$ s where R is a nuclear radius. They may be grouped into:

(*a*) Purely leptonic processes, e.g. muon decay

$$\mu^+ \to e^+ + \nu_e + \bar{\nu}_\mu \qquad (t_{1/2} = 1\cdot 5 \text{ μs}, \ Q = 105 \text{ MeV}) \quad (10.1)$$

(*b*) Semi-leptonic processes, e.g. nuclear beta-decay

$$^{24}\text{Na} \to {}^{24}\text{Mg} + e^- + \bar{\nu}_e \qquad (t_{1/2} = 15\cdot 4 \text{ h}, \ Q = 5\cdot 51 \text{ MeV}) \quad (10.2)$$

This category also includes atomic electron capture processes such as

$$^7\text{Be} + e^- \to {}^7\text{Li} + \nu_e \qquad (t_{1/2} = 53\cdot 4 \text{ d}, \ Q = 0\cdot 86 \text{ MeV}) \quad (10.3)$$

and the analogous capture of muons in muonic atoms (Sect. 3.3)

$$^{12}\text{C} + \mu^- \to {}^{12}\text{B} + \nu_\mu \qquad (t_{1/2} \approx 1\cdot 4 \times 10^{-6} \text{ s}, \ Q = 92 \text{ MeV}) \quad (10.4)$$

(*c*) Non-leptonic decays of strange particles, e.g.

$$\text{K}^0 \to \pi^+ + \pi^- \qquad (t_{1/2} \approx 10^{-8} \text{ s}, \ Q = 218 \text{ MeV}) \quad (10.5)$$

Experimental study of these processes has revealed new and basic properties of nature. First came the hypothesis of the neutrino; later, in a fruitful juxtaposition of events in nuclear and particle physics, the discovery of parity non-conservation in weak processes. There also came the important hypothesis of the non-conservation of strangeness in non-leptonic weak processes such as (10.5) and the recognition of the violation of other symmetries in kaon decay. More recently, and likely to be of outstanding importance for theoretical physics, are the discovery of neutral currents (Sect. 2.2.3) and the theories of Salam and Weinberg that unify the weak and the electromagnetic interactions.

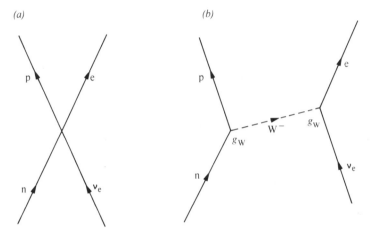

Fig. 10.1 The weak interaction. (*a*) Four-fermion point interaction, representing the beta decay of the neutron with absorption of a neutrino. (*b*) Beta decay of the neutron with an interaction of finite range, associated with exchange of a boson W⁻.

In processes such as (10.2), (10.3) and (10.4) the weak interaction converts a bound neutron into a bound proton or vice versa; the *free* neutron decays to a proton in an exactly similar way

$$n \rightarrow p + e^- + \bar{\nu}_e \qquad (t_{1/2} = 10{\cdot}8 \text{ min}, \ Q = 0{\cdot}782 \text{ MeV}) \qquad (10.6a)$$

This decay may be written more symmetrically by replacing the antineutrino *emission* by neutrino *absorption*, to which it is equivalent:

$$n + \nu_e \rightarrow p + e^- \qquad (10.6b)$$

Also, in processes such as (10.5), it is possible to introduce a virtual intermediate state in which a Λ-particle, a nucleon and a nucleon–antinucleon pair are present. It is thus possible to postulate that all weak interactions involve a four-fermion vertex, drawn for β-decay in Fig. 10.1*a*. If, however, the interaction is transmitted by an intermediate particle W the diagram is spread out (Fig. 10.1*b*) and then resembles the general interaction diagram drawn in Fig. 2.8. The strength of the weak interaction is represented by the magnitude g_W of the 'weak charge' associated with the vertices in Fig. 10.1*b*, but the intermediate particle W is not yet known experimentally, and until its mass M_W can be ascertained only a quantity proportional to g_W^2/M_W^2 can be determined by experiment. This quantity is known as the *weak interaction coupling constant g* and is analogous to the electromagnetic coupling constant $\alpha = (\mu_0 c^2/4\pi)e^2/\hbar c$. When expressed suitably in dimensionless form, the observed value of g is about $10^{-3}\alpha$. The theories that unify the weak and electromagnetic interactions predict a simple numerical relation between e and g_W that implies a boson mass $M_W \approx 80$ GeV.

10.2 Nuclear beta decay and lepton capture

10.2.1 *Experimental information*

Figure 9.7, which was presented in Section 9.3 in connection with the discussion of internal conversion, shows the spectrum of electron energies in the decay of 30-year ^{137}Cs, observed as a function of momentum p $(\propto B\rho)$ in a magnetic spectrometer. If the contributions to the counting rate due to the sharp conversion lines are removed, the remaining electron spectrum is as shown in Fig. 10.2*a*. This continuous distribution, with an upper limit, is the beta-particle spectrum of the ^{137}Cs nucleus, whose main decay may be written

$$^{137}\text{Cs} \rightarrow {}^{137}\text{Ba}^* + e^- + \bar{\nu} \qquad (t_{1/2} = 30 \text{ a}, \ Q = 0.512 \text{ MeV}) \quad (10.7)$$

Because of the particular angular momenta involved, the main beta decay of ^{137}Cs is to an excited state of the residual nucleus but there is also a weak ground-state decay that contributes electrons to the spectrum; these have also been removed in Fig. 10.2*a*.

The main characteristics of a beta-decay process, which are obtained by experiment and are to be elucidated by theory are, for both electron and positron emitters:

(*a*) *The lifetime*, as mean life τ or halflife $t_{1/2}$, which ranges from a fraction of a second to 10^6 a and is 30 a in the ^{137}Cs example.

(*b*) *The maximum electron or positron energy* in the decay, given by the upper limit of the spectrum. This is not accurately obtainable from the direct spectrum, Fig. 10.2*a*, but by use of a theoretical factor (Sect. 10.2.3) a linear plot known as the *Fermi–Kurie plot* (see Ex. 10.9) may be obtained. This extrapolates to the upper limit (Fig. 10.2*b*) and may be used to deduce the Q_β value for the decay. This in turn, together with the energy of any associated γ-radiation such as the 662 keV transition in ^{137}Ba, can be used to connect the masses of the initial

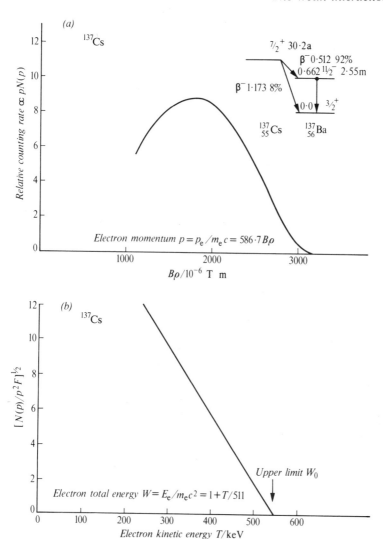

Fig. 10.2   The beta spectrum. (*a*) Sketch of electron distribution for ^{137}Cs obtained from Fig. 9.7 by removing internal conversion lines and the yield due to the high-energy transition. The ordinate is the counting rate at a spectrometer setting p, which is proportional to $pN(p)$ where $N(p)$ is the number of decay electrons per unit momentum at momentum p. (*b*) Fermi–Kurie plot to determine the end point of the spectrum. The abscissa gives electron kinetic energy (=electron total energy$-m_ec^2$). The ^{137}Cs decay is first-forbidden, and the FK plot in an accurate experiment is actually curved because of an energy-dependent shape factor additional to the Fermi factor F used in the ordinate. Omission of this factor means that an accurate end-point energy is not expected in the present case.

and final nuclei, on the assumption of zero mass for the neutrino. It is often convenient to express the upper limit in terms of
the *total* electron energy, i.e. including mass-energy $m_e c^2$ and it
is then written E_0, or in units of $m_e c^2$, W_0 $(=E_0/m_e c^2)$.

(c) *The detailed shape* of the electron or positron spectrum as a
function of total energy E_e $(W = E_e/m_e c^2)$ or momentum p_e
$(p = p_e/m_e c)$.

This information permits a broad classification of β-emitting
bodies and often gives an indication of spin and parity changes in
the decay. In addition, in recent years, and because of interest in the
fundamental nature of the weak interaction, there has been much
work on:

(d) The correlation between the spatial and spin directions of the
particles concerned in the decay (Sect. 10.4.2).

10.2.2 Classification

The extensive range of nuclear beta-decay lifetimes may be roughly
ordered by calculating a quantity known as the *comparative halflife*
$ft_{1/2}$ (often called the *ft*-value). The numerical factor f (Sect. 10.2.3,
eqn (10.21)) is determined mainly by the energy release E_0 in the
decay, and is strongly dependent on this quantity $(\approx E_0^5$ for high
energies). Use of $ft_{1/2}$ values rather than $t_{1/2}$ leads directly to the
basic quantity $1/(g^2 |M|^2)$ where g is the weak interaction coupling
constant and M is the nuclear matrix element. The number distribution for odd-mass beta-decaying nuclei with respect to $\log ft_{1/2}$ is
shown in Fig. 10.3.

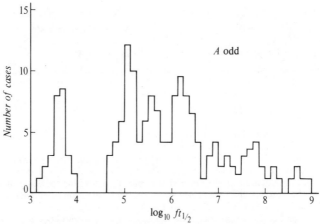

Fig 10.3 Comparative halflife for beta transitions between ground states of odd-
mass nuclei (Feenberg, E. and Trigg, G., *Rev. Mod. Phys.*, **22**, 399, 1950).

A group in the distribution shown in the figure suggests that for the corresponding decays the nuclear matrix element is independent of Z and of W_0. A clear case arises for $\log ft_{1/2} \approx 3\text{--}4$ which relates to especially probable transitions known as *superallowed*, in which it seems likely that there is excellent overlap between the initial and final nucleon wavefunctions. Decays of mirror nuclei (e.g. $^{13}\mathrm{N} \to$ $^{13}\mathrm{C}$) and decays within isobaric triplets (e.g. $^{26}\mathrm{Si} \to {}^{26}\mathrm{Al}^m \to {}^{26}\mathrm{Mg}$) fall into this class.

Other groupings perhaps exist for $\log ft_{1/2} \approx 4\text{--}5$ and for $\log ft_{1/2} >$ 6. For the former, the transitions may be *allowed* in the sense of the selection rules (Sect. 10.2.4) but are unfavoured with respect to the superallowed group because of change of nucleon wavefunction. The latter are the *forbidden transitions*, in which there is a more drastic change of structure. It will be seen in Section 10.2.3 that the $ft_{1/2}$ classification can equally well be described in terms of the orbital angular momentum transported by the leptons in the decay.

10.2.3 The neutrino and the Fermi theory

The starting point of the Fermi theory of beta decay is the Pauli hypothesis of the neutrino, a massless fermion that is created together with an electron in the beta-decay process just as a photon is created in a radiative transition between nuclear levels, but of course with a different probability. At the present point, neutrino is taken to embrace antineutrino and electron to include positron.

The neutrino was required because beta decay connects nuclear states with definite energy, angular momentum and parity. The continuous nature of the beta energy distribution (Fig. 10.2a) and the fact that no other familiar type of radiation could be seen in beta-decay processes (excepting, of course, radiation from the residual nucleus) meant that some new means of restoring an energy balance was required. Equally, it was necessary to ensure a similar balance of linear momentum in the decay. Moreover, since beta decay does not change mass number the initial and final nuclei must either both have integral or both have half-integral spin and the emission of only one fermion between these two states cannot conserve angular momentum.

Long after Pauli's hypothesis, the discovery of parity non-conservation in weak interactions (Sect. 10.4.2) led to the recognition of a specific connection between the momentum and spin vectors for a neutrino analogous to the circular polarization of a light quantum, but differing from it in that only one sense is permitted. For a massless fermion, it is useful to describe this property as *helicity*, and to speak of helicity states $+1$ and -1 corresponding with spin parallel or antiparallel with the direction of motion. If this is included, the properties required for the neutrino in a typical decay process are:

(a) zero charge;
(b) zero mass, since all experiments confirm the conclusion that the upper limit of the beta spectrum gives the energy change in the decay;
(c) half-integral angular momentum;
(d) extremely small interaction with matter because of the failure of conventional experiments to detect neutrino ionization;
(e) unit helicity ($\mathcal{H} = \pm 1$). This implies (b) and (c), i.e. that the neutrino is a massless fermion and consequently that it has a distinct *antiparticle*, the antineutrino $\bar{\nu}$, with opposite helicity.

The neutrino differs from a light quantum in respects (c) and (d) and in the fact that it is not associated with an electromagnetic field. It is, as shown in Fig. 10.1a, *not* the exchange particle of the weak interaction. Direct experimental evidence for the existence of the neutrino, apart from its success in making possible a theory of beta decay, is reviewed in Section 10.4.2.

The probability per unit time of a beta-decay process

$$(A, Z) \rightarrow (A, Z \pm 1) + e + \nu \qquad (10.8)$$

releasing total energy $E_0 = E_e + E_\nu$ is given by time-dependent perturbation theory as

$$T_{if} = (2\pi/\hbar) |H_{if}|^2 \rho(E) = \lambda, \quad \text{say} \qquad (10.9)$$

where H_{if} is the matrix element of the weak interaction Hamiltonian operator H_β between initial and final states i, f, and $\rho(E)$ is the number of momentum states in the volume V of the final system per unit energy at the total energy $E = E_0$. The hadronic wavefunctions to be used in the matrix element H_{if} are those of the initial and final nuclei, so that beta decay is basically a many-body problem. It is, however, possible to divide the calculation into a summation over single-particle transitions $n \xrightarrow{\beta, \nu} p$, plus small corrections due to many-body features. The latter will not be considered here; a discussion of such features will be found in Reference 11.4.

In its simplest version the theory of beta decay expresses a nucleon transition in the process (10.8) in the symmetrical form of equation (10.6b) in which two particles are absorbed and two created. The interaction Hamiltonian H_β cannot be given a specific form, as in the case of the electromagnetic interaction, but it clearly produces both a hadronic (nucleon) transition and the emission of leptons. If we indicate these processes by dimensionless operators $\mathbf{O}_h$ and $\mathbf{O}_l$ respectively then a possible form for the decay is

$$H_\beta = g\mathbf{O}_h[\phi_e^* \mathbf{O}_l \phi_\nu] \qquad (10.10)$$

where g is a coupling constant. The matrix element between initial and final nucleon states is then

$$H_{if} = g \int (\psi_f^* \boldsymbol{O}_h \psi_i)(\phi_e^* \boldsymbol{O}_l \phi_\nu) \, \mathrm{d}^3 r \qquad (10.11)$$

in which it is assumed that all the particles interact at a point. As in the electromagnetic case we also assume a normalization of wavefunctions such that the system volume V does not appear in T_{if}. If the matrix element is to be an energy the coupling constant g then has the dimensions energy × volume.

The nature of the operators $\boldsymbol{O}_h$ and $\boldsymbol{O}_l$ will be discussed in Section 10.4. For the present, we recall that the vector potential in the electromagnetic interaction could be written as an outgoing plane wave and note that in the nucleon weak decay there are lepton waves in the final state (10.6a) or in both initial and final state (10.6b). Assuming again a plane wave system and omitting the $\mathrm{e}^{i\omega t}$ term, these waves contribute a factor $\mathrm{e}^{i\boldsymbol{k} \cdot \boldsymbol{r}}$ to the matrix element, where $\boldsymbol{k} = \boldsymbol{k}_e + \boldsymbol{k}_\nu$ is the sum of the electron and neutrino wave vectors. For $r \approx R$, the nuclear radius, $\boldsymbol{k} \cdot \boldsymbol{r}$ is normally small ($\approx 0{\cdot}1$) and an expansion of the exponential may be made, either directly or in multipoles. Successive terms in this expansion correspond with the emission of the electron–neutrino pair in (10.8) with $l = 0, 1, 2, \ldots$ units of *orbital* angular momentum; strictly, it is not possible to separate orbital from total angular momentum for relativistic fermions but here a non-relativistic treatment is used. The transition probability associated with these terms decreases as $(kR)^2$ as l increases and this offers an explanation of the empirical classification of beta-decay processes into allowed ($l = 0$) and forbidden ($l \geq 1$) types. If the factor r^l is taken from the expansion into the interaction operator, leaving energy-dependent factors outside it, the classification is then related to the nuclear matrix element of the weak interaction and selection rules (Sect. 10.2.4) for changes of nuclear spin and parity may be inferred, as in the electromagnetic case.

We now consider specifically an *allowed transition* ($l = 0$), i.e. s-wave lepton emission. We replace the lepton bracket in (10.11) by unity although the operator $\boldsymbol{O}_l$ must still be considered to determine the fundamental nature of the lepton emission, as discussed in Section 10.4. The operator $\boldsymbol{O}_h$ could be simply the isospin-shifting operator τ^+ that converts a neutron into a proton. The matrix element for the transition (I_i, m_i) to (I_f, m_f) then reduces to

$$H_{if} = g \int \psi_f^* \tau^+ \psi_i \, \mathrm{d}^3 r \qquad (10.12)$$

If this is summed for all nucleons the isospin factor becomes T^+ and if the squared matrix element is averaged over the initial substates and summed over the final substates, allowing for lepton spins, we obtain the squared *nuclear matrix element* $|M|^2 = |H_{if}|^2/g^2$.

The density of states factor $\rho(E)$ in (10.9) refers to all possible distributions of energy between the electron and the neutrino and to a total energy E_0 within a range of width dE_0. For each electron energy E_e, the neutrino energy E_ν is $E_0 - E_e$ because the heavy recoil nucleus $(A, Z \pm 1)$ to which the nucleon is bound takes very little energy. The nucleus does, however, participate in the momentum balance so that the direction of the neutrino momentum is not determined by that of the electron. The factor $\rho(E)$ is, therefore, a product of phase-space factors for both electron and neutrino but with no statistical weight $2s + 1$ for intrinsic spins since this is implicit in the overall squared matrix element. For lepton momenta p_e, p_ν the transition probability for the selected energy range may therefore be written, omitting volume factors,

$$d\lambda = (2\pi/\hbar)g^2 |M|^2 \cdot \frac{4\pi p_e^2 \, dp_e}{h^3} \cdot \frac{4\pi p_\nu^2 \, dp_\nu}{h^3} \cdot \frac{1}{dE_0} \qquad (10.13)$$

using the standard form $4\pi p^2 \, dp/h^3$ for the number of states per unit volume at momentum p. Remembering that for the leptons, assuming a finite neutrino mass m_ν,

$$\left. \begin{array}{r} E_e^2 = p_e^2 c^2 + m_e^2 c^4 \\ E_\nu^2 = p_\nu^2 c^2 + m_\nu^2 c^4 \end{array} \right\} \qquad (10.14)$$

we have

$$\left. \begin{array}{r} c^2 p_e \, dp_e = E_e \, dE_e \\ c^2 p_\nu \, dp_\nu = E_\nu \, dE_\nu \end{array} \right\} \qquad (10.15)$$

Also, for a given electron energy $dE_\nu = dE_0$, so that (10.13) becomes

$$d\lambda = (1/2\pi^3)(g^2 |M|^2/\hbar^7 c^4) p_e p_\nu E_e E_\nu \, dE_e \qquad (10.16)$$

Equation (10.16) gives the energy distribution in the beta spectrum; for $m_\nu = m_e$ it would be symmetrical about the mean energy in the absence of Coulomb effects, but for $m_\nu = 0$ the maximum yield is shifted to a lower energy.

If the dimensionless variables W and p are used instead of E_e and p_e for the electron total energy and momentum, and if m_ν is set equal to zero, equation (10.16) becomes

$$d\lambda = (1/2\pi^3)(g^2 |M|^2 m_e^5 c^4/\hbar^7) p W (W_0 - W)^2 \, dW \qquad (10.17)$$

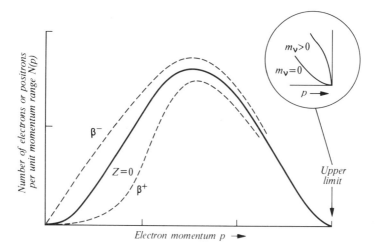

Fig 10.4 The beta spectrum. The full line marked $Z = 0$ shows the momentum distribution in beta decay without Coulomb correction. Introduction of the Fermi factor alters the statistical shape as shown schematically by the dotted lines for electrons and positrons. The inset is an enlarged version of the spectrum near the upper limit in the cases of zero and finite neutrino mass.

or, in terms of momenta rather than energies,

$$d\lambda = (1/2\pi^3)(g^2 |M|^2 m_e^5 c^4/\hbar^7) p^2 (W_0 - W)^2 \, dp \qquad (10.18)$$

The momentum distribution $N(p) = d\lambda/dp$ is plotted in Fig. 10.4. This spectrum approaches zero intensity at the upper limit with a horizontal tangent if $m_\nu = 0$ but with a vertical tangent if m_ν is finite as shown in the figure inset. Detailed experimental study of the spectrum for the tritium decay $^3H \rightarrow {}^3He + e^- + \bar{\nu}$ near the upper limit has set the present upper bound $m_\nu < (1/8500)m_e$, i.e. $<60 \, eV$.

If the nuclear matrix element is energy-independent it follows from (10.18) that if the number of electrons per unit momentum range in the spectrum $N(p) = d\lambda/dp$ is plotted as $[N(p)/p^2]^{1/2}$ against W, the graph should extrapolate to the total energy W_0 corresponding to the upper limit. This, however, is not precise enough because all electron and positron distributions are affected by the nuclear charge and the calculable Fermi factor $F(Z \pm 1, p)$ must be introduced to correct for this. The plot of $[N(p)/p^2 F]^{1/2}$ against W is the Fermi–Kurie plot, shown for ^{137}Cs in Fig. 10.2b. The factor F is really a penetration factor for the nuclear Coulomb barrier; its effect is to suppress low energies in positron spectra and to enhance them in the case of electrons; the spectral shape then changes as indicated by the dotted lines in Fig. 10.4.

The total decay probability λ for the transition (10.8) may now be obtained as

$$\lambda = \int d\lambda = (1/2\pi^3)(g^2 |M|^2 c^4 m_e^5/\hbar^7)$$
$$\times \int_0^{p_0} p^2 F(Z \pm 1, p)(W_0 - W)^2 \, dp \quad (10.19)$$

where p_0 is the electron momentum corresponding to the upper limit of the spectrum. This is normally written in terms of the halflife $t_{1/2}$ for the decay

$$(\ln 2)/t_{1/2} = (1/2\pi^3)(g^2 |M|^2 c^4 m_e^5/\hbar^7) f(Z \pm 1, p_0) \quad (10.20)$$

where

$$f(Z \pm 1, p_0) = \int_0^{p_0} p^2 F(Z \pm 1, p)(W_0 - W)^2 \, dp \quad (10.21a)$$

$$= \int_1^{W_0} pWF(Z \pm 1, W)(W_0 - W)^2 \, dW \quad (10.21b)$$

may be derived from tabulated values of the Fermi function F, and W and p are related by the expression, given by (10.14) for E_e,

$$W^2 = 1 + p^2 \quad (10.22)$$

The quantity $ft_{1/2}$ obtained from (10.20) is the comparative half-life (Sect. 10.2.2) for allowed decays, in which the leptons remove zero orbital angular momentum. Within the allowed class, groupings correspond to particular values of the structure-dependent nuclear matrix element M. If this quantity is known, as it is for some of the simplest nuclei, the $ft_{1/2}$-value yields the coupling constant g; this is found to be $1 \cdot 41 \times 10^{-62}$ J m^3 (Sect. 10.4.3).

The Fermi theory is immediately applicable to the nuclear capture of electrons or negative muons from an atomic orbit, processes (10.3) and (10.4). Only *two* particles are then involved in the actual final state, the residual nucleus and the neutrino, and the latter essentially has a unique energy if only one final nuclear state is involved. The capture probability is directly predictable from (10.9) using the $\rho(E)$ appropriate to a two-body final state and an initial lepton wavefunction representing a particle bound in the *original* atom rather than a plane wave. For an s-state, with Bohr radius a ($a_\mu = \frac{1}{207}a_e$) this is

$$\phi(r) = (1/\pi^{1/2})(Z/a)^{3/2} \exp(-Zr/a) \quad (10.23)$$

and the particle density at the nucleus is $|\phi(0)|^2$. The number of final neutrino states per unit energy range is as in (10.13) with $dE_0 = dE_\nu$

$$\rho(E) = \frac{4\pi p_\nu^2 \, dp_\nu}{h^3} \cdot \frac{1}{dE_0} = \frac{4\pi p_\nu^2}{h^3 c} \quad (10.24)$$

For electron capture, the stability conditions, equation (6.24), show that the overall energy release is just m_ec^2 (511 keV) greater than the *total* energy $W_0 m_e c^2$ available for the alternative process of positron decay between the same initial and final nuclei, neglecting nuclear recoil and atomic binding energy terms.

For an allowed nuclear transition, we obtain from (10.9), (10.23) and (10.24) the probability of electron capture averaged over the K-electrons as

$$\lambda_K = (1/2\pi^3)(g^2 |M|^2 m_e^2 c/\hbar^4)(Z/a)^3 2\pi(W_0 + 1)^2 \qquad (10.25)$$

and in the present approximation the ratio of the intensity of this transition λ_K to that of positron emission, λ_+, does not depend on the nuclear matrix element.

In some cases, e.g. $^7\text{Be} \rightarrow {}^7\text{Li}$ (eqn (10.3)) insufficient energy is available for positron decay and electron capture is the only decay mode. Generally, because of the additional energy available, the increased density of atomic electrons near the nucleus for heavy atoms, and the deterrent effect of the nuclear potential barrier on low-energy positron emission, electron capture increases in relative importance with Z. Capture is usually observed by detection of X-rays from the atomic vacancy; the decay constant λ_K can be altered by about 0·01 per cent by altering the electronic environment of the nucleus by chemical methods.

Muon capture, such as that shown in equation (10.4), may take place from the K-shell of a muonic atom. The basic process is

$$\mu^- + p \rightarrow n + \nu_\mu \qquad (10.26)$$

and this releases an energy of about 106 MeV so that many nuclear states may be excited. Gamma radiation from these states may be detected. Since the muon has to find a proton with which to interact, the capture probability includes an extra factor Z in comparison with electron-capture, i.e. $\lambda_{\mu c} \propto Z^4$. In practice, decay of the muon is more probable for $Z \approx 10$ or less. The intrinsic probability is still determined by the weak interaction coupling constant, but additional nuclear matrix elements are now important because of the high momentum transfer.

10.2.4 Selection rules (non-relativistic)

Among the most probable beta decays, namely the *superallowed* transitions with $ft_{1/2} \approx 3100$ s, are found the transitions

$$^{14}\text{O} \rightarrow {}^{14}\text{N}^* + e^+ + \nu \qquad (10.27)$$

in which the nuclear spin change is from 0^+ to 0^+, and

$$^6\text{He} \rightarrow {}^6\text{Li} + e^- + \bar{\nu} \qquad (10.28)$$

in which the spin change is from 0^+ to 1^+.

In the non-relativistic approximation used hitherto, which of course cannot be generally accurate for electron + neutrino emission, orbital and spin angular momenta have been considered separately. If the allowed character of (10.27) and (10.28) is attributed to s-wave emission of the lepton pair the total angular momentum change must be provided by appropriate alignment of the two lepton spins of $\frac{1}{2}\hbar$. In cases such as (10.27) with $\Delta I = 0$ the electron and neutrino are emitted with spins antiparallel; these are known as *Fermi transitions* (F). For decays such as (10.28) with $\Delta I = 1$ the lepton spins are parallel and these are *Gamow–Teller transitions* (GT).

Evidently the nuclear matrix elements for F and GT transitions will be different and in decays in which both types of transition are possible, e.g. for spin changes $\frac{1}{2}^{\pm} \rightarrow \frac{1}{2}^{\pm}$ or $1^{\pm} \rightarrow 1^{\pm}$ the matrix element is written

$$|M|^2 = C_F^2 |M_F|^2 + C_{GT}^2 |M_{GT}|^2 \qquad (10.29a)$$

where C_F and C_{GT} are coupling constants in units of g. The ratio $|C_{GT}/C_F|^2$ depends on the weak interaction itself. The ratio $|M_{GT}/M_F|^2$ is dependent on nuclear structure and can be predicted accurately only in simple cases, e.g. neutron decay. In the next section, evidence will be presented for the identification of C_F and C_{GT} with the vector (V) and axial vector (A) coupling constants C_V and C_A and if we write $g_V = gC_V$ and $g_A = gC_A$ then equation (10.20) takes the general form (after some rearrangement).

$$ft_{1/2} = \frac{K}{g_V^2 |M_F|^2 + g_A^2 |M_{GT}|^2} \qquad (10.29b)$$

where K is known, from the values of the fundamental constants appearing in equation (10.20), to be $1 \cdot 2308 \times 10^{-120} \, J^2 \, m^6 \, s$.

The two types of transition are clearly also possible in forbidden decays, as shown in the non-relativistic selection rules given in Table 10.1. These results apply to spin and parity changes for the initial and final nuclear states, for which isobaric spin is a good quantum number. The isospins actually observed show that this quantity is

TABLE 10.1 Selection rules for beta decay (non-relativistic)

Type of transition	Fermi		Gamow–Teller	
	Spin ch.	Parity ch.	Spin ch.	Parity ch.
Allowed	0	No	0 (except 0→0), ±1	No
First forbidden	0 (except 0→0), ±1	Yes	0, ±1, ±2	Yes

not conserved in the weak interaction, since allowed transitions with $\Delta T = 0, \pm 1$ are found. In the Fermi superallowed decays, identity of initial and final state wavefunctions implies $\Delta T = 0$ and since there must be at least two nuclei related by the transition, $T \geqslant \frac{1}{2}$. If $\Delta T \neq 0$, Fermi transitions are *isospin-forbidden* and then take place, with reduced probability, because of isospin impurities in the initial and final states.

10.3 Muon decay

Figure 10.5*a* is a spectrum of the electrons observed from the decay of muons at rest, process (10.1). The spectrum is continuous, suggesting that the final state contains three particles and it tends to a maximum intensity at the upper limit, suggesting that two particles can emerge directly opposite to the decay electron with high probability, as shown in Fig. 10.5*b*. The maximum electron energy is ≈ 54 MeV, i.e. about half the muon mass, and no high-energy radiation is observed so that processes such as $\mu \to e + \gamma$ and $\mu \to e + 2\gamma$ are ruled out.

(a)

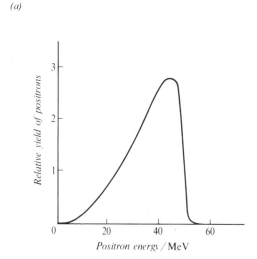

(b)

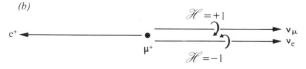

Fig. 10.5   Muon decay. (*a*) Spectrum of positrons from decay of a μ^+-particle at rest. (*b*) Helicities of neutrinos in μ^+-decay.

The purely leptonic processes that account for these facts are, as already indicated for μ^+ in (10.1),

$$\mu^- \rightarrow e^- + \bar{\nu}_e + \nu_\mu$$
$$\mu^+ \rightarrow e^+ + \nu_e + \bar{\nu}_\mu \tag{10.30}$$

Figure 10.5b shows that the two neutrinos emitted cannot be identical because these particles have a definite helicity and the highly probable collinear decay then implies that two identical fermions are emitted in the same state of motion. This is forbidden by the Pauli principle, and for identical neutrinos the electron intensity would tend to zero at the upper limit. It was originally thought that the neutrinos might be particle and antiparticle, but it is now known that there is an additional difference, namely that, as shown above, one is electron-associated and the other muon-associated. The form in which the processes (10.30) is written implies a law of lepton conservation for both (e, ν_e) and (μ, ν_μ) with μ^-, e^- counted as particles and μ^+, e^+ as antiparticles.

The fact that there are two distinct kinds of neutrino in nature, each with its own antiparticle, was established by experiments in which neutrinos from pion decay

$$\pi \rightarrow \mu + \nu \tag{10.31}$$

were allowed to pass through the plates of a large spark chamber. It was found that occasionally an interaction with a nuclear proton

$$\nu_\mu + p \rightarrow n + \mu \tag{10.32}$$

would produce a muon, but no electrons from the reaction

$$\nu_e + p \rightarrow n + e \tag{10.33}$$

were seen, although they would have been expected to produce a very obvious electromagnetic shower in the spark chamber.

The neutrino ν_e has a mass <60 eV. The mass of the neutrino ν_μ is known with less precision but certainly seems to be <650 keV. If it is assumed that both particles have zero mass then the muon-to-electron decay probability may be calculated using perturbation theory. The result, which assumes also that all the fermions have unit helicity if fully relativistic, is that

$$\lambda_\mu = 1/\tau_\mu = g_\mu^2 m_\mu^5 c^4 / 192 \pi^3 \hbar^7 \tag{10.34}$$

No nuclear matrix element, with attendant uncertainty of nuclear structure, appears in this expression, and since both the mean life of the free muon and its mass are known to high accuracy, the coupling constant g_μ may be obtained from these data with similar accuracy. Its value, after some small electromagnetic corrections, will be discussed in Section 10.4.3.

10.4 Nature of the weak interaction in nuclear beta decay

The discussion of beta processes in Section 10.3 was phenomeno-logical and presented no detailed account of the weak interaction matrix element. Such an account was possible in the electromagnetic case because of the existence of a full classical theory in which an energy could arise from the interaction of a current density and a vector potential and, therefore, also from the interaction of two currents. Fermi's original formulation of weak interaction theory proceeded in close analogy with electromagnetic theory and leads, in fact, to a similar current–current type of interaction, although the charges concerned are the 'weak charges' g_W of Fig. 10.1b rather than the electric charges e, and the currents interact only over a very small region, perhaps only at a point, rather than over all space. In the present section an outline is given of the way in which this formulation leads to a matrix element, starting with appropriate wavefunctions for the leptons and hadrons concerned and appealing directly to experiment to select the combinations that actually occur in nature. A more detailed discussion will be found in Reference 1.1a, Chapter 4.

10.4.1 Relativistic invariants

Electromagnetic theory, as described by Maxwell's equations, is invariant under a Lorentz transformation and it is reasonable to expect that the beta-decay matrix element (squared) shall also be a Lorentz invariant. Of the four wavefunctions appearing in H_{if}, equation (10.11), ϕ_e and ϕ_ν are solutions of the Dirac equation for fermions of finite and zero mass respectively, while ψ_f and ψ_i are solutions of the Schrödinger equation since the nucleons are non-relativistic. No discussion of the Dirac equation will be given in this book except to note that for a given four-momentum the equation has four independent solutions (corresponding in the non-relativistic limit to a particle with spin up or spin down with positive or negative energy, i.e. to particle and antiparticle).

The leptonic wavefunctions ϕ, therefore, have *four components* and the weak interaction operator $\boldsymbol{O}_l$ assembles these into sums of products. It is found that these products can be arranged into one or more of five different forms having well-defined properties under a Lorentz transformation and listed below:

(a) a *scalar* (S);
(b) a *four-vector* (V);
(c) an *antisymmetric tensor* (T);
(d) an *axial vector* (A);
(e) a *pseudoscalar* (P).

The four-vector and tensor are analogous to the corresponding quantities in electromagnetic theory. An axial vector describes quantities such as angular momentum and a pseudoscalar is, for instance, the scalar product of an axial vector with a velocity. It therefore has the important property that it changes sign under the parity, or mirror reflection, operation.

If the matrix element (10.11) is to be an energy, i.e. a scalar, a simple prescription is to allow the operator $O_l = O$ to act on the nucleonic wavefunctions as it does on the leptonic functions. The matrix element may then be written

$$H_{if} = g \int (\psi_f^* \boldsymbol{O} \psi_i)(\phi_e^* \boldsymbol{O} \phi_\nu) \, \mathrm{d}^3 r \tag{10.35}$$

for a particular operator, and if all suitable operators are included

$$H_{if} = g \int \sum_x C_x (\psi_f^* \boldsymbol{O}_x \psi_i)(\phi_e^* \boldsymbol{O}_x \phi_\nu) \, \mathrm{d}^3 r \tag{10.36}$$

where $x = $ S, V, T, A, P and the C_x are coupling coefficients, in units of g.

10.4.2 Basic experiments on the interaction

Studies of decay rate and spectrum shape do not by themselves determine the precise nature of the weak interaction matrix element. They do, however, provide a starting point by establishing properties (a)–(d) of the associated neutrino, as listed in Section 10.2.3. The evidence that the neutrino and antineutrino of beta decay are distinct particles is the failure to observe the reaction

$$^{37}_{17}\mathrm{Cl} + \bar{\nu}_e \rightarrow \, ^{37}_{18}\mathrm{A} + e^- \tag{10.37}$$

when antineutrinos from beta-decaying fission products in a nuclear reactor, e.g. reactions such as

$$^{140}_{57}\mathrm{La} \rightarrow \, ^{140}_{58}\mathrm{Ce} + e^- + \bar{\nu}_e \tag{10.38}$$

are allowed to pass through a volume of chlorine. No radioactive $^{37}\mathrm{A}$ decaying by the electron-capture process

$$^{37}_{18}\mathrm{A} + e^- \rightarrow \, ^{37}_{17}\mathrm{Cl} + \nu_e \tag{10.39}$$

was found. This experimental result verifies that (10.37) and (10.39) are *not* equivalent forward and backward reactions as they would be if the neutrino and antineutrino were identical particles. The same conclusion follows from the fact that double beta-decay processes such as

$$^{124}\mathrm{Sn} \rightarrow \, ^{124}\mathrm{Te} + 2e^- + 2\bar{\nu}_e \tag{10.40}$$

experimentally have a much lower probability than would be the case if the reaction could proceed in two stages

$$\left.\begin{array}{l} {}^{124}\text{Sn} \rightarrow {}^{124}\text{Sb} + e^- + \bar{\nu}_e \\ \bar{\nu}_e + {}^{124}\text{Sb} \rightarrow {}^{124}\text{Te} + e^- \end{array}\right\} \tag{10.41}$$

Lepton conservation, with distinct ν_e and $\bar{\nu}_e$, requires that the second stage should involve absorption of a neutrino. On the other hand, and now in a positive sense, antineutrino absorption has been demonstrated by Reines and Cowan for the 'inverse β' process

$$\bar{\nu}_e + p \rightarrow n + e^+ \tag{10.42}$$

which is formally equivalent to the neutron decay

$$n \rightarrow p + e^- + \bar{\nu}_e \tag{10.43}$$

The absorption was indicated by the appearance of a prompt positron e^+, with associated annihilation quanta, followed by radiations from the capture of the thermalized neutron in cadmium present in the reaction vessel.

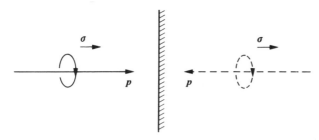

Fig. 10.6   Mirror reflection of a particle with definite helicity. The operation reverses the sign of the product $\boldsymbol{\sigma} \cdot \boldsymbol{p}$; the spin operator $\boldsymbol{\sigma}$ is used so that the eigenvalues of helicity are unity.

The property of helicity for the (zero-mass) neutrino was not acceptable while strict conservation of parity in the weak interaction was required. This may be seen in Fig. 10.6, which shows that the mirror reflection, to which the parity operation is equivalent, turns a right-handed ($\mathscr{H} = +1$) particle into a left-handed particle ($\mathscr{H} = -1$), e.g. $\bar{\nu} \rightarrow \nu$, so that the physical situation changes. The form of the weak interaction in beta-decay was studied before helicity considerations were recognized, by investigating the *correlation* between the direction of electron emission and of antineutrino emission in a suitable decay, such as ${}^6\text{He}(\beta, \bar{\nu}){}^6\text{Li}$. The neutrino direction had to be obtained from observation of the ${}^6\text{Li}$ nuclear recoil. Such experiments, after initial difficulties, indicated that Gamow–Teller transitions such as ${}^6\text{He}$ decay involved the axial vector (A) interaction ($C_{\text{GT}} = C_A$) and Fermi transitions involved the vector (V) type

($C_F = C_V$). Before this conclusion was final, however, these experiments were overtaken by the collapse of the parity-conservation requirement.

This major advance in our understanding of the weak interaction derives from the consideration by Lee and Yang in 1956 of experimental evidence on the decay of charged kaons. In brief, the 3π (τ-mode) and 2π (θ-mode) decay of the K^+-meson indicate identical lifetime ($\approx 10^{-8}$ s) and mass for the parent particle but apparently opposite parity, because of the nature of the 2π and 3π final states. The kaon lifetime indicates that the decay is a weak process, and it is non-leptonic. Lee and Yang suggested that the difficulty could be resolved by the simple but revolutionary assumption that parity is not conserved in a weak decay, and they went on to observe that whereas the validity of parity selection rules seemed well substantiated in electromagnetic and strong interactions, and indeed in the weak interaction as far as the nuclear levels are concerned, nevertheless existing experiments gave no information of this sort for the weak interaction itself. This was essentially because of experimental averaging over spin directions which concealed interference between amplitudes that behaved differently under the parity operation.

The appropriate parity-sensitive experiments were those in which the leptonic decay of a particle or nucleus could be examined with knowledge of the spin axis of the decaying state. The first results were obtained in 1957 for the $\pi \rightarrow \mu \rightarrow e$ decay and for the Gamow–Teller beta-decay of polarized ^{60}Co. In the former experiment the electron emission from muons was found to be asymmetric with respect to the direction of travel of the muon, and if the muon were assumed to be longitudinally polarized, this would indicate an asymmetry with respect to spin direction. In the ^{60}Co experiment the nuclear spin was aligned by cryogenic techniques and again the angular distribution of the electron emission in the decay was found to have the form $(a + b \cos \theta)$ with respect to the spin axis. Reflection of this asymmetric decay process in a mirror displays a different physical situation (Fig. 10.7), contrary to the symmetry expected if parity were conserved. In both cases the effect is seen by the observation of a *pseudoscalar* quantity which is not invariant in mirror reflection, namely the correlation of a vector (linear momentum of the electron) with an axial vector (angular momentum of the decaying particle). The effect is produced by interference of the amplitudes of opposite symmetry that are permitted if conservation of parity between the initial or final state and the emitted leptons is not required. The consequential effect that electrons emitted from unaligned nuclei should be longitudinally polarized (and muons from pions, as assumed above) was also observed. Similar considerations apply to hyperon decays, e.g. $\Lambda^0 \rightarrow p + \pi^-$.

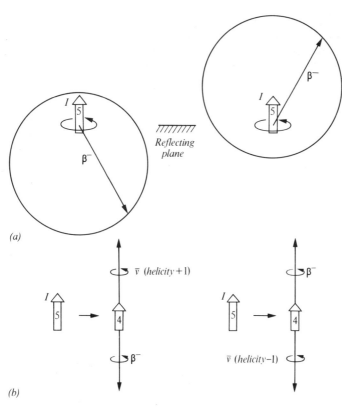

Fig. 10.7 Interpretation of ^{60}Co decay, an allowed Gamow–Teller process. (*a*) Reflection of the nuclear spin (5) and electron momentum in a horizontal mirror; the polar diagram is schematic only. (*b*) Explanation of the asymmetry shown in (*a*) in terms of neutrino theory. In order to change the nuclear spin by $1\hbar$, the spins of the light particles must point along the nuclear axis, but for a prescribed neutrino helicity only *one* of the two mechanisms shown is possible. Experiment indicates the left-hand diagram of (*b*) and hence $\mathcal{H}(\bar{\nu}) = +1$.

The longitudinal-polarization phenomenon, resulting from a pseudoscalar formed from the linear momentum of a particle and its own spin, suggested to Goldhaber, Grodzins and Sunyar (1958) an elegant way of determining the character of the nuclear beta interaction. It was essentially to measure the helicity of the neutrino emitted in the electron-capture process

$$^{152}\text{Eu} + \text{e}^- \rightarrow {}^{152}\text{Sm}^* + \nu_\text{e} \qquad (10.44)$$

The simple nature of this reaction associated this helicity immediately with the circular polarization of the photon from the excited nucleus ^{152}Sm* when it is emitted in a direction opposite to the neutrino. Such photons were selected by imposing a requirement for maximum recoil energy by including a resonant scattering process in the chain of observation and their polarization sense was found by transmission through magnetized iron. The helicity of the neutrino was unambiguously found to be *negative*, i.e. a left-handed screw. Other experiments relating to the longitudinal polarization of the charged particles in beta decays had already been done and in combination with the results of the ^{152}Eu experiment, indicated that the Gamow–Teller interaction was of axial-vector (A) character in agreement with evidence from the electron–neutrino correlation experiments. All 'parity' experiments were also consistent with the conclusions that Fermi-type decays were of vector (V) character and that the antineutrino had positive helicity, as for a right-handed screw. In both cases the helicity was the maximum possible, namely ± 1 for the massless leptons and $\pm v/c$ for electrons or positrons.

The *ft*-values for transitions in which both the Fermi and Gamow–Teller interactions are effective, and for which the nuclear matrix elements are known, show that for the hadronic term in the matrix element $|C_A/C_V| = |C_{GT}/C_F| = 1.23$. From experiments with polarized neutrons, for which the spin axis is known, it further follows that $C_A/C_V = -1.23$; the negative sign leads to the common description 'V–A *interaction*'.

10.4.3 *The weak matrix element*

Following Feynman and Gell-Mann, we now formulate the weak interaction matrix element, using as essential ingredients the non-conservation of parity, the similar strength of all non-strangeness-changing weak interactions, and the observation of left-handed neutrinos.

The matrix element (10.36) is a scalar, whatever the values of the operators. To describe parity non-conservation we need it to contain a pseudoscalar term as well, so that it ceases to be invariant under the parity operation, although it still represents an interaction energy. This is most simply achieved by adding an extra term to the lepton operator which ensures that the neutrino taking part in the weak interaction is left-handed (or right-handed for an antineutrino). Although this apparently ascribes parity non-conservation to the leptonic system it must be remembered that the effect is found also in non-leptonic processes, e.g. (10.5) and is therefore a property of the interaction as well as of the neutrinos. The lepton term then reads

$$\phi_e^* \boldsymbol{O}_x (1 + \gamma_5) \phi_\nu \qquad (10.45)$$

where γ_5 is a Dirac operator with the required property and $\gamma_5^2 = 1$; this quantity also occurs in the axial-vector operator. It has already been noted that the parity operation applied to such a neutrino leads to a non-physical state.

Introducing now the experimental result that only the V and A interactions are effective in nuclear beta decay and omitting all operators other than γ_5, the matrix element (10.36) can be put in the form (Ref. 1.1a, Appendix B)

$$H_{if} = g \int \psi_f^*(C_V - \gamma_5 C_A)\psi_i \cdot \phi_e^*(1+\gamma_5)\phi_\nu \, d^3r \qquad (10.46)$$

This shows that if C_V were equal and opposite to C_A the hadron and lepton terms would be exactly equivalent, each invoking left-handed particles. In fact, we have already noted that $C_A = -1.23 C_V$, probably because owing to strong interaction effects the nucleons are not strictly Dirac particles, and also because of virtual emission of pions in the axial vector interaction.

It might be thought that strong-interaction effects would be significant in vector processes such as the pure Fermi superallowed beta decays. However, the coupling constant $g_\beta^V = g C_V$ extracted from the ft-values for such processes is very close to that derived from the halflife of the muon according to the V–A theory. This agreement has been clarified by Feynman and Gell-Mann in the *conserved vector current* (CVC) *hypothesis*.

To see how this works, a further analogy is drawn to electromagnetism, in which the interaction of a proton with an electromagnetic field is the same, apart from sign, as that of an electron. The electric charge, in fact, is not *renormalized* by the emission of virtual pions. Now for beta decay, referring back to Fig. 10.1b, we may represent the process as a coupling of two particle *currents*, via an intermediate particle exchanged at the vertices. The hadronic current must contain a vector part and an axial-vector part, to account for the observed decays, but these parts are identical in the leptonic current. The CVC hypothesis asserts that the vector part of the hadronic current is *strictly* analogous to the electromagnetic current and is, therefore, free of divergence. Physically, the situation is as shown in Fig. 10.8 and means that although virtual emissions of pions may take place from a nucleon, the emitted pion must itself be capable of beta decay, so that the overall decay probability is unaltered. The predicted decay

$$\pi^+ \rightarrow \pi^0 + e^+ + \nu_e \qquad (\Delta I = 0, \, \Delta T = 0) \qquad (10.47)$$

has been observed with the correct basic rate, or $ft_{1/2}$-value, which of course is exactly that of superallowed beta decay. Other detailed consequences of the CVC hypothesis have been verified.

In terms of currents, the general weak interaction can usefully be described by regarding the Hamiltonian energy as due to the

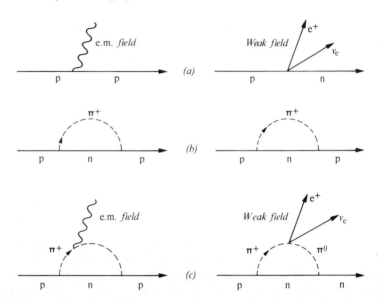

Fig. 10.8 Coupling of a proton to the electromagnetic and weak interaction fields. (*a*) Proton without meson cloud. (*b*) Proton and virtual meson, uncoupled. (*c*) Restoration of coupling by ascribing consequences of charge and weak decay to the pion.

interaction of a current J_W with itself (cf. eqn (9.6) for the electromagnetic interaction). The weak current will contain leptonic parts j_e, j_μ for the creation of electrons or muons, and hadronic parts S and J for strangeness-changing and non-changing processes respectively:

$$J_W = j_e + j_\mu + J + S \qquad (10.48)$$

Taking a product of two such currents we obtain terms such as $(j_e J)$ which applies to the nuclear beta decay just discussed, in which J breaks down into the vector and axial vector parts J_V, J_A. Other products describe muon decay and capture and strange-particle processes.

An important modification to this description of the weak interaction was made by Cabibbo, who proposed that the hadronic currents J and S should not be coupled-in equally. This is indeed necessary because strangeness-changing processes are slower than ordinary beta decays by at least a factor of 10 and apparently do not participate in the $\Delta S = 0$ universality. Specifically, Cabibbo wrote

$$J_W = j_e + j_\mu + J \cos \theta_C + S \sin \theta_C \qquad (10.49)$$

so that if the coupling constant for muon decay $(j_e j_\mu)$ which involves only the leptonic current is g_μ, then that for vector beta decay $(j_e J_V)$

312

is $g_\beta^V = g_\mu \cos \theta_C$. The $\Delta S = 1$ leptonic decays of the strange particles $(j_e S, j_\mu S)$ have a coupling constant $g_\mu \sin \theta_C$ and universality is redefined in terms of the Cabibbo angle.

From existing hyperon decay data the Cabibbo angle is found to be

$$\theta_C = \arcsin (0 \cdot 233 \pm 0 \cdot 005) \tag{10.50}$$

and from the observed muon coupling constant

$$g_\mu = 1 \cdot 4358 \pm 0 \cdot 0001 \times 10^{-62} \, \text{J m}^3 \tag{10.51}$$

application of the relation $g_\beta^V = g_\mu \cos \theta_C$ gives

$$g_\beta^V = 1 \cdot 3962 \pm 0 \cdot 0017 \times 10^{-62} \, \text{J m}^3 \tag{10.52}$$

The directly observed value of g_β^V is obtained from the pure Fermi superallowed transitions of the mass $(4n + 2)$ nuclei ^{14}O, ^{26}Al^m, ^{34}Cl, ^{38}K^m, ^{42}Sc, ^{46}V, ^{50}Mn, ^{54}Co. These decays are between $I^\pi = 0^+$ states and have $T = 1$, $\Delta T = 0$, for which $M_{GT} = 0$ and $M_F = \sqrt{2}$. The $ft_{1/2}$-values, after the introduction of small charge-dependent corrections, are closely similar, in accordance with the CVC hypotheses, and the mean value of 3088 s, inserted in equation (10.29b), gives

$$g_\beta^V = 1 \cdot 4116 \pm 0 \cdot 0008 \times 10^{-62} \, \text{J m}^3 \tag{10.53}$$

This differs from the value predicted from g_μ, but is sufficiently close to it to lend support to the CVC theory as modified by Cabibbo, which explains why beta decay is slower than expected from muon decay.

The residual discrepancy between the values (10.52) and (10.53) is almost certainly connected with the fact that certain model-dependent electromagnetic corrections are not included in the present calculation of g_β^V. These bring in not only the postulated exchange particle of the weak interaction but also the structure of the nucleon itself, namely its quark content. Difficulties in these calculations have now largely been removed by the theoretical advances that unify the weak and electromagnetic interactions and by the observation of the non-strangeness-changing neutral currents in neutrino interactions (Sect. 2.2.3) that support these theories (Ref. 2.4). Since the experimental accuracy of the coupling-constant measurements is quite high, the discrepancy may be used either to estimate the mass of the exchange particle or to predict the average charge of the quarks in the nucleon. In an analysis directed to the latter object, Wilkinson (Ref. 10.1) concludes that nuclear beta-decay data are consistent with quark charges of $\frac{2}{3}e$ and $-\frac{1}{3}e$, in agreement with the results of electron and neutrino scattering experiments (Ref. 2.3).

10.5 Summary

The weak interaction regulates the nuclear decays, such as beta decays, that are not dominated by radiative (electromagnetic) or hadronic (strong) processes. Throughout the history of nuclear physics it has been a fertile ground for new concepts. The chapter mentions Pauli's hypothesis of the neutrino, after which Fermi was able to formulate a theory of nuclear beta decay by analogy with radiative processes. The classification of beta-decaying nuclei in terms of Fermi's theory is described. No detailed formulation of the weak interaction matrix element can be given, since the interaction is not known, but it is constrained by relativistic invariance principles. Since 1957 it has been known also that it contains terms which interfere in such a way that parity is not conserved in weak processes. The experimental evidence for this discovery is outlined. The strength of the weak interaction is directly measurable by beta-decay experiments among which the vector-type Fermi decays have a special significance. It is shown that their decay rate may be related to that of the (structureless) muon if a parameter determined by strange-particle leptonic decay rates is introduced. Even more fundamentally, there is promise of a unification of the weak and electromagnetic interactions, at least at high energies.

Examples 10

10.1 The atomic mass of ^{137}Ba is given by $(M-A) = -87\,734$ keV. Assuming the decay scheme shown in Fig. 10.2 calculate the $(M-A)$ value for ^{137}Cs given that the electron kinetic energy at the upper limit of the main beta spectrum is 514 keV, and that the radiative transition energy is 661·6 keV. [−86 558 keV]

10.2 The atomic masses of $^{74}_{32}$Ge, $^{74}_{33}$As and $^{74}_{34}$Se are 73·921 18, 73·923 93 and 73·922 48. Calculate the Q-values for the possible decay schemes linking these nuclei. [$Q_{\beta^-} = 1·35$, $Q_{\beta^+} = 1·54$, $Q_{EC} = 2·56$ MeV].

10.3* Show that the relativistic expression for the ratio of nuclear recoil energy to maximum electron energy in β^--decay is $T_R/T_e = (Q + 2m_e)/(Q + 2M_R)$ where Q is the energy available for the decay products and M_R is the mass of the recoil nucleus.

Evaluate T_R for the ^{137}Cs case, Example 10.1. [0·003 keV]

10.4 Use the result of Example 10.3 to infer the nuclear recoil energy in the case of electron capture, releasing energy Q_{EC}. Assume that the neutrino mass is zero.

For the decay ^{37}A → ^{37}Cl, $Q_{EC} = 0·82$ MeV and $T_R = 9·7 \pm 0·8$ eV. Show that these figures are consistent with $m_\nu = 0$.

10.5 The reaction ^{34}S(p, n)^{34}Cl has a threshold at a laboratory proton energy of 6·45 MeV. Calculate non-relativistically the upper limit of the positron spectrum of ^{34}Cl assuming $m_e c^2 = 0·51$ MeV, $M_n - M_H = 0·78$ MeV. [4·47 MeV] (See Ex. 1.25 for the relativistic correction)

10.6 Write down the phase-space part of expression (10.16), in terms of the variables W and p for the electron (upper limit $= W_0$) and obtain the quantity $N(p) = d\lambda/dp$ *without* the assumption of zero neutrino mass.

By examination of the variation of $N(p)$ with p near the upper limit, verify the dependence on neutrino mass shown in Fig. 10.4. (Assume for this purpose that the main variation near W_0 is due to the neutrino momentum, given by $p_\nu^2 \propto (W_0 - W)$ for non-relativistic motion.)

Compute and plot the momentum spectrum (a) for $m_\nu = 0$, and (b) for $m_\nu = nm_e$. Compute and plot also the energy spectrum $d\lambda/dW$ given by equation (10.16) for the same two cases and verify that the spectrum is symmetrical for $m_\nu = m_e$.

10.7 In a very low-energy β-spectrum (e.g. ^{14}C with maximum kinetic energy $T_0 = 0\cdot156$ MeV), the energy distribution may be approximated by $d\lambda \propto T^{1/2}(T_0 - T)^2\, dT$.

Show that in this case the mean kinetic energy of the spectrum is one-third of the minimum energy.

10.8* In a high-energy β-spectrum, with the maximum energy $E_0 \gg m_e c^2$, show that the decay constant λ is approximately proportional to E_0^5.

10.9 In a typical experiment to determine the β$^-$-spectrum of ^{137}Cs, the following relative counting rates were observed at the indicated $B\rho$-values in a magnetic spectrometer

$B\rho$ (tesla m $\times 10^6$)	1370	1780	2230	2640	2866	2960	3080
Relative counting rate	726	855	760	305	152	40	30

Allowing for the momentum dependence of the spectrometer acceptance, make a Fermi–Kurie plot of these data, interpolating values of the function $p^2 F(Z, p)$ from the following table

$p = p_e/m_e c$	0·6	0·8	1·0	1·2	1·4	1·6	1·7	1·8	1·9
$p^2 F(Z, p)$	2·0	2·9	3·9	5·1	6·5	8·0	8·9	9·7	10·6

Deduce the end-point energy of the spectrum on the (incorrect) assumption that it has an allowed shape.

10.10* Show that the maximum total electron energy E_0 in the decay of a muon at rest is

$$E_0 = \tfrac{1}{2}m_\mu c^2\left[1 + \left(\frac{m_e}{m_\mu}\right)^2\right]$$

10.11* The maximum positron energy in the decay of ^{34}Cl ($t_{1/2} = 1\cdot57$ s) is 4·47 MeV and the Fermi factor $F(Z, p)$ is equal to 0·71 over most of the spectrum.

Evaluate the integrated Fermi function $f(Z, p_0)$ and the comparative halflife.

10.12* On the assumption that the nuclear matrix element for the ^{34}Cl decay is given by $|M_F|^2 = 2$ and $|M_{GT}|^2 = 0$, obtain the weak interaction coupling constant g_β ($= g_V$).

10.13 Use equation (10.25) to predict the decay rate by K-electron capture in the case of $^{34}_{17}$Cl, for which data are given in Example 10.11. [3×10^{-4} s^{-1}]

10.14 Repeat the calculation of Example 10.13 for the case of muon capture by $^{34}_{16}$S under the simplest possible assumptions. [$\approx 10^7$ s^{-1}]

10.15 The value of the weak interaction coupling constant is $g = 1\cdot44 \times 10^{-62}$ J m^3. Show that in natural units, with $\hbar = c = 1$, it has the value $G = 10^{-5}/m_p^2$ where m_p is the proton mass.

11 The strong interaction

11.1 General properties

The hadronic interaction is a central concern of particle physics, since it confines the quarks in the hadrons and therefore determines the spectrum of particle states and the course of particle reactions. In the physics of the complex nucleus the strong interaction is responsible for the main features of nuclear wavefunctions and, therefore, essentially determines both the properties of the complex nucleus and the nature of nuclear reactions. The link between the strong internucleon force as seen in the two-body problem and the models that can be used to provide wavefunctions for complex nuclei has been explored superficially in Chapters 5, 6, 7 and 8. In the present chapter some well-defined types of nuclear reaction will be briefly reviewed as examples of strong interaction processes, and as a dynamical complement to the calculation of static properties already outlined. The role of nuclear structure phenomena in clarifying ambiguities in the nucleon–nucleon interaction is an important one, but will not be dealt with.

The strong interaction between nucleons was first described by Yukawa in terms of one-meson exchange (Sect. 5.8 and Fig. 2.8); the coupling constant is so large that the perturbation methods used in the theory of the weak and electromagnetic interactions are no longer appropriate. Like all other known processes, the strong interaction conserves baryon number B, which is equivalent to the mass number A for complex nuclei. There is good evidence, from the successful operation of selection rules, that it conserves parity, except for a minor intervention by the weak interaction. The equivalence of cross-sections for forward and backward reactions is evidence for invariance under time reversal T and by Lorentz invariance we are then led to expect invariance under charge conjugation C, (Table 1.1) though this is mainly of significance in particle physics.

Evidence for isobaric-spin conservation is clearly seen in pion scattering by nucleons (Sect. 5.6.2) and in the existence of isospin multiplets and isospin selection rules in both nuclear and particle physics. Ordinary charge conservation (Q or Z) is a fact of experience and the phenomenon of associated production (Sect. 2.2.2) shows that the strong interaction also conserves strangeness.

Summarizing the evidence, the strong interaction is a short-range force between hadrons (range $\approx 10^{-15}$ m corresponding to one-pion exchange, but requiring other exchanges as well), a force that dominates all other interactions when not prevented from operating by trespass against the selection rules. These are

$$\Delta B = \Delta Q = \Delta S = 0$$

$$\Delta (\text{parity}) = 0$$

$$\Delta T = \Delta T_z = 0$$

for interacting hadrons and for hadronic decays, e.g. decays of resonant states both of particles and complex nuclei.

In reaction processes between complex nuclei the properties of the strong interaction become properties of the *S-matrix* which relates the amplitude of a given outgoing state to that of the initial system and is a generalization of the scattering amplitude defined in Section 1.2.7. An important property of the *S*-matrix is that it is *unitary*, and this leads to limits on cross-sections and to the optical theorem (Ex. 1.35). Another property is *reciprocity*, which is an expression of time-reversal invariance and leads to a connection between the cross-sections for forward and backward reactions if an average is taken over spin states. But the basic property of the *S*-matrix is *analyticity* with respect to the appropriate dynamical variables such as total energy and momentum transfer, which means that it can be defined for unphysical values of these variables as well as real values. In particular, the *S*-matrix will exhibit a singularity at a centre-of-mass energy corresponding to a resonant state, including the ground state, of a system of two particles. We, therefore, start by considering some phenomenological aspects of these resonances.

11.2 Nuclear and particle resonances $(\Gamma < D)$

11.2.1 Resonances and the level spectrum

Figure 11.1 gives the energy variation of the total cross-section for three nuclear interactions:

(*a*) positive pions with protons;
(*b*) fast neutrons with oxygen;
(*c*) slow neutrons with gadolinium.

The results are similar in showing well-defined peaks or *resonances* at certain energies near which one process is dominant (elastic scattering for (a), (b) and capture for (c)). In the resonance region of energies the full width Γ of the peaks at half-maximum intensity is of the order of, or less than, their spacing D. There is a tendency for such peaks to broaden as the incident energy increases and finally to merge into a *continuum* (Sect. 11.3). As indicated in the survey of nuclear spectra given in Section 8.8 the continuum is not featureless, but includes broad resonant features related to shell-model states, or to excitations across major shells (giant resonances). The isobaric analogue states (Sect. 5.6.1) are also seen clearly in the general continuum, partly because they are related to the ground state and low-lying levels in an analogue nucleus and therefore have a greater spacing than the surrounding background states.

For particle resonances such as that of Fig. 11.1a, the excitation energy is often a large fraction of the rest mass of the target particle, e.g. the proton, and the resonance is specified by the corresponding total c.m. energy E_0 in MeV, or the mass value M in MeV/c^2. The excitations of complex nuclei of interest to nuclear structure theory are normally very much less than the mass of the nucleus concerned. They may be specified by the actual energy E_{ex} of excitation above the ground state or by the laboratory energy T_a of an incident particle which may excite the resonance in a *formation process* such as

$$X + a \rightarrow C^* \qquad (11.1)$$

The relation between E_{ex} and T_a is (cf. (1.17))

$$E_{ex} = S_a + T_a M_X / (M_X + M_a) \qquad (11.2)$$

neglecting relativistic corrections, where S_a is the separation energy of particle a in the nucleus C and M_X, M_a are atomic masses. A c.m. correction must also be made to the level width Γ.

Particle resonances, both mesonic and baryonic, occur frequently in the energy range up to about 3 GeV. Nuclear resonances are found as discrete states in heavy nuclei for incident energies up to about 1 MeV and in light nuclei for energies up to about 10 MeV. They form the greater part of the nuclear spectrum, since bound states are limited in number, and because of this ready accessibility, their decay properties can be studied to good accuracy. They are, therefore, significant not only for reaction processes but also for theories of nuclear structure.

The level widths indicated in Fig. 11.1 differ considerably but each total width Γ is related to the mean lifetime τ for decay of the state by the relation

$$\Gamma\tau = \hbar = 6 \cdot 6 \times 10^{-16} \text{ eV s} \qquad (11.3)$$

The quantity $\Gamma/\hbar = 1/\tau$ thus represents the probability of decay per

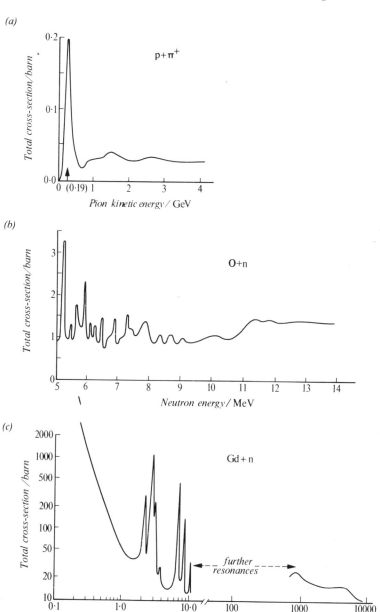

Fig. 11.1   Nuclear resonances shown in the interaction of: (a) positive pions with protons, indicating the Δ(1236) resonance at 190 MeV pion energy; (b) fast neutrons with oxygen; (c) slow neutrons with gadolinium.

319

unit time. Particle resonance widths (≈ 100 MeV) and fast-neutron widths in excited nuclei (≈ 100 keV) are determined by the strong interaction, but many slow-neutron resonance widths ($\approx 0 \cdot 1$ eV) are due to the emission of capture radiation and Γ is then equal to Γ_γ for radiation of about 8 MeV energy (Sect. 9.4.4).

All resonant states, like bound states, are characterized by a spin I, an isospin T and a parity π, as well as by measurable electric and magnetic moments. Mixing of different T-values due to the electromagnetic interaction is fairly common for highly excited states; small parity impurities due to the weak interaction may, in principle, be observed.

11.2.2 Particle resonances

The phenomenological description of particle resonances does not so far involve substructures and may be based on the general partial-wave formalism of Section 1.2.7. The scattering amplitude is, in general, a complex quantity and its magnitude is limited by conditions such as (1.83), modified if necessary by spin factors, that arise from the requirement of unitarity, or 'conservation of probability'. The mass M and total width Γ of a great many such resonances have been found by phase-shift analyses (see Ref. 1.1) of scattering and polarization data in which the elastic partial-wave amplitude is taken to have the *Breit–Wigner* form

$$f \propto \tfrac{1}{2}\Gamma_{\text{el}}/(E - M + \tfrac{1}{2}i\Gamma) \tag{11.4}$$

where E is the total c.m. energy, Γ is the total width and Γ_{el} is a partial width representing the part of Γ that is connected with elastic scattering. The $\Delta(1236)$ resonance of Fig. 11.1a is a well-known example of a p-wave resonance ($l = 1$, $J = \tfrac{3}{2}$) in which the detailed shape is affected by energy dependence of Γ and Γ_{el}. The elastic cross-section $\sigma_{\text{el}}(\propto |f|^2)$ with energy-independent Γ_{el} and Γ, is shown in Fig. 11.2.

The Breit–Wigner amplitude (11.4) may be understood by regarding the particle resonance as an oscillation that may be excited by a driving force, in analogy with a mechanical or electrical resonant system. The excitation is not a pure normal mode because of damping due to re-emission of particles and its wavefunction may be written

$$\psi = \psi_0 \exp\left[(-i/\hbar)(E_0 - i\Gamma/2)t\right]$$
$$= \psi_0 \exp\left(-\Gamma t/2\hbar\right) \exp\left(-iE_0 t/\hbar\right) \tag{11.5}$$

This represents a state decaying with a characteristic time, for $|\Psi|^2$, of $\tau = \hbar/\Gamma$. The expression (11.5) may be Fourier-transformed to a superposition of normal modes, or stationary states of energy E

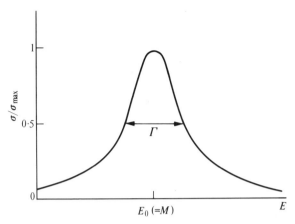

Fig. 11.2 A pure Breit–Wigner cross-section curve showing the resonance energy E_0 and total width Γ.

with an amplitude proportional to

$$1/(E - E_0 + i\Gamma/2) \tag{11.6}$$

which is of the Breit–Wigner form (11.4), and may reasonably be supposed to represent the probability amplitude for excitation of the system by particles of energy E. Equations (11.5) and (11.6) imply that a formation resonance may be taken to have a complex energy

$$E_0 - i\Gamma/2 \tag{11.7}$$

where Γ is the total width of the state. (The form $E_0 + i\Gamma/2$ would represent a density increasing with time and is not realistic.)

The detailed application of the Breit–Wigner amplitude to particle resonances is discussed in Reference 1.1; many of the results obtained in the following section (11.2.3) are directly applicable.

11.2.3 Nuclear resonances

The nuclear resonances shown in Fig. 11.1*b, c* also result from the existence of quasi-stationary states in the nuclear spectrum and should, therefore, be characterized by Breit–Wigner amplitudes as just discussed. Their external features may also be described by the partial-wave formalism, but their existence is determined by the internal nuclear structure. In this chapter, no attempt will be made to set up suitable internal wavefunctions, but the way in which they may be represented by the matching parameter ρ, introduced in equation (5.12), will be outlined.

We deal first with the narrow resonances, reserving mention of the broad (giant) resonance features of the continuum until Section

11.3. As pointed out in Section 8.1.1, slow-neutron data led Niels Bohr to the conclusion that C* in equation (11.1) must be a relatively long-lived state in which the energy of the incident particle is rapidly shared with nucleons by strong interactions within the many-body system. The probability of emission, or re-emission, of a particle is reduced by this sharing process and the compound nucleus may survive long enough for the radiative de-excitation

$$C^* \rightarrow C + \gamma \qquad (11.8)$$

with a lifetime $\approx 10^{-14}$ s, to take place. Alternatively, and especially at higher incident energies for both neutrons and charged particles, the compound nucleus C* formed in process (11.1) may decay by particle emission

$$C^* \rightarrow X + a \qquad (11.9a)$$
$$C^* \rightarrow X^* + a' \qquad (11.9b)$$
$$C^* \rightarrow Y + b \qquad (11.9c)$$

The first of these is *compound elastic scattering* X(a, a)X, the second is *compound inelastic scattering* X(a, a')X* and the third is a *reaction process* X(a, b)Y passing through the compound nucleus. Inelastic scattering is a particular case of a reaction. Each of these processes defines a particular channel with which is associated a specific Q-value given by the mass change and a probability factor or partial width Γ_a, $\Gamma_{a'}$, Γ_b, ... The total width is

$$\Gamma = \Gamma_a + \Gamma_{a'} + \Gamma_b + \cdots \qquad (11.10)$$

and if the cross-section for formation of the compound nucleus is σ_a then the cross-section Γ_{ab} for the reaction X(a, b)Y is

$$\sigma_{ab} = \sigma_a \Gamma_b / \Gamma \qquad (11.11)$$

This expression embodies Bohr's *independence hypothesis* namely, that the relative probability of different decays of the compound nucleus should be independent of its mode of formation; the partial widths are properties of the compound state itself and not (except for Γ_a) of the entrance channel. Good evidence for this proposition is found in reactions in which well-defined levels ($\Gamma \ll D$) may be excited in different ways; the hypothesis, in fact, also applies in many cases in which there is strong overlapping of levels (Sect. 11.3).

To give a formal description of the formation and decay of a compound nucleus resonance we confine attention to the (spinless) elastic scattering of s-wave neutrons X(a, a)X. The internal nuclear wavefunction is no longer determined by a simple potential well of the sort used in Chapter 5 to describe neutron–proton scattering. It will, however, contain both ingoing and outgoing components and

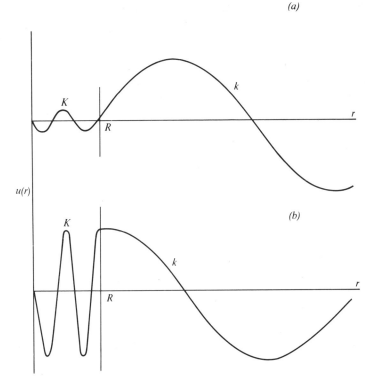

Fig. 11.3 Continuity of logarithmic derivative ρ of nuclear wavefunction $u(r)$ at a boundary $r = R$. In general, since the internal wavenumber $K \gg$ the incident wavenumber k, only a small internal amplitude results (a). At resonance, however, $\rho = 0$ and the internal amplitude is large (b).

will be characterized by a specific value of the matching parameter (eqn (5.12))

$$\rho = \left(\frac{r}{u} \frac{du}{dr} \right)_{r=R}$$

In general, as illustrated in Fig. 11.3, continuity of ρ at the boundary $r = R$ will reduce the amplitude of the wave within the nucleus. But if $\rho = 0$ at $r = R$, so that the wavefunctions have zero slope, they may be joined with equal amplitude. This is the condition for resonance, and leads to a phase shift of $(n + \frac{1}{2})\pi$ in the present simplified picture and to a cross-section, according to (1.79), of $4\pi/k^2$ if no inelastic processes occur ($\eta_0 = 1$). The general principle of *causality*, according to which a scattered wave must not originate before the incident wave arrives, requires that for a true resonance the phase shift must *increase* through the resonant value as the energy increases.

It may readily be shown (see Ex. 11.4) that the s-wave phase shift δ_0 is related to the parameter $\rho = \rho_0$ by the equation

$$\eta_0 \exp 2i\delta_0 = [(\rho_0 + ikR)/(\rho_0 - ikR)] \exp(-2ikR) \quad (11.12)$$

If ρ_0 is a real quantity, this requires that η_0 as defined in Section 1.2.7 is unity, and equation (1.81) confirms that the reaction cross-section vanishes. The expression (11.12) is then identical with (5.12) except that ρ_0 cannot be set equal to a known internal quantity. For the particular case $\rho_0 = 0$, when the derivative of the wavefunction vanishes at the nuclear surface, (11.12) gives $\delta_0 = (n + \frac{1}{2})\pi$ as already indicated. A further important case arises when the final wavefunction itself vanishes at the surface $r = R$, giving $\rho_0 = \infty$. This describes the reflection of a wave from a *hard sphere* of radius R and (11.12) gives $\delta_0 = -kR$. For kR small the elastic scattering cross-section according to (1.79) is $4\pi R^2$. *If ρ_0 is a complex quantity*, both scattering and reaction processes take place and $\eta_0 < 1$.

In a more general case, when ρ_0 is neither infinite nor zero, the elastic scattering amplitude contains two terms, one of the form of a hard-sphere contribution and the other resonant. The former, which by itself leads to scattering that is sometimes described as *potential* or *shape-elastic*, will interfere with the latter, which changes sign as the energy passes through resonance. For charged particles the potential scattering contains a Coulomb term and this also interferes with the nuclear amplitude. The characteristic pattern in the yield of scattered particles in this case is shown in Fig. 11.4*a*. The reaction amplitude contains only the resonant term and the cross-section near resonance has the Breit–Wigner shape, as shown in Fig. 11.4*b*.

If only compound elastic scattering is significant, substitution of the expression (11.12) in the equations (1.79) and (1.81), as illustrated in Example 11.5, leads to the Breit–Wigner formulae for the $l = 0$ cross-section for the elastic scattering X(a, a)X of spinless uncharged particles near resonance:

$$\sigma_{el}^0 = (\pi/k^2)\Gamma_a^2/[(T - T_0)^2 + \Gamma^2/4] \quad (11.13)$$

and for the reaction process X(a, b)Y:

$$\sigma_{ab}^0 = (\pi/k^2)\Gamma_a\Gamma_b/[(T - T_0)^2 + \Gamma^2/4] \quad (11.14)$$

where T is the c.m. energy and T_0 is its value at resonance. In practice, laboratory energies are frequently used in these equations.

The cross-section for the formation of the compound nucleus is the sum of all partial cross-sections and follows immediately from (11.11), or by adding all such equations as (11.14):

$$\sigma_a = (\pi/k^2)\Gamma_a\Gamma/[(T - T_0)^2 + \Gamma^2/4] \quad (11.15)$$

These equations pay no regard to possible energy dependence of the widths and, therefore, only apply near a specific energy, e.g. that of

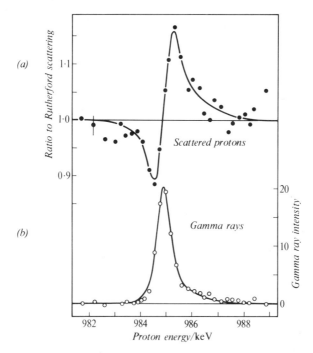

Fig. 11.4   Scattering and absorption of protons at the 985-keV resonance in the reaction ^{27}Al+p. (a) Yield of scattered protons. (b) Yield of capture radiation (Bender, R. S., *et al.*, *Phys. Rev.*, **76**, 273, 1949). The interference in (a) is between the nuclear and Coulomb amplitudes.

the resonance $T = T_0$. Away from a dominant compound elastic scattering resonance the background may be due to the potential scattering with cross-section $4\pi R^2$ for neutrons.

Formulae (11.13) and (11.14) may be used to describe processes involving photons, as has been done in Section 9.4.4.

11.2.4 Introduction of spin

The cross-sections (11.13) and (11.14) may be generalized to include single orbital momenta other than $l = 0$ and intrinsic spins by adding a statistical factor

$$(2I+1)/(2s_1+1)(2s_2+1)(2l+1)$$

where I is the spin of the (pure) compound state and s_1, s_2 the spins of the colliding particles; for photons $(2s+1) = 2$ because only two independent polarization states exist. The statistical factor is obtained by averaging over initial and summing over compound substates. If the intensity $(\pi/k^2)(2l+1)$ of the lth component of the

325

incident plane-wave is also recognized the factor becomes

$$g(I) = (2I + 1)/(2s_1 + 1)(2s_2 + 1) \tag{11.16}$$

Examples of the application of the cross-section formula including the spin factor to the determination of radiative width Γ_γ were given in Section 9.4.4. As a further example we may consider the scattering of pions ($s_1 = 0$) by protons ($s_2 = \frac{1}{2}$). The resonant elastic cross-section is

$$\sigma_{el} = (4\pi/k^2)g(\Gamma_\pi/\Gamma)^2 = (4\pi/k^2)(I + \tfrac{1}{2})x^2 \tag{11.17}$$

where $x = \Gamma_\pi/\Gamma$ is the elastic branching ratio or *elasticity*. The inelastic cross-section is similarly

$$\sigma_{inel} = (4\pi/k^2)g[\Gamma_\pi(\Gamma - \Gamma_\pi)/\Gamma^2] = (4\pi/k^2)(I + \tfrac{1}{2})x(1 - x) \tag{11.18}$$

and the total cross-section is

$$\sigma_{tot} = (4\pi/k^2)(I + \tfrac{1}{2})x \tag{11.19}$$

Anomalies in total cross-section curves as a function of bombarding energy permit the product $(I + \frac{1}{2})x$ to be obtained, although a full partial-wave analysis, with both scattering and polarization cross-sections as input data, is more informative. For the (π^+p) resonance shown in Fig. 11.1a the maximum cross-section is about $8\pi/k^2$, corresponding with $I = \frac{3}{2}$, $x = 1$.

The energy dependence of partial widths, so far neglected, is determined by centrifugal and Coulomb barrier effects (Sect. 11.6.2). For a neutron of momentum q and angular momentum $l\hbar$

$$\Gamma_b \propto q^{2l+1} \propto (T - T_0)^{(2l+1)/2} \tag{11.20}$$

in the non-relativistic case (cf. Ref. 9.1, p. 361).

The angular distribution of elastic scattering is governed by the general considerations of Section 1.2.7 and is determined by the orbital momentum required to form the compound state. In general, more than one l-value may be possible and the partial-scattering amplitudes must then be summed coherently to give the final differential cross-section.

11.3 The continuum ($\Gamma > D$)

No treatment will be given here of the continuum region of the elementary-particle spectrum. The continuum of compound-nucleus states may be studied by techniques of low resolution, and indeed experimental averaging may be desirable in order to conceal features due to the fine-structure resonances. When this is done, the broad virtual levels of a potential well may be seen; at energies for which Γ is not yet $\gg D$, the broad resonance is split into fine-structure components (Fig. 11.5).

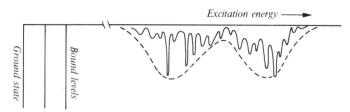

Fig. 11.5 Schematic of nuclear levels when $\Gamma \approx D$. The fine-structure levels are of such a density that their envelope builds up the broad resonances shown by the dotted line. The level contours may be taken to represent the yield of a particular nuclear reaction.

If it is assumed that an individual resonance may still be represented by a Breit–Wigner expression with energy-independent widths, then in the continuum the *cross-section for formation of the compound nucleus* may be obtained by averaging the single-level formation cross-section σ_a. For the simple case of spinless particles and the s-wave, without Coulomb barrier, we take an energy interval W small enough for k to be regarded as constant and obtain, using equation (11.15)

$$\overline{\sigma_a} = 1/W \int_{T-1/2W}^{T+1/2W} (\pi/k^2) \sum_s \Gamma^s \Gamma_a^s / [(T - T_s)^2 + \tfrac{1}{4}(\Gamma^s)^2] \, . \, dT \tag{11.21}$$

where s enumerates the individual level considered within the range W. This gives

$$\bar{\sigma}_a = (\pi/k^2)(2\pi/W) \sum_s \Gamma_a^s = (\pi/k^2) 2\pi (\bar{\Gamma}_a/D) \tag{11.22}$$

where $\bar{\Gamma}_a$ is the mean width and D the mean spacing over the range W. The quantity $\bar{\Gamma}_a/D$ is the s-wave *strength function* and is a parameter representing (theoretically) the complexity of nuclear motion in the compound state and (experimentally) the average formation cross-section.

The strength function may be related formally to the internal nuclear motion in the way adopted for the discrete resonances, using the parameter ρ_0. Since the number of reaction channels in the continuum region is normally large, the incident wave is heavily damped in the nuclear interior (*black nucleus*) and there is little compound elastic scattering. The internal wavefunction may then be assumed to be ingoing only and of the form e^{-iKr} where K is the internal wavenumber. This is in contrast with the resonance case discussed in Section 11.2.3 in which the internal wavefunction must include an outgoing component to give the compound elastic scattering. In the present case we obtain, for the s-wave,

$$\rho_0 = -iKR \tag{11.23}$$

327

The resulting formation cross-section, which is, of course, now just equal to the inelastic cross-section, is found from (11.12) and (1.81) to be

$$\sigma_a = (\pi/k^2)4kK/(k+K)^2 \approx 4\pi/kK \tag{11.24}$$

since $k \ll K$. This $l = 0$ cross-section has no resonant features; it predicts a $1/v$ decrease with increasing energy. Such a variation is well known experimentally for slow neutrons and may be seen at the lowest energies in Fig. 11.1c. Comparing (11.24) with (11.22) gives

$$\overline{\Gamma_a}/D = 2k/\pi K \tag{11.25}$$

which is the required relation.

It is normally thought that under the so-called *statistical assumption* of random motion in the compound state, and with adequate averaging over fine-structure levels (i.e. an energy interval $W \gg \overline{\Gamma}$), the Bohr independence hypothesis applies to the continuum. As noted in Section 11.2.3 there is good experimental evidence for this, in the comparison of branching ratios to distinct channels from compound nuclei formed in the same state but through different incident channels. The cross-section for a process $X(a, b)Y$ (Fig. 11.6) passing through continuum intermediate states is then written in the form (11.11), with the formation cross-section given for s-waves by (11.24).

At higher energies when more l-values are possible, subject of course to the condition $[l(l+1)]^{1/2} \leq kR$, the formulae of Section 1.2.7 may be used to predict a limiting cross-section. Thus, for a black nucleus we have, neglecting barrier effects and approximating $[l(l+1)]^{1/2}$ by l,

$$\left.\begin{array}{ll} \eta_l = 0 & \text{for} \quad l \leq kR, \quad \text{and} \\ \eta_l \exp 2i\delta_l = 1 & \text{for} \quad l > kR \end{array}\right\} \tag{11.26}$$

so that from (1.81) the compound-nucleus formation cross-section is

$$\sigma_{\text{inel}} = (\pi/k^2) \sum_0^{kR} (2l+1) = \pi(R + 1/k)^2 \tag{11.27}$$

which shows a monotonic decrease to the limiting geometrical value πR^2. The accompanying shadow scattering has an equal cross-section if the inelastic cross-section has its maximum value so that the *total* black-nucleus cross-section for energy of say ≈ 50 MeV is

$$\sigma_{\text{tot}} = 2\pi(R + 1/k)^2 \tag{11.28}$$

The decay width Γ_b in continuum theory contains external factors (cf. (11.20)) but must also be evaluated for the whole accessible range of levels of the final nucleus Y. The width is usually expressed in terms of the *level density* ω_Y in the residual nucleus and this is

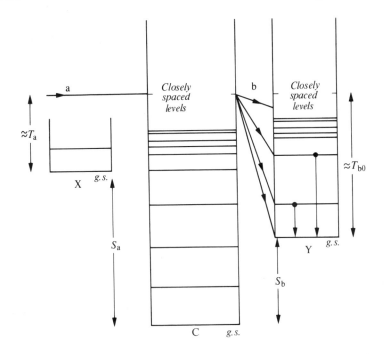

g.s. = *ground state*

Fig. 11.6   Nuclear reaction X(a, b)Y passing through continuum levels of compound nucleus C. The incident energy is T_a and the product particles b have energies T_{b0}, T_{b1}, . . . leaving the final nucleus in excited states which may emit radiation to the ground state. S_a and S_b are the separation energies of a, b in the compound nucleus C, and the Q-value for the reaction is $Q = S_a - S_b$. Centre-of-mass corrections are disregarded in this illustration.

obtained from a suitable model. For a Fermi-gas model, and with an excitation E above the ground state of Y (Fig. 11.6)

$$\omega_Y(E) \propto \exp 2(aE)^{1/2} \qquad (11.29)$$

where a is a parameter proportional to the single-particle level spacing near the Fermi energy. From the expressions for σ_a and Γ_b the *excitation function* or variation of yield with energy for the reaction X(a, b)Y may be calculated, with due allowance for barrier penetration factors for the incident particle and for competition between alternative channels b, c, ... The form of an excitation function for the *charged-particle* reaction $(\alpha, n) + (\alpha, 2n) + (\alpha, 3n)$ is shown in Fig. 11.7.

The intensity of particles b falls off at high energies because it is not likely that a large amount of energy will be concentrated on one

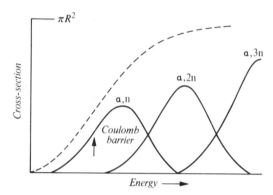

Fig. 11.7 Cross-sections in the continuum theory. The figure shows the cross-sections for the X(α, xn) reactions and the theoretical cross-section for formation of the compound nucleus. The rise is due partly to barrier penetration and partly to the onset of new processes.

particle in the compound state. It also diminishes at low energies, partly at least because of barrier transmission effects for the low-energy emitted particle. The resulting spectrum has similarities with the Maxwell distribution for the energy of the molecules of a gas. The analogy is often used to suggest that a quantity with the dimensions of energy may be defined as the *nuclear temperature* and that the reaction particles b may be regarded as evaporation products. The temperature, however, is simply an alternative expression of the level density of Y.

The concepts of the continuum region discussed in this section all suggest that the emergent particles should have an isotropic or at least symmetric angular distribution in the c.m. system, because of the many incident *l*-waves involved. Experiment confirms this but also shows that for transitions to the lower states of Y the angular distributions have a marked forward peak. These are associated with non-compound nucleus processes known as *direct interactions* (Sect. 11.5).

11.4 The optical model

The experimental study of nuclear states over a wide range of energy that has been surveyed in Chapters 7 and 8 and in Sections 11.2 and 11.3 suggests that the theory of nuclear reactions must include both the narrow fine-structure states and the broad single-particle or giant resonances. The optical model, which treats a nuclear reaction in analogy with the propagation of light through a partially absorbing medium, has proved successful in describing these apparently conflicting aspects of nuclear behaviour in one formalism. The model is mainly a theory of elastic scattering and this aspect has already been described in Section 7.2.4 in connection with the determination of the shell-model potential. In the present section the origin of the model and its treatment of inelastic processes will be examined.

In the continuum region of the nuclear spectrum the total cross-section for a neutron-induced reaction at high energy should be given by the black-nucleus formula (11.28) which predicts a smooth dependence of σ_{tot} on incident energy T (through k) and on target mass number (through R). At low energies the black-nucleus assumption $\eta_l = 0$ may not be valid, but as the energy increases, so does the number of reaction channels and the nucleus should then become 'blacker'. Experiments with neutrons of energy ≈ 100 MeV for which $\lambda \ll R$, however, already in 1949 showed that, contrary to expectation, nuclei seemed to become more *transparent* as the incident particle energy increased; the radius R obtained from (11.28) was smaller than expected. This was explained by assuming that interactions tended to occur with individual target nucleons rather than with the target nucleus as a whole, thus reducing the amount of compound-nucleus formation.

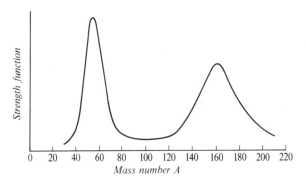

Fig. 11.8 Low-energy neutron strength function. For a black nucleus, no peaks would be seen.

At about the same time, the anomalies in slow-neutron scattering length described in Section 7.2.4 and shown in Fig. 7.7 were established. Moreover, the s-wave neutron strength function (eqn (11.22)) that measures the probability of compound-nucleus formation, and is deduced from total cross-section measurements, also showed peaks, as in Fig. 11.8. In both cases the anomalies occurred at mass numbers at which an s-wave neutron would form a resonance at zero energy (infinite scattering length) in a potential well whose radius was increasing with mass number A, i.e. at mass numbers at which, for instance, the 2s, 3s, ... neutrons just become bound. For a given A, broad resonances in total cross-section are seen (notably in the work of Barschall) as the neutron energy increases up to, say 5 MeV and for a given energy the inferred nuclear radius deviates from a strict $A^{1/3}$ dependence in a way consistent with some passage of the incident wave through the target nucleus. As with the higher-energy experiments these observations lead to a clear requirement that the incident particles shall have a long mean free path in the target nucleus. Precisely this requirement is necessary (Ch. 7) for the validity of the shell-model description of bound levels; it is met by invoking the Pauli principle as discussed in Section 6.4 in connection with nuclear matter.

The optical model proposes that a nuclear reaction should be determined by the solution of the Schrödinger equation for a one-body rather than a many-body problem, namely, the motion of an incident particle of given energy in a limited region of complex attractive potential

$$-[V(r) + iW(r)] \qquad (11.30)$$

A detailed prescription for this potential has already been given in equation (7.13) and $V(r)$ has been related to the shell-model potential. Elastic scattering produced by V is evidently of single-particle type and will show broad, size-dependent resonances. The imaginary part W of the potential leads to a scattering amplitude that decreases with time (the sign is chosen to ensure this, otherwise a non-physical situation of continually increasing amplitude results). This term of the complex potential, therefore, represents an attenuation of the incident wave in the nucleus and determines the mean free path; it represents all reaction processes that are not associated with the prompt elastic scattering arising from the real potential V, and in particular W must parameterize any *compound elastic scattering*.

The course of a nuclear reaction according to these ideas may be represented as in Fig. 11.9. There is a preliminary or single-particle stage in which the interaction of the incident wave with the potential V leads to *shape-elastic* or *potential scattering*. Part of the wave is absorbed (potential W) to form a compound system, towards which the first step may be excitation of a nucleon from the target nucleus. This creates a two particle–one hole (2p, 1h) state which has sometimes been called a *doorway state* since it leads both to direct interactions (Sect. 11.5) and to further particle–hole excitations (3p, 2h; 4p, 3h; etc.) which can finally create the strongly interacting many-body state that is the compound nucleus. Both the doorway state and the compound nucleus may decay back into the initial system, contributing to the total elastic scattering.

Optical-model analysis works best at energies that are high enough (say, ≈ 10 MeV or above) for *compound* elastic scattering to be disregarded in comparison with inelastic processes and certainly

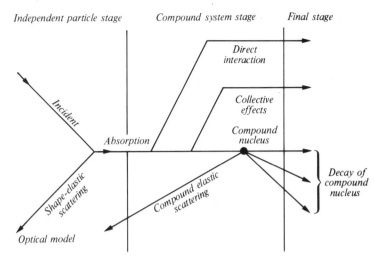

Fig. 11.9 Nuclear reaction scheme according to V. F. Weisskopf (*Rev. Mod. Phys.*, **29**, 174, 1957).

with shape-elastic scattering. It should not be used for light nuclei, for which sharp resonance effects may be seen. Often both scattering and polarization differential cross-sections are available experimentally and the model provides the corresponding amplitudes. From W, the *total* reaction cross-section may be obtained. A model fit to 30·3-MeV proton scattering is shown in Fig. 7.8 and a set of potential parameters is given in Table 7.2. The theoretical curves in the figure are actually provided by an improved version of the optical model for nucleons due to Greenlees, Pyle and Tang, in which the number of arbitrary parameters in the potential is reduced by relating specific parts of the potential to the fundamental nucleon–nucleon interaction on the basis of a reasonable nuclear matter distribution using a folding procedure.

The optical model has enjoyed continuing success in its analysis of experimental data and the optical potential, like the shell-model potential to which it is so closely related, must be regarded as a property of nuclear matter. This success would be less pleasing if it were not possible to formulate some connection between the broad resonances and the underlying fine-structure states. This can be done by recognizing that a single-particle strength may, in fact, be split up between a number of states of the same spin and parity because of the residual interactions already invoked in the shell model. The degree to which the strength is distributed among the fine-structure states depends on the strength of the residual forces; in general, there will be some clustering about the unperturbed single-particle energy (cf. Fig. 11.5) and low-resolution experiments will show the single-particle features.

11.5 Direct interaction processes

11.5.1 Experimental characteristics

Figure 11.10 shows the angular distribution of two nuclear reactions that are typical of a very large number of similar processes observed at incident energies of the order of 10 MeV. The first is *inelastic scattering*

$$^{24}\text{Mg} + \alpha \rightarrow {}^{24}\text{Mg}^* + \alpha' \tag{11.31}$$

in which the target nucleus is excited and the second is *deuteron stripping*

$$^{16}\text{O} + \text{d} \rightarrow {}^{17}\text{O} + \text{p} \tag{11.32}$$

which is an example of a large subset of processes (*transfer reactions*) in which a particle rather than excitation energy is transferred to (or from) the target nucleus.

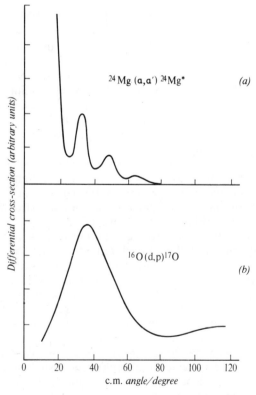

Fig. 11.10 Angular distributions of direct reactions. (*a*) ^{24}Mg (α, α')$^{24}\text{Mg}^*$ with 30-MeV α-particles, exciting the 1·37 MeV level. (*b*) $^{16}\text{O}(\text{d}, \text{p})^{17}\text{O}$ with 8-MeV deuterons. In each case the angular distribution is that of the light product particle.

These processes are taking place at energies which would produce continuum states in any compound nucleus, but the spectrum of particles emitted at forward angles shows many more particles of high energy, corresponding to the production of low-lying states of the residual nucleus, than would be expected from evaporation theory. The particles emitted at large angles, however, do tend to show an evaporation spectrum, indicating that compound-nucleus formation takes place for central collisions. The high-energy particles corresponding with the non-compound nucleus reactions show oscillatory angular distributions often with a maximum at small angles, in contrast with the symmetric type of angular distribution expected for a compound nucleus process. Finally, the energy dependence of yield is monotonic, in contrast with the behaviour of resonance reactions (Sect. 11.2.3). These characteristics strongly suggest that, as for scattering in the optical model, the basic interaction at these energies is not with the target nucleus as a whole but with just a part of it, e.g. a single nucleon, perhaps peripherally in the nuclear surface and effectively with a mean free path of the order of nuclear dimensions. The compound nucleus two-stage picture

$$X+a\rightarrow C\rightarrow Y+b \qquad (11.33)$$

is then inappropriate except for central collisions and is replaced by the single-stage process or *direct reaction*

$$X+a\rightarrow Y+b \qquad (11.34)$$

in which the final products are the same but the reaction mechanism is different. Both direct and compound-nucleus modes may occur together as different channels of the compound system indicated in Fig. 11.9, but they evolve on different time scales.

Spectroscopically, direct reactions are important because inelastic scattering picks out *collective modes* of excitation and nucleon-transfer reactions select *single-particle* or single-hole states preferentially. The experimental spectrum (Fig. 11.11) shows an enhanced yield to such states, or to states closely connected with them structurally, in comparison with the yield to more complex states. In each case the energy of the state with respect to the ground state is obtained directly from the spectrum. A special case that has already been discussed is the (p, 2p) reaction (Sect. 7.1.4) although here the transferred particle does not form a bound state with the projectile and we speak of a *knock-out* process.

For each state that is excited directly, some indication of the total angular momentum I may be derived from the angular distribution, and the absolute value of the differential cross-section contains information on the structure (i.e. the wavefunctions) of the complex nuclei concerned.

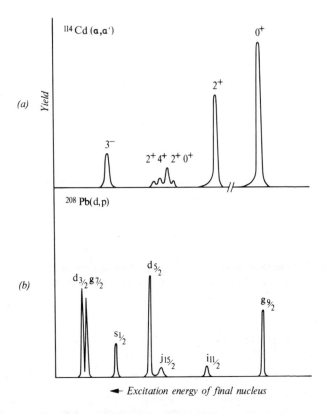

Fig. 11.11 Selectivity of direct processes shown in reaction yields. (*a*) Inelastic scattering, showing quadrupole phonon (Fig. 8.11) and octupole phonon states. (*b*) Transfer reaction, showing single-particle states in ^{209}Pb above the $N = 126$ neutron shell. The yields are representative, schematically, of a number of experiments reported in the literature for incident particles of energy about 20 MeV.

11.5.2 Theoretical treatment

Only a very brief outline of the theory of direct interactions will be given here; detailed numerical calculations for comparison with theory are simple in principle but complicated in practice, requiring elaborate computer codes. A full discussion will be found in Reference 11.1.

A basic consideration is that the occurrence of forward-peaked angular distributions means that high angular momenta of the *incident* particle are involved, since low-order partial waves give much more symmetrical distributions. This is good evidence for the *peripheral* (or surface) nature of the reactions, which has also been established as an important mechanism in the collisions of elementary particles at high energies, e.g.

$$p + p \rightarrow p + Y + K \tag{11.35}$$

in which the incident proton is considered to collide with a virtual pion associated with the target proton (Fig. 11.12). The matrix element for the process then includes the pion-nucleon coupling amplitude for one vertex and the $\pi p \rightarrow YK$ reaction amplitude for the other. The differential cross-section for the production of one of the particles in the final state is obtained from the matrix element by time-dependent perturbation theory and is in principle the product of a quantity relating to the target structure with the cross-section for the peripheral process. For direct nuclear reactions, it is the structure-dependent term that is of interest.

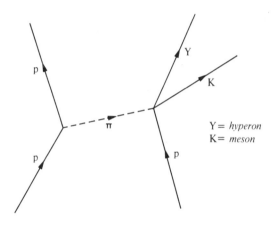

Fig. 11.12   A peripheral collision, in which an incident proton (on the right) collides with a virtual pion from a target proton, producing a kaon and hyperon by the (πp) interaction.

(i) *Direct inelastic scattering.*   The most primitive approach is to represent a reaction such as $X(p, p')X^*$ by the diagram of Fig. 11.13. This indicates that the peripherally struck proton is transferred from a state with wavefunction ψ_i to a state ψ_f of different energy and possibly different angular momentum, leaving the rest of the nucleus unaffected in this approximation. The linear momentum transferred to the nucleus is $q\hbar$ where in terms of the initial and final wavenumbers of the incident particle

$$\mathbf{q} = \mathbf{k}_i - \mathbf{k}_f \tag{11.36}$$
$$q^2 = k_i^2 + k_f^2 - 2k_i k_f \cos\theta$$
$$= (k_i - k_f)^2 + 4k_i k_f \sin^2\tfrac{1}{2}\theta \tag{11.37}$$

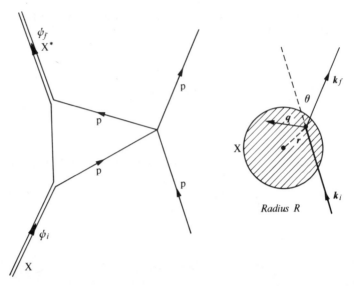

Fig. 11.13 A direct inelastic scattering $X(p, p')X^*$ in which a proton collides peripherally with a proton in the target nucleus and transfers it from an initial state ψ_i to a final state ψ_f. The momentum vectors are shown on the right.

and θ is the c.m. angle of scattering. For nuclear problems q is usually expressed in fm^{-1}.

If ψ_i and ψ_f in this simple case represent single-particle states with orbital quantum numbers l_i and l_f and if there is no change of spin orientation in the collision, the angular momentum transfer to X is $l\hbar$ where l is restricted by the inequality

$$l_f + l_i \geqslant l \geqslant |l_f - l_i| \tag{11.38}$$

by the need to conserve parity between initial and final states, and also by the geometrical condition

$$|\mathbf{r} \times \mathbf{q}| = [l(l+1)]^{1/2} \tag{11.39}$$

where $\mathbf{r}$ is the position vector of the struck nucleon with respect to the nuclear centre. For peripheral reactions $r \approx R$, the nuclear radius, since collisions with smaller r are more likely to result in compound-nucleus formation. The waves representing the inelastically scattered particles, therefore originate on the nuclear surface at points for which the condition (11.39) is fulfilled. Interference between them creates the observed oscillatory angular distribution (cf. Fig. 11.10a), with an angular periodicity characteristic of nuclear dimensions and of the l-value, and with a principal maximum at an angle θ given by

$$q(\theta) \approx [l(l+1)]^{1/2}/R \tag{11.40}$$

using the known properties of the spherical Bessel function $j_l(qR)$ that describes the angular distribution.

These considerations determine particular l-values but say nothing about the selectivity of the reaction for individual final states. If, however, the *plane-wave impulse approximation* (PWIA) is used, the matrix element between initial and final states is just the product of the matrix element for incident particle–*free nucleon* scattering (e.g. (pp) scattering Fig. 11.13) with the structure-dependent quantity

$$\int \exp{(i\boldsymbol{q} \cdot \boldsymbol{r})} \psi_f^*(r) \psi_i(r) \, \mathrm{d}^3 r \qquad (11.41)$$

The direct inelastic scattering thus picks out states ψ_f whose wavefunction overlaps strongly with ψ_i, e.g. the members of a rotational band or the two-phonon states, and displays them as a spectrum. The cross-section for the excitation of a given state is thus a model-dependent quantity, although equation (11.41) indicates that it is likely to be proportional to the radiative transition probability between the two states (cf. eqn (9.5)).

The PWIA matrix element may be applied to describe direct inelastic scattering processes such as (e, e'), (p, p'), (d, d') and (α, α'). Charge exchange reactions such as (n, p) or (p, n) may be included in the formalism and show a high resonance-type selectivity for the analogue state (Sect. 5.6.1) in which $\psi_f(r)$ and $\psi_i(r)$ differ only in that a neutron has been exchanged for a proton or vice versa (with consequent displacement of energy). In other respects the process is effectively an elastic scattering. Finally, if ψ_f is an *unbound* state, knock-out reactions of the type (p, 2p) (Sect. 7.1.4) may also be described.

The plane-wave impulse approximation may be improved by the use of distorted waves which take account of the refracting effect of the nuclear potential. These are, in fact, superpositions of plane-wave states and give the *distorted-wave impulse approximation* (DWIA). At the energies of the order of 10 MeV at which much spectroscopic work has been conducted, distorted waves are certainly required, but it is also necessary to discard the impulse approximation itself because it is not reasonable, except perhaps at high incident energies, to think of free particle–particle collisions, with disregard of the rest of the nucleus. Since, however, it is the essence of a direct reaction that it does not much modify the character of the incident wave, the *Born approximation* of quantum mechanics may be used to calculate the cross-section, with a suitable effective interaction for particle–nucleus scattering in the matrix element. As for weak and electromagnetic processes, the interaction may be expanded in multipoles and the collision cross-section (which may also be obtained directly by time-dependent perturbation theory) is a sum of terms corresponding with different orbital momenta. In practice, most calculations are now based on the *distorted-wave Born approximation* (DWBA), with distorting potentials provided by the optical-model analysis of elastic scattering. The DWBA analysis of inelastic data for a given *l*-value determines parameters of the nuclear model, e.g. the collective model, used to describe the excitation.

(ii) *Transfer reactions.* Reactions such as (d, p), (d, n), (t, d) (stripping) and (p, d), (n, d), (d, t) (pickup) and others of this general type have angular distributions for incident energies of the order of 10 MeV or more of the type shown in Fig. 11.10*b*. The process X(d, p)Y may be represented diagrammatically as in Fig. 11.14, and the other reactions can be shown similarly. Their essential feature is the transfer of a particle (or even of a group of particles in more complex cases, Sect. 11.8.2) between the projectile and the target nucleus.

As with direct inelastic scattering, we have for the linear momentum transfer

$$q^2 = (k_p - k_d)^2 + 4k_p k_d \sin^2 \tfrac{1}{2}\theta \tag{11.42}$$

together with the angular momentum condition

$$|\mathbf{r} \times \mathbf{q}| = [l(l+1)]^{1/2} \tag{11.43}$$

The *l*-value again determines the parity change between the initial and final nuclei, and limits the total angular momentum change. If the reaction amplitude arises mainly in the nuclear surface then in (11.43) *r* is set equal to the nuclear radius *R* and the angular distribution of the observed final particle has a principal maximum at the angle for which $q \approx [l(l+1)]^{1/2}/R$. Theoretical angular distributions for the (d, p) reaction are shown in Fig. 11.14*b*. These characteristics of transfer reactions have provided much information on the quantum numbers of nuclear levels since the development of the theory by Butler in 1951.

Nuclear structure information is contained in the differential cross-section for the reaction, e.g. at the principal maximum. The transition matrix element is now more complicated than that for direct scattering since it must allow for the structure of the complex bombarding or emitted particle. The simplest approach is the *plane-wave Born approximation* (PWBA) in which the matrix element contains the probability of combining the incident momentum with the momentum in the projectile of the particle or particles x to be transferred in such a way that the resultant matches the momentum of a state in the final nucleus Y (= x + X). The interaction effective in this process is that which exists between x and the remainder of the projectile, e.g. the (np) interaction in (d, p) stripping. The

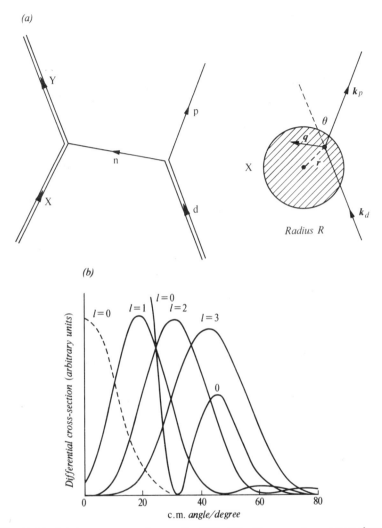

Fig. 11.14   (a) A neutron-transfer reaction X(d, p)Y, with momentum vectors shown on the right. (b) Theoretical proton angular distributions for different angular momentum transfers $l\hbar$ in the (d, p) reaction with 8·8-MeV deuterons (Butler, S. T., *Proc. Roy. Soc.*, **A208,** 559, 1951, Fig. 10).

overlap between initial and final states (cf. 11.41) will tend to be large when single-particle states are involved, since these have relatively simple wavefunctions, and transfer reactions show a high selectivity for such states. Other more complicated states can, of course, also be excited but for these the cross-section is smaller, because of their particular structure. If the differential cross-section

341

for a single-particle state can be calculated, then the ratio of the observed cross-section for any state to the single-particle value is known as the *spectroscopic factor;* it measures the probability that in the initial state the configuration, except for the particle to be transferred, is just that of the final state. The spectroscopic factor for the transfer of a nucleon to a state outside a closed shell, without core excitation, would be unity.

Although the PWBA theory of transfer processes has been effective in determining quantum numbers, perhaps fortuitously, it generally underestimates cross-sections and the sophisticated procedures of the DWBA must be used for more meaningful analysis.

11.6 Alpha decay and barrier penetration

Because of the relatively low atomic mass of the α-particle, resulting from the high binding energy of 28·3 MeV (7 MeV per nucleon) most naturally occurring nuclei with $A > 150$ are unstable against α-emission. The fact that this form of spontaneous *decay* and indeed the similar decay process known as spontaneous fission (Sect. 11.7) are relatively rare phenomena compared with the corresponding *instability* is due to an exponential dependence of decay probability on the available energy. This in turn is due to the necessity for penetration of the nuclear potential barrier in the emission process. Such barrier effects have already been noted in connection with beta decay (Sect. 10.2.3) where they are responsible for the difference in shape between electron and positron spectra (Fig. 10.4). Alpha and fission decay processes, however, differ from weak interaction and electromagnetic decays because the emitted particle is not created by the interaction itself. The constituent nucleons of the α-particle or the fission fragment are part of the strongly interacting structure of the parent nucleus and only a detailed nuclear model can predict the probability that they will assume the configuration that leads, after barrier penetration, to the observed decay. The assembly of four nucleons into an α-particle, however, does release 28 MeV of energy and the resulting excitation places the particle in a state from which barrier penetration is possible. A further difference from β-decay is that α-particles are emitted in groups of uniform energy, characteristic of a two-body process

$$C \rightarrow Y + \alpha \tag{11.44}$$

The emission of α-particles from a radioactive nucleus such as ^{212}Bi (ThC, $E_\alpha = 6·21$ MeV, $t_{1/2} = 60·5$ min) is not, in principle, different from the decay of a compound nucleus C^* formed in a nuclear reaction through the channel $Y + \alpha$. It follows that the barrier effects that so profoundly affect lifetimes for α-decay are also present in both the formation and decay stages of a nuclear reaction. In fact, narrow resonances in many low-energy nuclear reactions would not be observed as such were not the particle widths Γ_p, Γ_α,... reduced by a barrier penetration factor. Although Coulomb factors do not occur for neutrons there is still some retardation due to the momentum change at the nuclear boundary (cf. eqn (11.24)).

In this section emphasis will be placed on the α-decay of fairly long-lived nuclei with a view to illuminating the part played by the nuclear potential barrier in the course of nuclear reactions in general. The barrier has a crucial role in the design of thermonuclear power sources and, of course, in the understanding of energy production in stars.

11.6.1 Experimental information on α-decay

The regularities of the ground state α-decay energies for $A > 200$ are shown in Fig. 11.15. In this region of A the occurrence of the *radioactive series* has been known since the early days of studies of radioactivity. In such a series, of which the decays following that of ^{232}Th, ^{235}U and ^{238}U are well-known examples, a sequence of α-emissions mainly between ground states leads finally to a β-unstable nucleus and one or two β-decays then occur before α-transformations are resumed; the series terminates when a nucleus

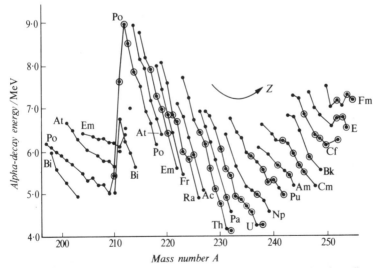

Fig. 11.15 Energy release in the α-decay of the heavy elements, showing effect of neutron shell closure at $N = 126$. The nuclei ringed are stable against beta decay (adapted from Hanna, G. C., in *Experimental Nuclear Physics III*, Wiley 1959).

(^{208}Pb, ^{209}Bi or ^{206}Pb) is reached which is stable in the sense of having both beta stability and an unobservably long life for α-decay.

Figure 11.15 shows clearly the effect on α-decay energies of the closure of the $N = 126$ neutron shell. Extra energy is available just beyond the shell closure, since two loosely bound neutrons are then removed by the α-emission. The same effect occurs for $N = 82$ and results in the appearance of α-activity among the rare earths.

The connection between α-particle energy and mean lifetime for decay was first investigated in 1911 by Geiger and Nuttall, who found a linear relationship between the logarithm of the decay constant λ_α and the range of the α-particle group in air. Figure 11.16 is a modern version of the Geiger–Nuttall plot; it relates to even–even nuclei and shows a linear dependence of $\ln \tau$ (or $\log_{10} t_{1/2}$) on $T_\alpha^{-1/2}$ where T_α is the α-particle kinetic energy. This is the basic fact of α-decay that needs interpretation. An additional fact for theory is that similar plots for odd-mass, and odd-Z–odd-N nuclei show a retardation of about a factor of 100 compared with even-Z–even-N α-decays.

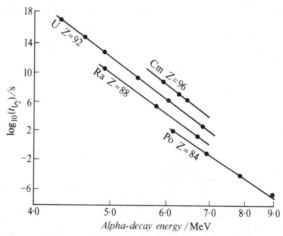

Fig. 11.16   Energy–lifetime relation for even–even α-emitting nuclei of indicated Z. The energy scale is linear in $1/T_\alpha^{1/2}$ (Hanna, G. C., *loc. cit.*).

The α-particle spectra of the heavy elements show two types of structure, historically known as '*fine structure*' and '*long-range α-particles*'. These are no more than examples of phenomena that are now extremely familiar in nuclear reactions in which (*a*) transitions take place to a series of states of the final nucleus Y in (11.44), or (*b*) transitions take place to the ground state of Y from excited states of the nucleus C. In α-decay the nucleus C* in the case of long-range particle emission will, in general, have been formed by a preceding beta decay, since the decay rate for this process (Sect.

10.2.3) does not have an exponential energy dependence and excited states can be reached with observable yield. Such *beta-delayed* emission is known for neutrons (Sect. 11.7.1) and for protons as well as for α-particles. The actual yield of long-range α-particles is only about 10^{-5} of that for the α-transition between ground states because the radiative transition $C^* \rightarrow C + \gamma$ competes in the decay of the state C^*. The intensity of the fine-structure groups in an α-particle spectrum, however, is determined by the α-decay process itself. The experimental fact that most of the fine-structure components have an intensity of about 1 per cent of the most probable transition, which is not necessarily that to the ground state, means that there is a sharp dependence of decay probability on energy and probably on angular momentum.

11.6.2 *Barrier penetration*

(i) *The penetration factor for α-decay.* In α-decay the barrier factor, although an external and calculable quantity, essentially determines the probability of decay and accounts for its very wide spread of values.

The Coulomb potential surrounding a point nucleus Y of charge $Z_2 e$ is given by

$$V_C(r) = Z_1 Z_2 e^2 / 4\pi\varepsilon_0 r \qquad (11.45)$$

where Z_1 is the charge number of a point-like particle of positive charge at a distance r from the nuclear centre. In the very simple and unrealistic model used in Section 6.2 this potential is cut off at a distance R_C from the centre and replaced abruptly by an attractive nuclear well of depth U, Fig. 11.17. The nuclear radius R is then taken to be equal to R_C and the barrier height is

$$B = Z_1 Z_2 e^2 / 4\pi\varepsilon_0 R_C \qquad (11.46)$$

This has the value $1 \cdot 2 Z_1 Z_2 / A^{1/3}$ MeV for $R_C = 1 \cdot 2 \times 10^{-15} A^{1/3}$ fm. In a more realistic approach the nuclear well is that indicated by an optical-model analysis of α-particle elastic scattering with nucleus Y as a target. More fundamentally still, the potential may be constructed from a detailed nuclear model. Such calculations show that the nuclear well is not sharp, as drawn in Fig. 11.17, and that the barrier B is rather lower in value than is indicated by (11.46).

We assume without discussion that inside the nucleus the operation of the strong interaction between nucleons leads to a quasi-stationary state in which there is a preformed α-particle of kinetic energy $U + T_\alpha$; the way in which this configuration arises is the proper business of the theory of α-decay. If this particle emerges spontaneously its kinetic energy at infinite distance would be T_α, assuming no loss of energy by recoil. Such spontaneous emission is

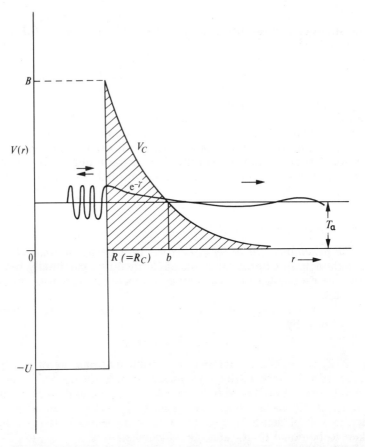

Fig. 11.17 Theory of α-decay. The wave representing the α-particle has large amplitude and short wavelength within the nuclear well and is attenuated exponentially in the region of negative kinetic energy ($R < r < b$). The total energy of the emitted particle is T_α.

not possible classically because between the points $r = R$ and $r = b$, where

$$T_\alpha = 2Z_2 e^2 / 4\pi\varepsilon_0 b \qquad (11.47)$$

the kinetic energy is negative. However, as first pointed out by Gamow, and Gurney and Condon, quantum mechanical tunnelling through the potential barrier can take place and the α-particle will then come out with zero kinetic energy at the radius $r = b$. It gains its final kinetic energy by repulsion as it moves away from the residual nucleus Y.

Solution of the Schrödinger equation for this problem provides wavefunctions that are oscillatory for $r < R$ and for $r > b$ as in the case of a potential-well problem such as neutron–proton scattering (Sect. 5.2.2). A difference in the present case is that the assumed potential boundary at $r = R$ causes strong reflection of the internal wave and the transmission for $T_\alpha < B$ is normally very small. The potential discontinuity or matching factor that arises in this way (cf. (11.24)) is part of the overall *transmission coefficient*. It may be calculated by imposing the normal boundary conditions on the wavefunction at $r = R$; the special cases of resonance that are predicted will be disregarded.

In the intermediate, under-the-barrier region $R < r < b$, the wavenumber is imaginary and this leads to a non-oscillatory, exponentially decreasing amplitude. The decrease over the zone is given by the factor $e^{-\gamma}$ where

$$\gamma = [(2\mu_\alpha)^{1/2}/\hbar] \int_R^b [V(r) - T_\alpha]^{1/2} \, dr \qquad (11.48)$$

where μ_α is the reduced mass of the α-particle. We define the *barrier penetration factor* as the corresponding intensity ratio

$$P_0 = e^{-2\gamma} \qquad (11.49)$$

The subscript indicates that only s-wave emission is considered. The factor P_0, multiplied by the potential discontinuity factor, gives the overall transmission coefficient.

Evaluation of (11.48) (see Ex. 11.19) gives

$$\gamma = [Z_1 Z_2 e^2/2\pi\varepsilon_0\hbar v]\{\cos^{-1}(T_\alpha/B)^{1/2} - [(T_\alpha/B)(1 - T_\alpha/B)]^{1/2}\} \qquad (11.50)$$

where v is the α-particle velocity. In the limit of high barriers $R \to 0$ and $B \to \infty$ so that the bracket in (11.50) becomes $\pi/2$ and

$$P_0 \to \exp(-Z_1 Z_2 e^2/2\varepsilon_0\hbar v) = \exp(-2\pi\eta) \qquad (11.51)$$

where the dimensionless Coulomb parameter η is given by

$$\eta = Z_1 Z_2 e^2/4\pi\varepsilon_0\hbar v \text{ (positive for a repulsive force)} \qquad (11.52)$$

The expression (11.51) is known as the *Gamow factor*; the Coulomb parameter is large and the Gamow factor is small for particles of low velocity v in the field of nuclei of high charge $Z_2 e$.

Use of the Gamow factor in α-decay systematics provides an explanation of the observed dependence of lifetime on energy, since the decay constant λ_α will be proportional to P_0 and hence dependent on v, or more strictly on T_α/B. For the heavy nuclei an increase in T_α by 1 MeV decreases $\tau_\alpha = 1/\lambda_\alpha$ by a factor of about 10^5 while a 10 per cent decrease in R (increase in B) increases τ_α by a factor of 150. From (11.51) it can also be seen that for a given Z_1, i.e. for a given radioactive element rather than a series,

$$\ln \tau_\alpha = -\ln \lambda_\alpha = a + bT_\alpha^{-1/2} \qquad (11.53)$$

where a and b are constants if the nuclear radius is regarded as a constant. This is of the form of the plot shown in Fig. 11.16.

(ii) *Transmission coefficients for nuclear reactions.* Coulomb forces have not so far been introduced explicitly into the discussion of nuclear reactions, except that the optical potential was written (Sect. 11.4) with a Coulomb term that is important for small-angle scattering. Barrier penetration factors are certainly required in all cases when charged particles of energy comparable with nuclear Coulomb potential barrier form or emerge from a compound nucleus. As already noted, they are responsible for the sharpness of many nuclear resonance levels. For neutrons, although no Coulomb factors are necessary, potential discontinuity factors arise, and for all particles we may write for the limiting total inelastic cross-section (neglecting spin)

$$\sigma^l_{\text{inel}} = (\pi/k^2)(2l + 1)T_l \tag{11.54}$$

in contrast with (1.83), where T_l is the transmission coefficient in the incident channel for orbital angular momentum $l\hbar$. In the final stage of the reaction similar factors are required for each individual decay channel, and these factors are included in experimentally observed widths Γ.

A calculation similar to that just described for α-decay will give the order of magnitude of the barrier penetration factor in a particular case. For accurate work, however, it is necessary to use wavefunctions for particles moving in the Coulomb field of the target nucleus with angular momentum l that creates an additional centrifugal barrier. These functions cannot be given generally in simple form, but are tabulated. From them, T_l values can be obtained, e.g. for $l = 0$ and $R \to 0$ the s-wave barrier penetration part of T_0 is unity for neutrons and is given by

$$P_0 = 2\pi\eta/(\exp(2\pi\eta) - 1) \tag{11.55}$$

for charged particles. This reduces essentially to the Gamow factor (11.51) when $\eta \gg 1$.

The trend of the calculated transmission coefficients for both neutrons and protons in a particular case is shown in Fig. 11.18.

11.7 Fission

11.7.1 Discovery of fission

The factors that conspire to make α-decay an observable property of the ground states of many heavy nuclei (instability, potential barrier, shell effects) can, of course, be evaluated for other forms of decay, using the semi-empirical mass formula as a guide. Proton and neutron radioactivity of ground states can only be expected for

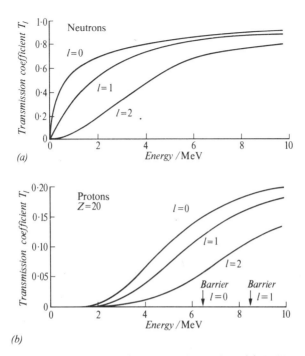

Fig. 11.18 Transmission coefficients for potential barriers. (*a*) Incident neutrons, $R = 5$ fm. (*b*) Incident protons, $R = 4.5$ fm and $Z = 20$ (Ref. 9.1, Ch. 8, Sect. 5).

nuclei very far removed from the line of stability and the same applies to the emission of other light fragments, most of which are less tightly bound than the α-particle, and all of which suffer a severe barrier penetration effect. The binding energy per nucleon (Fig. 6.3), however, reaches a maximum at $A \approx 60$ and is still about 8·5 MeV at $A \approx 120$, whereas because of the Coulomb energy it has dropped to about 7·5 MeV at $A = 240$. The division of a mass-240 nucleus into two equal parts would thus release 240 MeV in all, and if barrier transmission for this energy is finite the *spontaneous-fission* decay of ground states

$$C \rightarrow A + B + Q \qquad (11.56)$$

($Q \approx 200$ MeV) becomes possible. In fact, all stable nuclei with $A > 90$ (about) are unstable energetically against division into two nuclei of approximately half the original mass number but only for $A > 230$ does the halflife with barrier ˙become short enough to permit observation of this decay as a spontaneous mode.

Spontaneous fission decay was, in fact, not observed until after neutron-induced fission, i.e. the fission of a compound nucleus C^*, had been discovered and understood. This contrasts with α-decay, for which the ground-state decay was first observed, and, when accelerated beams of protons and deuterons became available, α-decay as a reaction channel, e.g. in the (p, α) or (d, α) reactions. Neutron-induced fission was established as an unexpected result of the search for transuranic elements in the bombardment of uranium by slow neutrons over the years 1934–39. It had been hoped to produce these elements (now familiar as neptunium and plutonium) by the succession of processes

$$\left. \begin{array}{l} {}^{238}_{92}U + n \rightarrow {}^{239}_{92}U + \gamma \\ {}^{239}_{92}U \rightarrow {}^{239}_{93}Np + e^- + \bar{\nu} \\ {}^{239}_{93}Np \rightarrow {}^{239}_{94}Pu + e^- + \bar{\nu} \end{array} \right\} \tag{11.57}$$

and indeed the production of 23-min ${}^{239}U$ was recognized, but accompanied by a great number of other activities. As in the early days of radioactivity, the elucidation of this complex of activities fell to the chemists, and despite the expectation that the elements produced would be either transuranics such as Np and Pu, or more familiar bodies derived from them by α-decay, e.g. radium, Hahn and Strassmann following Curie and Savitch were forced to conclude that there was a production of *barium* isotopes in the U+n reaction.

In 1938 Meitner and Frisch accepted these conclusions and proposed that the uranium nucleus, on absorption of a slow neutron, could assume a shape sufficiently deformed to promote *fission* into two approximately equal masses, e.g.

$$_{92}U + n \rightarrow {}_{92}U^* \rightarrow {}_{56}Ba + {}_{36}Kr \tag{11.58}$$

or to many other similar pairs. The two product nuclei, or *fission fragments* would each have energy ≈ 75 MeV and an estimated range of about 30 mm of air; they would recoil after the actual fission event (or *scission* of the compound nucleus U*) in opposite directions (Fig. 11.19). Their energy is converted into available heat as they slow down in matter with a yield of 1 joule for $3 \cdot 2 \times 10^{10}$ fission events.

Since the number of neutrons N in a stable nucleus increases faster than the number of protons Z, the fission fragments are initially neutron-rich, and a small number ($\approx 2 \cdot 5$) of *prompt* neutrons may be emitted from the rapidly moving fragments. These neutrons are those that permit the development of chain-reacting systems. The main mechanism by which the N/Z ratio is restored to normal is sequential beta decay, e.g. for a xenon fragment

$${}^{140}_{54}Xe \xrightarrow{e^-} {}^{140}_{55}Cs \xrightarrow{e^-} {}^{140}_{56}Ba \xrightarrow{e^-} {}^{140}_{57}La \xrightarrow{e^-} {}^{140}_{58}Ce \text{ (stable)} \tag{11.59}$$

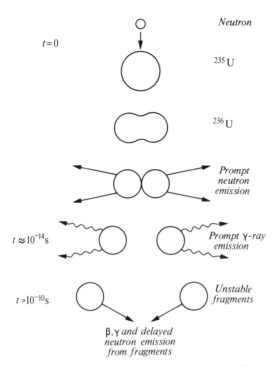

Fig. 11.19 Schematic representation of the fission process in uranium. The time scale gives orders of magnitude only.

Such chains are responsible for the intense antineutrino production in nuclear reactors.

In some of the fission decay chains a β-process may leave a product nucleus such as ^{87}Kr so highly excited that neutron emission is a predominant alternative to gamma-ray emission and to further β-decay. The resulting *delayed neutrons* have a halflife corresponding with that of the parent β-emitter; they are β-delayed in the same way as are groups of long-range α-particles (Sect. 11.6.1) and groups of protons in certain nuclear reactions. Delayed neutrons are important technically in the control of nuclear reactors.

These and other properties of the fission process were coordinated early in 1939 in a paper by N. Bohr and Wheeler, which gave a detailed theory based on the likeness of the fissioning compound nucleus to a liquid drop. An outstanding conclusion of this paper, rapidly verified experimentally, was that the slow-neutron fission of uranium (11.58) took place in the rare isotope ^{235}U. This was indicated because the complex of fission products did not show the particular resonance behaviour known to be associated with the production of ^{239}U* as a compound nucleus.

11.7.2 Theory of fission

The basic experimental facts, most of which were known to Bohr and Wheeler or anticipated by them, are:

(a) Thermal-neutron induced fission is observed for a number of even-Z–odd-N nuclei such as ^{233}U, ^{235}U and ^{239}Pu. The cross-section for the process, as assessed by counting the fragments, varies with energy as shown in Fig. 11.20a which indicates a (velocity)$^{-1}$ dependence at very low energies and a resonance behaviour in the 1–10^3 eV range. The fragment pairs have an asymmetric mass distribution (Fig. 11.21) in which a given fission event generally yields a high mass (H) and a low mass (L) product rather than two equal masses. Conservation of linear momentum then requires the energy distribution also to be asymmetric so that the mean energies of the two groups are $\bar{E}_{\text{H}}$ – 60 MeV, $\bar{E}_{\text{L}} = 95$ MeV.

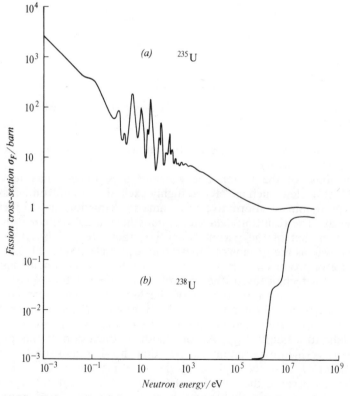

Fig. 11.20 Fission cross-sections, schematic, as a function of neutron energy. (a) ^{235}U, (b) ^{238}U.

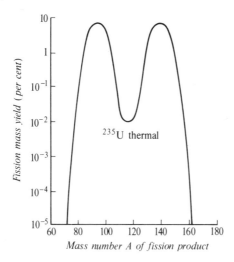

Fig. 11.21   Mass distribution of fission fragments.

(b) Fast-neutron induced fission is found for a number of even–
even nuclei such as ^{232}Th, ^{238}U. There is a 'threshold' energy for
the process, as shown in Fig. 11.20b, whereas in the thermal
region the main interactions are the (n, γ) capture process with a
cross-section at various resonances $\approx \pi \lambdabar^2$ and elastic scattering
between resonances with the hard-sphere cross-section $\approx 4\pi R^2$.
Below the threshold energy fission is still possible energetically,
but the cross-section decreases exponentially, as expected of a
barrier penetration effect. Fission is produced in many nuclei by
energetic particles, e.g. p, d, α, and also by γ-radiation. The
fragment distribution for fast fission is generally more symmetric
than for thermal fission.

(c) Spontaneous-fission halflives for even-Z–even-N nuclei vary
between 10^{16} a for ^{238}U to a few hours or less for transuranic
elements such as fermium ($Z = 100$). The lifetimes show a
general decrease with increasing value of a parameter Z^2/A
although locally for a given Z the lifetimes pass through a
maximum as N varies. Since 1967 it has been known that some
of the low-lying excited states of heavy nuclei, e.g. americium,
have unexpectedly short halflives for spontaneous fission,
≈ 1 ms, and this has led to an extension of earlier ideas about
the potential barrier in fission.

The difference between thermally fissile and fast-fissile nuclei is
that in the former case an even-Z–even-N compound nucleus is
formed, e.g.

$$^{235}\text{U} + \text{n} \rightarrow {}^{236}\text{U}^* \tag{11.60}$$

and in the latter case an even $-Z$ odd-N system, e.g.

$$^{238}U + n \rightarrow ^{239}U^* \tag{11.61}$$

The form of the pairing energy term in the semi-empirical mass formula, equation (6.10), indicates that the neutron binding energy for ^{236}U is greater than that for ^{239}U and the excitation of the compound nucleus for an incident thermal neutron in (11.60) is therefore greater than in (11.61). Any barrier resisting fission can therefore be more readily overcome in the former case.

The nature of the fission barrier was discussed by Bohr and Wheeler in terms of an incompressible liquid-drop model of the nucleus, in which energies (and hence masses) are determined by the physical effects already discussed in setting up the semi-empirical mass formula. Stability against fission, when energetically possible, depends critically on the relative importance of the short-range nuclear force and the long-range Coulomb force. If a spherical, constant-density nucleus is slightly deformed in a symmetrical way, the surface energy increases, since for a sphere the ratio of surface area to volume is a minimum, and the Coulomb energy decreases. For a given A, these changes are in first-order proportion to their undeformed values $\beta A^{2/3}$ and $\varepsilon Z^2/A^{1/3}$ (eqn (6.10)). If the surface energy increase exceeds the electrostatic energy decrease, the nucleus is stable against small deformations. When the two energy changes are equal fission becomes a possibility and may be characterized by the dimensionless *fissility* parameter proportional to the ratio of the two energies, i.e. to Z^2/A, a quantity that has already been introduced in connection with spontaneous-fission halflives. In fact, as pointed out by Aage Bohr, fissile nuclei are not spherical but deformed in their ground state, but the argument still applies with a different value for the fissility.

Further insight into the general nature of the fission process may be obtained by considering the inverse heavy-ion reaction, in which two heavy fragments with their final observed energies approach to form a compound system (Fig. 11.22). In all cases the potential energy of the system increases as the fragments approach to the point at which the influence of nuclear forces is felt, or in classical terms to the point of contact; at this point the separation of the centres is known as the *strong-absorption radius* R_{SA} (eqn. 6.9b and Sect. 11.8.1). In case (c), for Z^2/A small, the nuclei fuse together with the emission of energy as radiation, or as neutrons, to form an effectively stable system. The final nucleus has a high barrier E_f for the reverse process of spontaneous fission. (If Z^2/A is small enough, the final nucleus is absolutely stable with a ground-state energy below the zero corresponding with infinite fragment separation.) In case (a), for Z^2/A large, the Coulomb energy is high and cannot be compensated by the nuclear force to form even a quasi-stable

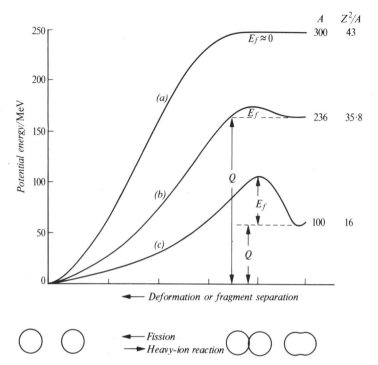

Fig. 11.22   Schematic representation of fission barriers for nuclei of mass number $A \approx 100$, 236 and 300. The energy release Q is defined in equation (11.56).

system, so that the final heavy nucleus is spontaneously fissionable with a very short halflife ($E_f \approx 0$). In case (b), for intermediate values of Z^2/A, the nuclear force is sufficient to bind the two particles together with a fission barrier E_f of a few MeV; the final nucleus is then spontaneously fissionable but with barrier penetration, which accounts for the dependence of halflife for this process on Z^2/A. In cases (c) and (b) the potential-energy curve is drawn to indicate a finite stable deformation in the ground state and the normal collective states will exist (cf. Fig. 8.6b).

From a detailed consideration of the liquid-drop model, and from the experimental observations on the fission of ^{238}U, Niels Bohr and Wheeler concluded that case (a), in which a nucleus is unstable with respect to the smallest deformation, is reached for $Z^2/A = 47{\cdot}8$. They also estimated the height of the fission barrier E_f in case (b) with $Z^2/A < 47{\cdot}8$. It is about 6 MeV for $A \approx 240$ and varies with A (or Z^2/A) in the manner indicated in Fig. 11.22. In induced fission an energy equal to or greater than the fission barrier E_f must exist in the compound nucleus as a result of the primary process. This is absorbed in creating the deformation that brings the nucleus to the point of scission. After scission the two fragments gain their final kinetic energy (Q in Fig. 11.22) by Coulomb repulsion. The sequence of configurations is shown at the bottom of the figure.

It remains to account for the anomalously short spontaneous-fission halflives ($\tau \approx 10^{-2}-10^{-8}$ s) found for excited states of the nuclei plutonium ($Z = 94$), americium ($Z = 95$) and curium ($Z = 96$). The states concerned are known from studies of the primary process to have an excitation of 2–3 MeV and a low spin, so that it is not obvious why they do not decay preferentially by a radiative transition, as in the case of many known isomers in lighter nuclei. Since this anomaly appears significantly in a certain range of Z-values it is reasonable to ascribe it to a shell-model effect. The spherical shell model has no magic numbers in the required range, but in a deformed potential these numbers may change and lead to a bunching in energy of the shell-model orbits in the transuranic region. From the calculations of Strutinsky and of Nilsson, it is now known that this effect can lead to a *double hump* in the potential-energy curve, and consequently in the fission barrier. Figure 11.22*b* must then be replaced by Fig. 11.23, in which *two* stable deformations are shown, each with a potential well (I and II) and with a set of energy levels.

If a nucleus like americium is excited to the level of well II, spontaneous fission (i.e. fission with barrier penetration) is impeded by a much thinner barrier than with the single-humped curve of Fig. 11.22*b*. The observed shorter fission halflife can then be understood.

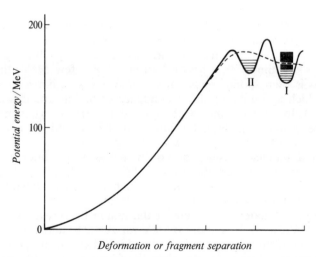

Deformation or fragment separation

Fig. 11.23 The double-humped fission barrier, arising because of shell structure. The dotted line is the potential energy curve of Fig. 11.22*b*, but the two potential wells I and II are not on the same energy scale.

For fission to occur at all of course, radiative decay to the ground state with $t_{1/2} \approx 10^{-14}$ s must be inhibited, and this is so because a transition to well I requires the nucleus to undergo another barrier penetration process (II→I).

The important discovery of the double hump, which is not necessarily confined to fissionable nuclei, also explains some remarkable features of fission excitation functions. The first surprise is that induced fission shows sharp compound-nucleus resonances in the eV–keV energy range, corresponding to a compound nucleus excitation approximately equal to the height of the fission barrier. The explanation of this is that most of the energy is expended in producing the critical deformation, so that very few fission channels are open and the resonances of well I are not broadened. The second feature is that the excitation function for fission of some nuclei such as ^{240}Pu (Fig. 11.24) shows that 'sub-threshold' fission occurs and is enhanced for certain regularly spaced bands of the fine-structure levels of well I. This is due to overlap of the more widely (and nearly equally) spaced levels of well II (whose lowest state is 2–3 MeV above the bottom of well I) with those of well I and is a convincing piece of evidence for the validity of the concept of the double hump.

The phenomena connected with the double hump are sometimes described as due to *shape isomerism* since in contrast with normal isomerism the increased lifetime for radiative decay, which permits fission to occur, has nothing to do with small energy or large spin

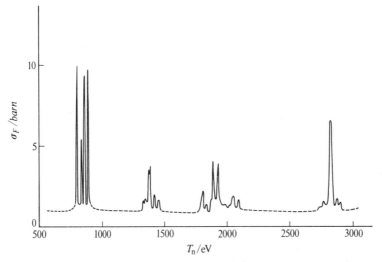

Fig. 11.24 'Sub-threshold' fission cross-section of ^{240}Pu (adapted from Migneco, E. and Theobald, J. P., *Nuclear Physics*, **A112**, 603, 1968).

changes. The shell-model calculations that predict the double hump also suggest a region of increased nuclear stability for $Z \approx 114$, $N \approx 184$–196. It is here that a possibly profitable search for superheavy elements, already existing in nature or produced synthetically in heavy-ion reactions may be made. Further regions of stability, for proton numbers of 126 and 164, have also been suggested.

11.8 The reactions between complex nuclei (heavy ions)

11.8.1 *General survey* (Ref. 11.2)

All nuclei with $A \geqslant 4$ are today regarded as heavy ions. There is no really sharp distinction between the nuclear reactions produced by particles such as ^{12}C, ^{40}Ca or ^{84}Kr for instance, and those produced by neutrons or hydrogen isotopes, but such ions have characteristics which emphasize particular features of the whole range of nuclear phenomena from elastic scattering to fission and fusion.

Heavy ions have been accelerated in tandem electrostatic generators, linear accelerators and cyclotrons to energies of 5–10 MeV per nucleon, and recently in a synchrotron to 2·1 GeV per nucleon. The complexity of the reactions in which these particles engage has stimulated the development of detectors with high resolution with discrimination between different particles by mass and charge. Magnetic spectrographs, time-of-flight techniques and solid-state counter telescopes have all been used in achieving clear identification of the particle products of heavy ion reactions with targets of even the heaviest nuclei. Using picosecond timing a sensitivity of 0·3 mass unit at $A = 16$ has been obtained. Gamma radiation from residual nuclei may, of course, be studied with high-resolution Ge(Li) detectors.

The special features of heavy-ion reactions arise from the short de Broglie wavelength of the incident particle (1·02 fm for 50-MeV ^{16}O ions, which is much less than the nuclear radius) and from the importance of Coulomb field effects due to the high ionic charge Z. Because of the short wavelength it is possible to discuss heavy-ion collisions in terms of classical trajectories, as shown in Fig. 11.25, which defines three general types of collision that will be of interest (although, in general, all three types will be present in a given interaction of sufficiently high energy). In Fig. 11.25a the initial impact parameter is so large that the colliding particles do not approach within the extent of their mutual short-range nuclear force. The trajectory is then the familiar hyperbola of the Coulomb field and there is a closest distance of approach (see Ex. 11.28)

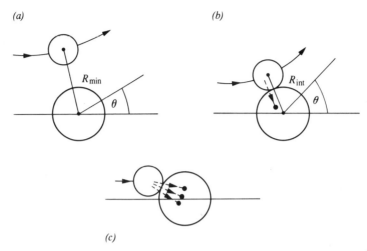

Fig. 11.25 Collisions between heavy ions. (*a*) Distant collision, (*b*) Grazing trajectory, (*c*) Formation of compound state. In (*b*) the possibility of nucleon transfer is indicated and in (*c*) partial or complete amalgamation of projectile and target takes place.

given by

$$R_{min} = (Z_1 Z_2 e^2/4\pi\varepsilon_0 2T)(1 + \text{cosec } \theta/2) = (\eta/k)(1 + \text{cosec } \theta/2)$$
(11.62)

where T is the total kinetic energy in the c.m. system and θ is the c.m. angle of scattering. The Coulomb parameter $\eta = Z_1 Z_2 e^2/4\pi\varepsilon_0\hbar v$ has already been introduced in Section 5.3.2 in connection with proton–proton scattering and in Section 11.6.2 in respect of barrier penetration. From (11.62) η can be seen to be half the closest distance of approach for $\theta = 180°$ in units of the reduced de Broglie wavelength and the condition for the validity of the classical trajectory is simply $\eta \gg 1$.

The initial impact parameter of the classical trajectory determines the relative orbital angular momentum L in the collision, and this in turn adds a centrifugal term to the total potential experienced by the incident ion. The quantum number L is given by

$$L(L+1) = \eta^2 \cot^2 (\theta/2)$$
(11.63)

and the total potential energy at interparticle separation R is

$$V(R) = Z_1 Z_2 e^2/4\pi\varepsilon_0 R + L(L+1)\hbar^2/2\mu R^2$$
(11.64)

where μ is the reduced mass. In terms of L, (11.62) may be rewritten

$$kR_{min} = \eta + [\eta^2 + L(L+1)]^{1/2}$$
(11.65)

a relation which also follows directly from equations (11.63) and (11.64) by the substitution $V(R_{min}) = T$. For collisions with c.m. energies less than the potential barrier determined by taking $R = R_{int}$ to be somewhat greater than the sum of the nuclear radii, i.e.

$$R_{int} = r_0(A_1^{1/3} + A_2^{1/3}) \tag{11.66}$$

with $r_0 \approx 1 \cdot 5$ fm, all interactions are of this type. They result only in Rutherford scattering or in inelastic scattering with excitation of the target or incident nucleus.

In Fig. 11.25*b*, for smaller impact parameters and sufficiently high energy the separation between the nuclear surfaces is comparable with the range of nuclear forces and the strong nuclear interaction can result in the transfer of particles between the colliding nuclei. It is not always necessary that the incident energy shall exceed the Coulomb barrier: particles, and especially neutrons, may tunnel through the barrier separating the potential wells of the two nuclei when these nuclei are in interaction through the outer parts of their wavefunctions. Coulomb processes as described for case (*a*) of Fig. 11.25 can also take place and in addition there may be nuclear inelastic scattering. For a given incident energy the critical trajectory shown defines a *grazing angle*. The separation R_{int} for the two nuclei in a grazing collision is the strong-absorption radius R_{SA} (eqn. 6.9*b*), at which distance particles may be transferred to or from the nuclear interior.

In Fig. 11.25*c*, for smaller impact parameters still, and energies above the Coulomb barrier, the nuclei actually collide and a complex set of processes including fragmentation of both partners and compound-nucleus formation or *fusion* may occur. At energies above the barrier, but still comparable with it, this latter process is dominant over inelastic scattering and transfer processes, although these will still take place.

In all of these reaction types the special features due to the high mass and charge of the incident ion are:

(*a*) A high linear momentum transfer for a given energy, so that experiments requiring high recoil velocities such as Doppler shift lifetime determinations are aided. A ^{16}O nucleus of 160 MeV energy fusing with a ^{40}Ca nucleus imparts a velocity of $0 \cdot 04c$ to the compound system while if it were scattered back the momentum transfer, divided by $\hbar$, would be $q = 15$ fm^{-1}.

(*b*) A high angular momentum transfer to a compound or residual nucleus. For a 160-MeV ^{16}O nucleus approaching a ^{40}Ca nucleus with an impact parameter of 5 fm, the total c.m. angular momentum calculated classically is $40\hbar$.

(*c*) An enhancement of all Coulomb excitation processes in comparison with those for light ions, so that multiple excitations may be observed.

A notable feature arising from the complexity of the incident ion is the possibility of using heavy ions for new modes of nuclear excitation, e.g. the production of shape isomers. The attainment of the charge numbers thought likely for superheavy nuclei from naturally occurring target nuclei seems impossible without the use of such ions, and within the more familiar range of Z-values many new neutron-deficient nuclei may be produced. From the point of view of nuclear spectroscopy of a more conventional type, however, heavy ions have both advantages and disadvantages; these will be briefly examined in the following sections.

11.8.2 Direct reactions of heavy ions

Because of the extended, complex and strongly interacting structure of a heavy ion, it must be expected that its mean free path in nuclear matter will be small. Reactions or scatterings which do not greatly disturb the ion must therefore be sharply located near the surface of a target nucleus. Deeper penetration will lead to compound-nucleus formation or fragmentation.

The *elastic scattering* of heavy ions, therefore, shows the phenomena illustrated in Fig. 11.26. In all cases shown, the scattering parameter kR, where k is the c.m. wavenumber and R is the interaction distance, is considerably greater than unity. Whether or not nuclear size effects leading to diffraction peaks in the angular distribution can be seen depends on the value of the Coulomb parameter η.

For large η, e.g. high charges and low energy, pure Coulomb scattering is seen, as in Fig. 11.26a, whereas in the opposite extreme of low η, e.g. moderate charges and high energy, the Fraunhofer diffraction pattern of a black disc of radius R is found, Fig. 11.26c. The minima of this pattern correspond with the zeros of the diffraction function $[J_1(kR \sin \theta)/kR \sin \theta]^2$. The interaction distance that fits this type of distribution is that shown in Fig. 11.25b and is given by (11.66).

In intermediate cases, in which both Coulomb and nuclear effects are seen, a typical pattern is as shown in Fig. 11.26b. For small angles Rutherford scattering is found but at the grazing angle, at which the nuclear interaction is felt, the elastic scattering falls off rapidly. Within the range of Rutherford-scattering angles, oscillations reminiscent of Fresnel-type diffraction phenomena appear; these differ from the Fraunhofer fringes because of the deflections produced by the Coulomb field.

Theories of the elastic scattering of heavy ions are necessarily more complicated than the optical-model analyses used for lighter particles. There is very little contribution from compound elastic scattering because of the many channels open once a compound nucleus

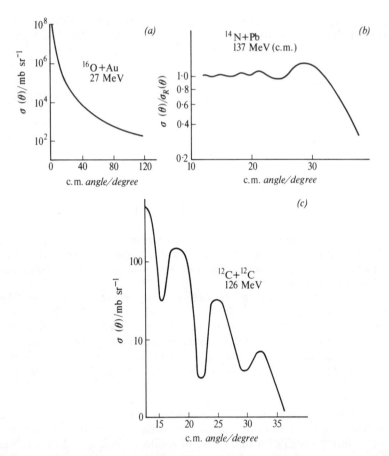

Fig. 11.26 Heavy-ion elastic scattering (examples adapted from Bromley, D. A., 'Heavy Ion Interactions', *Yale Univ. Report* 3223–32, 1965). The η values are (a) 73, (b) 27, (c) 1·8.

has been formed, but the absorption processes react back strongly on the elastic amplitude, causing the diffraction phenomena. Optical potentials that fit the data can be set up, but the main result of such analyses is a determination of the interaction radii (11.66). Prediction of the optical potentials themselves is a formidable problem but the empirical potentials are useful in distorted-wave calculations for inelastic processes.

Although compound elastic scattering is not expected for heavy ions, striking resonance peaks have been observed in the yield of elastic scattering and reactions for the $^{12}C + ^{12}C$, $^{12}C + ^{16}O$ and $^{16}O + ^{16}O$ system over the c.m. energy range 0–50 MeV. These have been attributed to the formation of 'molecular' systems with the two nuclei acting as 'atoms', and the resonances correspond with the molecular energy levels. Formally, this may be regarded as a type of shape-elastic scattering.

Inelastic scattering, if adequate resolution is available, is more informative than the elastic process since heavy-ion reactions show a useful preference for low-lying states of high angular momentum and this is easily supplied by the incident ion. The states that are closely coupled with the ground state, e.g. the members of a rotational band, are strongly excited and the differential cross-section for such states yields nuclear deformation parameters as with lighter ions (Sect. 11.5.2). For energies less than the barrier height, the mechanism is that of Coulomb excitation; at higher energies the nuclear interaction contributes as well, and will interfere with the Coulomb amplitude.

The most useful reactions of heavy ions for spectroscopic purposes are those of the *transfer type* (Fig. 11.25b), again providing that adequate resolution is available for identification of final states. In these reactions there is also high *selectivity*; single-nucleon transfer preferentially populates single-particle or single-hole states while multinucleon transfer leads to more complex excitations connected perhaps with the existence of clusters. In all cases, however, the number of states of a given type excited is fairly small. This may be understood in terms of kinematic constraints that are especially marked for heavy-ion reactions, because of the localization of both the trajectories and the reaction sites. Thus, from Fig. 11.25b the incoming and outgoing trajectories must be such that $R_i \approx R_f$ for the reaction to take place. Surface localization then allows only a small number of angular momentum transfers l for a given energy. The reaction probability is thus a maximum when the matching condition

$$l = L_i - L_f \qquad (11.67)$$

is satisfied, where the quantum number L is given by the formula (11.63). For small angles at least equations (11.62) and (11.63) show that it is a good approximation to write $L_i = kR_i$ and $L_f = kR_f$ and these are each large numbers in the general case. This matching is one reason for the selectivity of the heavy-ion reaction. Physically, it means that the velocity of the transferred nucleon in the incident ion, modified by the incident ionic velocity itself, must approximately equal its velocity in the residual nuclear state. Unfortunately, because $L_i \approx L_f \gg l$ the angular distribution of the process is mainly governed by the classical trajectories and is not primarily characteristic of the transferred l-value as for light-ion processes of this type (Sect. 11.5.2).

The angular momentum matching also leads to an optimum Q-value for the transfer reaction. Thus, from (11.62), if $R_i \approx R_f$ we find

$$\left(\frac{Z_{1i}Z_{2i}}{T}\right)(1+\operatorname{cosec}\theta_i/2) = \left(\frac{Z_{1f}Z_{2f}}{(T+Q)}\right)(1+\operatorname{cosec}\theta_f/2) \quad (11.68)$$

and if each θ is equal to the grazing angle then

$$Q_{opt} = [(Z_{if}Z_{2f}/Z_{1i}Z_{2i})-1]T \quad (11.69)$$

For incident energies above the Coulomb barrier, Q_{opt} corresponds with the optimum angular momentum transfer given by (11.67). If the reaction has $Q < Q_{opt}$, there will be some inhibition by the Coulomb barrier while if $Q > Q_{opt}$, the reaction will tend to take place with high angular momenta for which the overlap between the projectile and target wavefunctions is not strong.

The effect of these restraints is seen in the 'bell-shaped' angular distributions found for heavy-ion stripping processes at energies close to the barrier height, Fig. 11.27. The transfer cross-section first of all increases from zero scattering angle because of increasing overlap between the two nuclei and this continues until the grazing angle is reached, but then it falls as absorption of the incident particle as a whole, or in large part, sets in. At higher energies diffraction features begin to appear in the angular distribution.

Spectroscopic information is extracted from heavy-ion transfer cross-sections through DWBA calculations. These are necessarily more elaborate than for light ions because of the structure of the

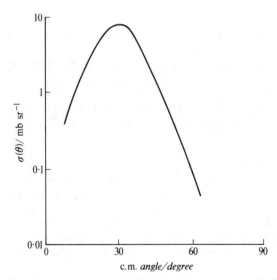

Fig. 11.27 Bell-type angular distribution for heavy-ion stripping. The curve corresponds approximately with data for the ^{48}Ca(^{16}O, ^{15}N)^{49}Sc reaction at 56 MeV.

incident particle and of the transferred cluster (if other than a single nucleon) and also because of the importance of recoil effects. As with light ions, the transfer cross-sections are expected to show a useful magnitude dependence on the j-value of a transferred single nucleon and on the configuration of the final state in the case of transfer of a group of nucleons.

The semi-classical picture given above has been successful in interpreting many of the reactions of ions such as ^{12}C, ^{16}O, ^{19}F (with energies comparable with the Coulomb barrier) on nuclei throughout the periodic table. Under these conditions it is still valid to regard the transfer as a small perturbation to an essentially Coulomb trajectory, and to regard the transferred angular momentum as small. As the incident energy increases to $\approx 10\,\mathrm{MeV/A}$, the transferred angular momentum tends to increase and there is increasing selectivity for states in which the transferred group has the maximum angular momentum allowed by the Pauli principle. This is demonstrated in Fig. 11.28 for three-nucleon (n2p) transfer in the reaction $^{12}C(^{12}C, {}^9Be)^{15}O$ in which known states with spin $\frac{5}{2}$ and $\frac{7}{2}$ are excited together with states of likely spin $\frac{11}{2}$ and $\frac{13}{2}$; the latter are strongly favoured.

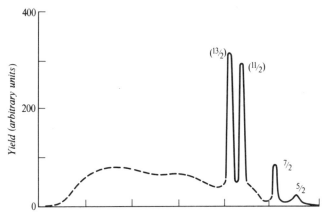

Fig. 11.28 Energy spectrum for the three-nucleon transfer reaction $^{12}C(^{12}C,$ $^9Be)^{15}O$ showing known or expected spin values of the states excited (adapted from Scott, D. K., *et al.*, *Phys. Rev. Letters*, **28**, 1659, 1972).

11.8.3 Compound-nucleus reactions of heavy ions

When the classical trajectory in a heavy-ion collision would indicate a deflection greater than the grazing angle, formation of a compound nucleus, or *fusion* of the two ions, can ensue if the incident energy is greater than the barrier height (Fig. 11.25c).

For projectiles such as ^{12}C, ^{16}O and ^{19}F complete fusion for targets with A up to about 75 yields a compound nucleus with an angular momentum of about $50\hbar$ and an excitation energy of perhaps 50 MeV, corresponding with the energy necessary to overcome the Coulomb and centrifugal barrier. Fusion has also been seen in other systems of mass up to $A \approx 160$. In the decay of such a highly excited nucleus, neutron emission is favoured above that of charged particles because of the Coulomb barrier. The resulting reaction A(HI, xn)B shows selectivity, since by choice of bombarding energy the average number x of successive neutron emissions may be controlled and the production of the particular residual nucleus B can be optimized. From its mode of production B will be a neutron-deficient nucleus.

The yield of the compound-nucleus (xn) reaction is limited by competition from fission for the target nuclei with $A \approx 200$, and there is a critical angular momentum l_{cr} above which fused nuclei rapidly fly apart into fission fragments. For fused nuclei with $I_0 < l_{cr}$, though still with a large angular momentum, neutron evaporation takes place but this cannot easily remove large amounts of angular momentum because of the centrifugal barrier. There is then the phenomenon of energetic photon emission between the levels of the compound nucleus in the form of a so-called '*statistical cascade*' of electric-dipole transitions. These remove angular momentum until a level of spin I, which is the lowest-lying level of this spin in the nucleus (*yrast level*), is reached. There then follows a sequence of transitions between yrast levels, which are in some cases just the levels of the ground-state rotational band, connected perhaps by E2 transitions. States with I as high as 30 have been excited in this way and provide experimentally a highly characteristic series of transition energies (cf. Fig. 8.6b). Because any neutron emission that precedes the gamma-ray cascade does not remove much angular momentum, the nucleus retains rather closely the spin alignment with which it was formed in the fusion reaction. The gamma-ray emission is then strongly anisotropic and this can be useful for spectroscopic purposes. Gamma-ray observations are normally made with pulsed accelerators, and both 'in-beam' and 'out-of-beam' data are obtained, so that the prompt radiations can be distinguished.

Unfortunately for the prospects of superheavy-element formation, fusion is not easily achieved as the mass of the projectile is increased towards $A \approx 200$. The target nucleus must have a similar mass in order to provide a reasonable N/Z ratio for the compound system. The angular momentum in this case is large but in fact fission is no longer the dominant process, as it is for lighter systems. Examination of the energy distribution of the products of such collisions shows that in addition to nearly-elastic direct processes involving little energy transfer, there is a comparable yield of groups of particles of mass near to one or other of the colliding nuclei but which have much smaller energies. The incident kinetic energy and angular momentum have been transferred to internal excitations of these products by a new mechanism known as *deep inelastic scattering*. This process, for which explanations of a classical nature, e.g. friction or viscosity, have been advanced, may lead to a deeper understanding of the interactions of highly excited nuclear systems in which interaction times intermediate between those for a direct reaction and for a compound nucleus are significant. It is a field which offers many new possibilities for the study of what may be described as new forms of strongly interacting complex systems.

11.9 Summary

The strong interaction determines the spectrum of elementary particle states and the binding energy and level spectrum of complex nuclei. It also regulates the course of particle and nuclear reactions. Resonant states both in particles and nuclei are often well-described by a Breit–Wigner type formula and the origin of this formula is outlined; it yields reaction cross-sections near a well-defined resonance. Beyond the resonance region of the nuclear or particle spectrum a continuum is found and reactions passing through continuum states often show the characteristic forward-peaked angular distributions of 'direct' processes.

In heavy complex nuclei the interplay between the strong forces and Coulomb repulsion has a special significance. This dominates not only α-decay, but also fission and the reactions of heavy ions. Corrections arising from shell-structure may exert an important effect on nuclear stability in this high-Z region.

The present Chapter, together with Chapters 8, 9 and 10, touches on many of the current experimental research topics in nuclear structure. A summary of such a wide field is difficult but for high resolution experiments at least it is offered by D. A. Bromley in Reference 11.3. And if, having traversed much of the well-charted territory of nuclear physics, one wishes to take stock of what has been achieved in our understanding of the strong interaction in the nucleus, a survey by D. H. Wilkinson, in Reference 11.4, may provide an answer.

Examples 11

Sections 11.1–11.3

11.1 What might be the isobaric spin of a resonant state producing a peak in the $(\pi^- p)$ total cross-section?

If the yield of the $(\pi^0 n)$ final state at the resonance is double that of the $(\pi^- p)$, show that one of the possible isospins is excluded.

11.2 Resonances are observed in the following reactions at the stated laboratory energies:

$$(a) \ \pi^+ + p \qquad\qquad T_0 = 815 \text{ MeV}$$
$$(b) \ ^{19}\text{F} + p \rightarrow {}^{20}\text{Ne}^* \rightarrow {}^{16}\text{O}^* + \alpha \quad T_0 = 340 \cdot 5 \text{ keV}$$
$$(c) \ ^{50}\text{V} + n \rightarrow {}^{51}\text{V} \qquad\qquad T_0 = 165 \text{ eV}$$

If the masses concerned are π^+, 139·6; p, 938 MeV/c^2; ^{1}H, 1·0078; ^{19}F, 18·9984; ^{20}Ne, 19·9924; ^{50}V, 49·9472; n, 1·0087; ^{51}V = 50·9440 a.m.u.: calculate the energy of the compound system in each case, in MeV. [1640, 13·16, 11·1 MeV]

11.3 In reaction (b) of Example 11.2 a resonance at $T_0 = 873$ keV (lab) has a total width of 5·2 keV and a width for proton re-emission of 1·1 keV. What is the resonant cross-section for α-particle emission, assuming that this is the only competitive process, and neglecting spin factors? [0·55b]

11.4 Write down the final-state wavefunction for elastically scattered neutrons of wavenumber k in terms of the parameters η_0 and δ_0 (as used in equation (1.72)), and deduce equation (11.12).

11.5* Starting with equations (1.78), (1.79) and (1.81) obtain the elastic and inelastic cross-sections in terms of the real and imaginary parts of ρ. Show that the potential (hard-sphere) scattering amplitude is

$$f_0^{\text{pot}} = \frac{\exp(2ikR) - 1}{2i}$$

and that the amplitude for compound scattering is

$$f_0^{\text{nucl}} = -\frac{1}{2i} \frac{2ikR}{\text{Re}\,\rho_0 + i(\text{Im}\,\rho_0 - kR)}$$

Show further that in the neighbourhood Re $\rho_0 = 0$ (resonance) the assumption

$$\rho_0 = -\alpha(T - T_0) - i\beta$$

leads to the Breit–Wigner equation (11.13).

11.6* Show that for the elastic scattering of neutrons the s-wave scattering amplitude when this is the only process possible may be written in the form $\exp(i\delta_0)\sin\delta_0$ where

$$\delta_0 = \beta_0 + \phi_0,$$

β_0 is a nuclear phase shift and $\phi_0 \approx kR$.

11.7 Prove equations (11.23) and (11.24).

11.8* The total energy of a degenerate Fermi gas may be written $E = a(kT)^2$ where T is the absolute temperature. The entropy of the gas, apart from a constant, is given by $S(E) = k \ln \omega(E)$.

Use these formulae to show that the density of states in an excited nucleus is proportional to $\exp[2(aE)^{1/2}]$.

11.9 Cadmium has a resonance for neutrons of energy 0·178 eV and the peak value of the total cross-section is about 7000 b. Estimate the contribution of scattering to the resonance. [13·5 b].

Sections 11.4–11.5

11.10* Use equation (5.12) to obtain the exact s-wave phase shift δ_0 in terms of kR and KR. Examine the variation of δ_0 (which determines the scattering cross-section) as KR varies from 0 to 2π in the case of 1-eV neutrons, with $R = 4\cdot55$ fm.

If $R = r_0 A^{1/3}$ with $r_0 = 1\cdot2$ fm, at what mass numbers would scattering anomalies be seen? (Assume $K = 0\cdot6$ fm^{-1}.)

11.11* The radial part of an $l = 1$ wavefunction is proportional to

$$[\sin kr/(kr)^2] - \cos kr/kr$$

(This agrees with the asymptotic form used in Chapter 1.) Use this expression to obtain the formula analogous to (5.12) for the phase shift δ_1 in terms of kR and KR.

11.12 Show that in the scattering of a particle M_1 by a target nucleus M_2 the momentum transfer q to the nucleus M_2 is the same in both the laboratory and c.m. systems of coordinates.

11.13* By applying the matrix element (11.41) in the case of two s-states, $l_f = l_i = 0$, verify that the angular distribution of direct inelastic scattering has a peak at $\theta = 0°$.

11.14 The neutron separation energy for the nucleus ^{132}Xe is $8\cdot9$ MeV (Table 6.1) and the binding energy of the deuteron is $2\cdot22$ MeV. What is the Q-value for the reaction $^{131}_{54}$Xe(d, p)$^{142}_{54}$Xe? [$6\cdot68$ MeV]

Using equation (1.11) find the proton energy for the ground-state reaction at a laboratory angle of 90° for deuterons of energy 10 MeV. Find also the difference in energy between protons observed at 0° and 180°. [$16\cdot41$ MeV, $0\cdot55$ MeV]

11.15 Suppose that the reaction of Example 11.14 takes place in the surface of the target nucleus, for which $R = 1\cdot2A^{1/3}$ fm. Using the single-particle shell model (Ch. 7) to indicate spin and parity values, find the angle at which the maximum yield of stripped protons would be expected (neglect c.m. corrections and dependence of proton energy on angle). [29°]

Sections 11.6–11.8

11.16 From equation (1.50) we may infer that the differential cross-section for producing particles b with velocity v_b by the reaction X(a, b)Y with particles a of velocity v_a is proportional to v_b/v_a. Use this result and the Gamow factor (11.51) to discuss the variation of the cross-section for the following general types of reaction either near threshold, or with low-energy incident particles:

(a) neutron elastic and inelastic scattering;
(b) the (n, α) reaction, Q positive;
(c) the (p, n) reaction, Q negative;
(d) the (p, α) reaction, Q negative.

11.17 Investigate the stability against α-decay of ^{80}Kr and ^{176}Hf, given the atomic masses ^{80}Kr, $79\cdot9164$; ^{76}Se, $75\cdot9192$; ^{176}Hf, $175\cdot9414$; ^{172}Yb, $171\cdot9364$; ^{4}He, $4\cdot0026$.

11.18 The Q-value for α-decay of RaC' (^{214}Po) is $7\cdot83$ MeV. What is the energy of the α-particles emitted? [$7\cdot68$ MeV]

11.19* Write down the Schrödinger equation for the α-particle penetration problem discussed in Section 11.6.2. Assuming a solution $\psi \approx \exp -\gamma(r)$ and neglecting $d^2\gamma/dr^2$, deduce equation (11.48).

Integrate this equation between $r = R$ and $r = b$ using the substitution $r = b \cos^2 \theta$ and verify equation (11.50).

11.20* Obtain from (11.50) a better approximation than equation (11.51) for the case $T_\alpha \ll B$, showing dependence of decay constant on nuclear radius.

11.21 If two α-emitting nuclei with the same mass number, one with $Z = 84$ and the other with $Z = 82$, have the same decay constant, and if the first emits α-particles of energy $5\cdot3$ MeV, estimate the energy of the α-particles emitted by the second. [$5\cdot05$ MeV]

11.22 Calculate the energy release in:
(a) the spontaneous fission of $^{232}_{92}$U to $^{145}_{57}$La and $^{87}_{35}$Br;
(b) the neutron-induced fission of $^{232}_{92}$U to $^{146}_{57}$La and $^{87}_{35}$Br;
(c) the neutron-induced fission of $^{231}_{92}$U to $^{145}_{57}$La and $^{87}_{35}$Br.
The masses concerned are
^{231}U, 231·0363; ^{232}U, 232·0372; ^{146}La, 145·9255; ^{145}La, 144·9217; ^{87}Br, 86·9203; ^{1}n, 1·0087. [182, 186, 189 MeV]

11.23* For a nucleus described by the semi-empirical mass formula (6.10) show that, neglecting pairing and asymmetry terms, the energy released in fission into two fragments is a maximum for equal division of the charge and mass.
Calculate the value of Z^2/A at which this division just becomes possible energetically.

11.24* Suppose that the fusion of two identical nuclei takes place when the energy change is just that needed to overcome the Coulomb repulsion of the two when they approach from the infinite separation to the point of contact. Then the reverse process of two-body fission will also be possible.
Calculate the corresponding Z^2/A, using the constants of the semi-empirical mass formula with $r_0 = 1\cdot25$ fm and again neglecting asymmetry and pairing terms.

11.25 Oxygen-16 ions of energy 100 MeV are timed over a distance of $0\cdot3$ m. What timing accuracy is necessary if the mass number determination shall be accurate to $\pm0\cdot3$ a.m.u.? [81 ps]

11.26 Verify the values of η shown in Fig. 11.26, caption.

11.27* Prove the formula for the impact parameter in Rutherford scattering:

$$b = [Z_1 Z_2 e^2 / 4\pi\varepsilon_0 \mu v_1^2] \cot(\theta/2)$$

where Z_1, Z_2 are the charge numbers of the colliding particles, μ is the reduced mass and v_1 is the laboratory velocity of the incident particle and θ is the c.m. angle of scattering.

11.28* Use the formula given in Example 11.27 to prove equation (11.62).

11.29* In the collision of $^{56}_{26}$Fe ions of energy 10 MeV per nucleon with $^{238}_{92}$U nuclei the grazing angle was found to be 39° (lab).
Calculate (a) the η value, (b) the angular momentum number for the grazing orbit, (c) the interaction distance, and (d) the reaction cross-section (using eqn (11.62)).

11.30 Compare the Coulomb and centrifugal barriers in the collision of oxygen-16 ions with $^{50}_{23}$V. Assume that the incident energy is just equal to the barrier height and that the impact parameter is the sum of the nuclear radii. [$34\cdot1$ and $25\cdot8$ MeV for $r_0 = 1\cdot25$ fm]

11.31 What is the optimum Q-value for
(a) a heavy ion neutron-transfer reaction [0]
(b) the proton transfer reactions ^{48}Ca (^{16}O, ^{15}N)^{49}Sc and ^{48}Ca (^{16}O, ^{17}F)^{47}K at $T_{\text{lab}} = 56$ MeV (Fig. 11.27). [$-3\cdot4$, $2\cdot9$ MeV]
Compare this Q_{opt} with the ground state Q-value given by the masses (^{48}Ca, 47·9525; ^{49}Sc, 48·9557; ^{16}O, 15·9949; ^{15}N, 15·0001; ^{47}K, 46·9617; ^{17}F, 17·0021). [$-7\cdot8$, $-15\cdot3$ MeV]

References

1.1 (a) Perkins, D. H., *Introduction to High Energy Physics*, Addison-Wesley, 1972.
(b) Tassie, L. J., *The Physics of Elementary Particles*, Longman, 1973.
1.2 (a,b) Feynman, R. P., *Lectures on Physics* Vols 1 and 2, Addison-Wesley. (a) 1963, (b) 1964.
1.3 (a) Dicke, R. H. and Wittke, J. P., *Introduction to Quantum Mechanics*, Addison-Wesley, 1960.
(b) Willmott, J. C., *Atomic Physics*, Wiley, 1975.
2.1 Evans, R. D., *The Atomic Nucleus*, McGraw-Hill, 1955.
2.2 Powell, C. F., 'Mesons', *Rep. Prog. Phys.*, **13**, 350, 1950.
2.3 Perkins, D. H., 'Neutrinos and Nucleon Structure', *Contemp. Phys.*, **16**, 173, 1975.
2.4 Dalitz, R. H. *et al.*, *Proc. Roy. Soc.*, **A355**, 441–637, 1977.
3.1 Sternheimer, R. M., 'Interaction of Radiation with Matter', in *Methods of Experimental Physics 5 (Part A)*, ed. Yuan, L. C. L. and Wu, C. S., Academic Press, 1961.
3.2 Dyson, N. A., *X-rays in atomic and nuclear physics*, Longman, 1973.
4.1 (a) Allen, K. W., 'Electrostatic Accelerators', in *Nuclear Spectroscopy and Reactions (Part A)*, Academic Press, 1974.
(b) Skorka, S. J., *Nucl. Instrum. Meth.*, **146**, 67, 1977.
4.2 Fry, D. W. and Walkinshaw, W., 'Linear Accelerators', *Rep. Prog. Phys.*, **12**, 102, 1949.
4.3 Livingood, J. J., *Principles of Cyclic Particle Accelerators*, van Nostrand, 1961.
4.4 England, J. B. A., *Techniques in Nuclear Structure Physics*, Macmillan, 1964.
England, J. B. A., 'Detection of ionizing radiations', *Jour. Phys.*, **E9**, 233, 1976.
4.5 Shutt, R. P. (ed.) *Bubble and Spark Chambers*, Academic Press, 1967.
5.1 Brown, G. E. and Jackson, A. D., *The nucleon–nucleon interaction*, North Holland, 1976.
6.1 Wapstra, A. H. and Gove, N. B., 'The 1971 Atomic Mass Evaluation', *Nuclear Data Tables*, **A9**, 265, 1971; Wapstra, A. H. and Bos, K., 'A 1975 Midstream Atomic Mass Evaluation', *Atomic Data and Nuclear Data Tables*, **17**, 474, 1976.
6.2 Bohr, A. and Mottelson, B., *Nuclear Structure I*, Benjamin, 1969.
6.3 Brown, G. E., *The Unified Theory of Nuclear Models and Forces*, North Holland, 1967.
6.4 Preston, M. A. and Bhadhuri, R. K., *Structure of the Nucleus*, Addison-Wesley, 1975.
6.5 (a) Clayton, D. D. and Woosley, S. E., 'Thermonuclear Astrophysics', *Rev. Mod. Phys.*, **46**, 755, 1974.
(b) Williams, P. M., 'The Evolution of the Elements', *Contemp. Phys.*, **19**, 1, 1978.

371

Elements of nuclear physics

7.1 Hodgson, P. E., 'Nucleon removal and re-arrangement energies', *Rep. Prog. Phys.*, **38**, 847, 1975.

8.1 Bohr, A. and Mottelson, B., *Nuclear Structure II*, Benjamin, 1975.

8.2 Kuo, T. T. S. and Brown, G. E., *Nuclear Phys.*, **A85**, 40, 1966.

8.3 (a) Satchler, G. R., 'New giant resonances in nuclei', *Physics Reports*, **14C**, 98, 1974.
   (b) Bertrand, F. E., 'Excitation of giant multipole resonances through inelastic electron scattering' *Ann. Rev. Nucl. Sci.*, **26**, 475, 1975.

9.1 (a) Blatt, J. B. and Weisskopf, V. F., *Theoretical Nuclear Physics*, Wiley, 1952, Appendix B.
   (b) Reference 6.4, Appendix E.

9.2 (a) Wertheim, G. K., *Mössbauer Effect; principles and applications*, Academic Press, New York, 1964.
   (b) Dale, B. W., 'Mössbauer Spectroscopy', *Contemp. Phys.* **16**, 127, 1975.

9.3 Siegbahn, K. (ed.), *Alpha, Beta and Gamma Ray Spectroscopy*, North Holland, 1965, p. 1687.

10.1 Wilkinson, D. H., *Nature*, **257**, 189, 1975.

11.1 (a) Jackson, D. F., *Nuclear Reactions*, Methuen, 1970.
   (b) Satchler, G. R., *Introduction to Nuclear Reactions*, Macmillan, 1980.

11.2 Zeidman, B., *Nucl. Instrum. Meth.*, **146**, 199, 1977 ('Heavy Ion Reactions').

11.3 Bromley, D. A., *Nucl. Instrum. Meth.*, **146**, 1, 1977 ('The Physics of Large Tandems').

11.4 Wilkinson, D. H., *Nucl. Instrum. Meth.*, **146**, 143, 1977 ('Where is the atomic nucleus going?').

General reference

Matthews, P. T., *Introduction to Quantum Mechanics*, 3rd edition, McGraw Hill 1974.

Definitions and data tabulations

nucleus, a general term for the finite structure of neutrons and protons constituting the centre of force in an atom.

nuclide, a specific nucleus, with given proton number Z and neutron number N.

isotope, one of a group of nuclides each having the same proton number Z.

isotone, one of a group of nuclides each having the same neutron number N.

isobar, one of a group of nuclides each having the same mass (baryon) number A $(=B) = Z + N$.

isomer, a nuclide excited to a long-lived state from which beta or gamma decay ensues.

curie (Ci), one curie of a radioactive nuclide is that quantity in which the number of disintegrations per second is $3 \cdot 700 \times 10^{10}$. This unit is being superseded by the

becquerel (Bq), as a unit of activity. One becquerel is a rate of one nuclear transmutation per second.

The symbol for a *chemical element* is:

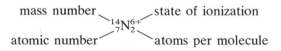

$$\text{mass number} \searrow \quad \nearrow \text{state of ionization}$$
$$^{14}_{7}\text{N}^{6+}_{2}$$
$$\text{atomic number} \nearrow \quad \searrow \text{atoms per molecule}$$

The following data tabulations are useful:

(Abbreviated references are: NDT, *Atomic Data and Nuclear Data Tables*; RPP, 'Review of Particle Properties' *Rev. Mod. Phys.*, **48**, No. 2 (Part II), 1976; αβγ, Siegbahn, Ref. 9.3.)

angular correlation coefficients, αβγ, p. 1687

atomic masses, Ref. 6.1

beta decay, 'Tables for analysis of beta spectra', National Bureau of Standards 1952; Landolt–Bornstein Tables, Group I, Vol. 4 (Behrens–Janecke), 1969; NDT **A10**, 205, 1971

Clebsch–Gordan coefficients, RPP p. S36 and Appendix 4

electromagnetic processes, Refs. 2.1 and 3.2

elementary particles, RPP p. S3 and Appendix 2

fundamental constants, Appendix 6

gamma-ray spectra, *Applied Gamma Ray Spectrometry*, Adams, F. and Dams, R., Pergamon Press, Oxford 1970

internal conversion coefficients, αβγ p. 1639

neutron cross-sections (*activation*), NDT **A11,** 621, 1973

nuclear level schemes, *Nuclear Data Sheets* (Academic Press, New York); *Table of Isotopes*, (7th Ed) by Lederer, C. M., and Shirley, V. S. (Wiley 1978)

nuclear spins and moments, NDT **A5,** 433, 1969; NDT **A7,** 495, 1970; αβγ p. 1621

particle detectors, Ref. 4.4

photon cross-sections, NDT **A7,** 565, 1970

Q-values, NDT **A11,** 127, 1972

radioactive radiations in order of energy and nuclide, NDT **13,** 89, 1974; *The Radiochemical Manual*, Radiochemical Centre, Amersham 1966

radioactive nuclides in order of halflife, NDT **A7,** 477, 1970

radioactivity and radiation protection, *The Radiochemical Manual*, Radiochemical Centre, Amersham 1966

ranges and stopping powers, NDT **A7,** 233, 1970

X-rays general and X-ray absorption edges, Ref. 3.2

Appendices 1–6

Appendix 1 Spherical harmonics

The spherical harmonic functions $Y_l^m(\theta, \phi)$ are of frequent occurrence in atomic and nuclear physics. They provide a convenient means of describing any deformation of an initially spherical surface and can be used to represent the angular 'shape' of wavefunctions. They are eigenfunctions of the operators for the square of the total angular momentum of a particle moving in a central field of force and for its resolved part (Sect. 1.4). Explicitly

$$Y_l^m(\theta, \phi) = (-1)^{(m+|m|)/2} \sqrt{\frac{2l+1}{4\pi} \frac{(l-|m|)!}{(l+|m|)!}} \times P_l^m(\cos \theta) \times e^{im\phi}$$

where l and m are integers and $|m| \leq l$. The function $P_l^m(\cos \theta)$ is an *associated Legendre function*, which may be derived from the corresponding *Legendre function* $P_l(\cos \theta)$ by a process of repeated differentiation.

From this expression it may be seen that the complex conjugate of a spherical harmonic is given by

$$Y_l^m(\theta, \phi)^* = (-1)^m Y_l^{-m}(\theta, \phi)$$

and that under the parity operation $(\theta \to \pi - \theta,\ \phi \to \phi + \pi)$

$$Y_l^m(\theta, \phi) \to (-1)^l Y_l^m(\theta, \phi)$$

so that the parity is even or odd according to l. The spherical harmonics are orthogonal and their square is normalized to unity over a sphere. This is expressed by the equation

$$\int_0^\pi \int_0^{2\pi} Y_{l'}^{m'}(\theta, \phi)^* Y_l^m(\theta, \phi) \sin \theta\, d\theta\, d\phi = 1 \quad \text{if } l = l',\ m = m'$$
$$= 0 \quad \text{otherwise}$$

For the Legendre functions:

$$\int_0^\pi \int_0^{2\pi} [P_l(\cos \theta)]^2 \sin \theta\, d\theta\, d\phi = 4\pi/(2l+1)$$

375

and for the associated Legendre functions:

$$\int_0^\pi \int_0^{2\pi} [P_l^m(\cos\theta)]^2 \sin\theta \, d\theta \, d\phi = \frac{4\pi}{2l+1} \frac{(l+|m|)!}{(l-|m|)!}$$

For a given l

$$\sum_m |Y_l^m(\theta,\phi)|^2 = (2l+1)/4\pi$$

and for cases in which $m=0$ (Sect. 1.2.7)

$$Y_l^0 = [(2l+1)/4\pi]^{1/2} P_l(\cos\theta)$$

The Legendre functions P_l also form an orthogonal normalized set and can be used to expand a scattering amplitude (Sect. 1.2.7) in terms of partial-wave amplitudes for a series of l-values. Expressions for the first few spherical harmonics and Legendre functions are given in the following table:

Legendre function	Associated Legendre function	Spherical harmonic
$P_0(\cos\theta) = 1$	$P_0^0(\cos\theta) = 1$	$Y_0^0(\theta,\phi) = \sqrt{1/4\pi}$
$P_1(\cos\theta) = \cos\theta$	$P_1^0(\cos\theta) = \cos\theta$	$Y_1^0(\theta,\phi) = \sqrt{3/4\pi}\cos\theta$
	$P_1^{\pm 1}(\cos\theta) = (1-\cos^2\theta)^{1/2}$	$Y_1^{\pm 1}(\theta,\phi) = \mp\sqrt{3/8\pi}(1-\cos^2\theta)^{1/2}e^{\pm i\phi}$
$P_2(\cos\theta)$ $=\frac{1}{2}(3\cos^2\theta - 1)$	$P_2^0(\cos\theta)$ $=\frac{1}{2}(3\cos^2\theta - 1)$	$Y_2^0(\theta,\phi) = \sqrt{5/16\pi}(3\cos^2\theta - 1)$
	$P_2^{\pm 1}(\cos\theta)$ $= 3\cos\theta(1-\cos^2\theta)^{1/2}$	$Y_2^{\pm 1}(\theta,\phi) = \mp\sqrt{15/8\pi}\cos\theta(1-\cos^2\theta)^{1/2}$
	$P_2^{\pm 2}(\cos\theta)$ $= 3(1-\cos^2\theta)$	$Y_2^{\pm 2}(\theta,\phi) = \sqrt{15/32\pi}(1-\cos^2\theta)e^{\pm 2i\phi}$

Appendix 2 Stable elementary particles

Stable particles are those that do not decay by the strong interaction. The information in this table has been taken from the regularly-appearing 'Review of Particle Properties' by the Particle Data Group, *Rev. Mod. Phys.* No. 2, Part II, April 1976. In this article the new resonances such as J/ψ are classified as mesonic; a full listing is given in *Phys. Letters* **68B**, 1, 1977.

The stable hadrons have a strangeness quantum number S and are all uncharmed. Parity is not uniquely defined for fermions although the relative intrinsic parity of a fermion–antifermion pair is odd. Conventionally, the nucleon and the Λ are assigned even parity and the Λ is given strangeness $S = -1$. Antiparticles are included in the table; their mass and lifetime are assumed to be those of the corresponding particle. The magnetic moment of a particle and antiparticle (with $J > 0$) are equal and opposite with respect to the direction of intrinsic spin.

Group	Name	Spin-parity J^π	Isospin T, T_z	Symbol	Mass MeV/c²	Strangeness	Mean life s	Main decay modes
Leptons $B=0$	Photon	1^-	—	γ	0	—	Stable	
	Neutrinos	$\frac{1}{2}$	—	ν_e, ν_μ $\bar\nu_e, \bar\nu_\mu$	0 0	— —	Stable Stable	
	Electrons		—	e^- e^+	0.51	—	Stable	
	Muons		—	μ^- μ^+	105.7	—	2.2×10^{-6}	$e\nu\bar\nu$
Mesons $B=0$	Pions	0^-	$1, 0$ $1, -1$	π^0 π^+ π^-	135.0 139.6	0 0	0.83×10^{-16} 2.6×10^{-8}	$\gamma\gamma,\ \gamma e^+e^-$ $\mu^+\nu,\ \pi\pi,\ \pi\pi\pi$ $\mu^-\bar\nu,\ \pi\pi\pi,\ \pi\pi\pi\pi$
	Kaons	$\frac{1}{2}, \frac{1}{2}$ $-\frac{1}{2}$ $\frac{1}{2}, -\frac{1}{2}$	K^+ K^- K^0 $\overline{K}_0$	493.7 497.7	1 -1 1 -1	1.24×10^{-8} $K_S^0\,0.89\times10^{-10}$ $K_L^0\,5.18\times10^{-8}$	$\pi\pi$ $\pi\pi\pi,\ \pi\mu\nu,\ \pi e\nu$
	Eta		$0, 0$	η	548.8	0	2.5×10^{-19}	$\gamma\gamma,\ \pi\pi\pi$
Baryons $B=\pm1$	Nucleons	$\frac{1}{2}$	$\frac{1}{2}, \pm\frac{1}{2}$ $\frac{1}{2}, \mp\frac{1}{2}$	$p^+, \bar{p}^-$ $n, \bar{n}$	938.3 939.6	0, 0 0, 0	Stable 932	$pe\bar\nu$
	Lambda hyperon		$0, 0$	$\Lambda^0, \bar\Lambda^0$	1115.6	$-1, +1$	2.6×10^{-10}	$p\pi, n\pi$
	Sigma hyperons		$1, \pm1$ $1, 0$ $1, \mp1$	$\Sigma^+\bar\Sigma^-$ $\Sigma^0\bar\Sigma^0$ $\Sigma^-\bar\Sigma^+$	1189.4 1192.5 1197.4	$-1, +1$ $-1, +1$ $-1, +1$	0.8×10^{-10} $<10^{-14}$ 1.48×10^{-10}	$p\pi, n\pi$ $\Lambda\gamma$ $n\pi$
	Cascade hyperons		$\frac{1}{2}, \pm\frac{1}{2}$ $\frac{1}{2}, \mp\frac{1}{2}$	$\Xi^0\bar\Xi^0$ $\Xi^-\bar\Xi^+$	1314.9 1321.3	$-2, +2$ $-2, +2$	2.96×10^{-10} 1.65×10^{-10}	$\Lambda\pi$ $\Lambda\pi$
	Omega hyperon		$0, 0$	$\Omega^-\bar\Omega^+$	1672.2	$-3, +3$	1.3×10^{-10}	$\Xi\pi,\ \Lambda K$

Appendix 3   Effective range

The definition of scattering length given in Section 5.2.2 equation (5.15), is not confined to the case of a square-well potential. It is in fact a property of the final-state wavefunction for the interacting system at zero energy, as is illustrated in Fig. A3.1 for two short-ranged potentials of different shape.

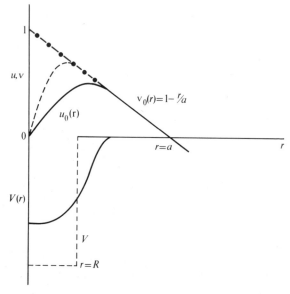

Fig. A3.1   Effective range theory for (np) scattering. The full line shows zero-energy wavefunctions u_0 for a potential $V(r)$ and v_0 for no interaction. The extrapolation of v_0 to $r = 0$ is shown as $\cdot - \cdot - \cdot -$. The dashed (−−) line shows the wavefunction u_0 for the square-well potential that gives the same scattering length a. The effective range is determined by the difference between u_0 and v_0 within the actual range of the force.

Outside the range of the nuclear force, which need not be specified, the final-state wavefunction is a solution of the Schrödinger equation

$$\frac{d^2(r\psi)}{dr^2} + k^2(r\psi) = 0 \qquad (A3.1)$$

where $k^2 = (2\mu/\hbar^2)\tfrac{1}{2}T_n$ and T_n is the incident neutron energy. We will consider two solutions to this equation, one called u relating to the actual case in which a short-range interaction exists and the other, v, for a case in which there is no such interaction. Each of these wavefunctions has the form (cf. (5.8))

$$\sin(kr + \delta_0)/\sin\delta_0 \qquad (A3.2)$$

where δ_0 is the s-wave phase shift and the factor $1/\sin \delta_0$ has been introduced to normalize the wavefunction $v(r)$ to unity at $r = 0$.

For low energies $(k \to 0)$ (A3.2) takes the form

$$u(r) = v(r) \approx 1 + kr \cot \delta_0 \qquad (A3.3)$$

which is the straight line shown in Fig. A3.1 intersecting the axis of r at the point $r = a$ where

$$k \cot \delta_0 = -1/a \qquad (k \to 0) \qquad (A3.4)$$

and a is, of course, the scattering length. We can, therefore, write the no-potential solution v for zero energy in the form

$$v_0(r) = 1 - r/a \qquad (A3.5)$$

and this solution is shown in Fig. A3.1 extrapolated back to $r = 0$.

Let us now consider the realistic solution u both for zero energy (u_0) and for an energy small compared with the depth V of the potential well, so that $u \approx u_0$. This wavefunction is sketched for the two wells in Fig. A3.1 and obviously differs greatly from v within the range of the force. The difference between u_0 and v_0, suitably expressed, determines the effective range.

To make this explicit, we note that within the range of the force the Schrödinger equation takes the form

$$\frac{d^2 u}{dr^2} + k^2 u = \frac{2\mu}{\hbar^2} V(r) u \qquad (A3.6)$$

both for k finite (u) and k zero (u_0). To eliminate $V(r)$, we write the equation also for u_0, cross-multiply the two equations by u_0 and u and subtract, finding

$$u_0 \frac{d^2 u}{dr^2} - u \frac{d^2 u_0}{dr^2} + k^2 u u_0 = 0$$

or

$$\frac{d}{dr}\left(u \frac{du_0}{dr} - u_0 \frac{du}{dr}\right) = k^2 u u_0 \approx k^2 u_0^2 \qquad (A3.7)$$

Since this expression does not involve any assumption about $V(r)$ it will also apply to the no-potential case so that we may also write

$$\frac{d}{dr}\left(v \frac{dv_0}{dr} - v_0 \frac{dv}{dr}\right) = k^2 v v_0 \approx k^2 v_0^2 \qquad (A3.8)$$

If (A3.7) is subtracted from (A3.8) and if an integration over r from 0 to ∞ is performed we obtain

$$\left[\left(v \frac{dv_0}{dr} - v_0 \frac{dv}{dr}\right) - \left(u \frac{du_0}{dr} - u_0 \frac{du}{dr}\right)\right]_0^\infty = k^2 \int_0^\infty (v_0^2 - u_0^2)\, dr \qquad (A3.9)$$

Because u and v are identical outside the range of the force, the quantity in the square bracket vanishes at the upper limit. At $r = 0$, $v(0) = v_0(0) = 1$ from (A3.5) and $u(0) = u_0(0) = 0$ since the real wavefunction ψ must be finite at the origin. Also $dv_0/dr = -1/a$ from (A3.5) and $dv/dr = k \cot \delta_0$ from (A3.3).

The expression (A3.9) thus reduces to the shape-independent formula (5.19)

$$k \cot \delta_0 = -1/a + \tfrac{1}{2} r_0 k^2 \qquad (A3.10)$$

if the effective range is defined to be

$$r_0 = 2 \int_0^\infty (v_0^2 - u_0^2) \, dr \qquad (A3.11)$$

This quantity depends on the actual range (although it is not equal to it even for a square well) and on the depth of the potential, but it does not require a particular potential shape. It is possible to fit the observed s-wave phase shift with any reasonable shape, so long as the potential is of short range; the depth and shape are connected through the requirement that $1/a$ and r_0 must be correctly predicted.

Appendix 4 Clebsch–Gordan coefficients

The addition of angular momenta, or of isobaric spins, is described in Reference 1.1a, Appendix C. Briefly, the total angular momentum $\mathbf{J}$ of two particles of angular momentum $\mathbf{j}_1$ and $\mathbf{j}_2$ is given by the vector equation

$$\mathbf{J} = \mathbf{j}_1 + \mathbf{j}_2 \qquad (A4.1)$$

together with the similar equation for the z-components

$$M = m_1 + m_2 \qquad (A4.2)$$

The final quantum number J can take any value between $j_1 + j_2$ and $|j_1 - j_2|$. To find the probability of a given J we write

$$\phi_1(j_1 m_1) \phi_2(j_2 m_2) = \sum_J C_J \psi(J, M) \qquad (A4.3)$$

where $\psi(J, M)$ is a normalized angular momentum eigenfunction and C_J is a Clebsch–Gordan coefficient, sometimes denoted by $(j_1 j_2 m_1 m_2 \,|\, JM)$. The same coefficients permit the state $\psi(J, M)$ to be expressed in terms of a superposition of the product states $\phi_1 \times \phi_2$ for the different values of j_1 and j_2 allowed by equations (A4.1) and (A4.2).

The following two tables cover several simple cases, and further tables will be found in Reference 1.1a.

$j_1 = \frac{1}{2}, \; j_2 = \frac{1}{2}$

m_1	m_2	$\begin{matrix} J \\ M \end{matrix}$	$\begin{matrix} 1 \\ 1 \end{matrix}$	$\begin{matrix} 1 \\ 0 \end{matrix}$	$\begin{matrix} 0 \\ 0 \end{matrix}$	$\begin{matrix} 1 \\ -1 \end{matrix}$
$\frac{1}{2}$	$\frac{1}{2}$		1			
$\frac{1}{2}$	$-\frac{1}{2}$			$\sqrt{\frac{1}{2}}$	$\sqrt{\frac{1}{2}}$	
$-\frac{1}{2}$	$\frac{1}{2}$			$\sqrt{\frac{1}{2}}$	$-\sqrt{\frac{1}{2}}$	
$-\frac{1}{2}$	$-\frac{1}{2}$					1

$j_1 = 1, \; j_2 = \frac{1}{2}$

m_1	m_2	$\begin{matrix} J \\ M \end{matrix}$	$\begin{matrix} \frac{3}{2} \\ \frac{3}{2} \end{matrix}$	$\begin{matrix} \frac{3}{2} \\ \frac{1}{2} \end{matrix}$	$\begin{matrix} \frac{1}{2} \\ \frac{1}{2} \end{matrix}$	$\begin{matrix} \frac{3}{2} \\ -\frac{1}{2} \end{matrix}$	$\begin{matrix} \frac{1}{2} \\ -\frac{1}{2} \end{matrix}$	$\begin{matrix} \frac{3}{2} \\ -\frac{3}{2} \end{matrix}$
1	$\frac{1}{2}$		1					
1	$-\frac{1}{2}$			$\sqrt{\frac{1}{3}}$	$\sqrt{\frac{2}{3}}$			
0	$\frac{1}{2}$			$\sqrt{\frac{2}{3}}$	$-\sqrt{\frac{1}{3}}$			
0	$-\frac{1}{2}$					$\sqrt{\frac{2}{3}}$	$\sqrt{\frac{1}{3}}$	
-1	$\frac{1}{2}$					$\sqrt{\frac{1}{3}}$	$-\sqrt{\frac{2}{3}}$	
-1	$-\frac{1}{2}$							1

The Clebsch–Gordan coefficients satisfy the symmetry relations

$$(j_1 j_2 m_1 m_2 \mid JM) = (-1)^{j_1 + j_2 - J}(j_2 j_1 m_2 m_1 \mid JM)$$
$$= (-1)^{j_1 + j_2 - J}(j_1 j_2 - m_1 - m_2 \mid J - M) \quad \text{(A4.4)}$$

It will be noted in the tables that the expansions of $\phi_1 \times \phi_2$ for different values of m_1, m_2 but the same M are orthogonal.

Appendix 5 Solutions to selected examples

The examples selected are marked by an asterisk.

Chapter 1

1.1 $\boldsymbol{p}_1' = \boldsymbol{p}_1 - \boldsymbol{q}$ and the scalar product of each side with itself is

$$p_1'^2 = p_1^2 + q^2 - 2p_1 q \cos \phi_L$$

Substituting for p^2 in terms of T gives

$$T_1' = T_1 + M_2 T_2 / M_1 - p_1 q \cos \phi_L / M_1$$

and substituting for T_1' from (1.2) and putting $T_2 = q^2/2M_2$ gives the result.

1.3   Conservation of energy and momentum give

$$T_4 = Q + T_1 - T_3 \quad \text{and} \quad p_4 = p_1 - p_3$$

Forming scalar products gives

$$p_4^2 = p_1^2 + p_3^2 - 2p_1 p_3 \cos \theta_{3L}$$

and substituting $p^2 = 2MT$ and the value of T_4 gives the result.

1.4   Write (1.11) in the quadratic form

$$a(T_3^{1/2})^2 + b(T_3^{1/2}) + c = 0$$

For one value of $T_3^{1/2}$, which must be positive since $T_3^{1/2} \propto |\mathbf{p}_3|$, we find $\sqrt{b^2 - 4ac} \geq -b$, i.e. $-4ac \geq 0$. For two solutions $b^2 \geq 4ac$. Substitution of the values of a, b, c gives the results.

1.11   In the centre-of-mass system, the component of momentum along the direction of motion of the c.m. is $(p_L \cos \theta_L - M_3 v_c)$ and that perpendicular to this direction is $p_L \sin \theta_L$ where p_L is the laboratory momentum. The resultant c.m. momentum p is then given by

$$p^2 = p_L^2(\cos^2 \theta_L + \sin^2 \theta_L) + M_3^2 v_c^2 - 2M_3 v_c p_L \cos \theta_L$$

which gives the required expression when p^2 is replaced by $2M_3 T$ and $\frac{1}{2}M_3 v_c^2$ by a. Similarly the second expression follows by transforming from the c.m. system.

Since the component of momentum perpendicular to v_c is $p_L \sin \theta_L$ in both c.m. and laboratory frames, we have $p_L \sin \theta_L = p \sin \theta$ i.e. $\sin \theta_L / \sin \theta_L = p/p_L = (T/T_L)^{1/2}$.

1.13   The Lorentz transformation connects the momentum in the c.m. system with momentum–energy in the laboratory system. The relation is

$$p_c = \gamma_c(p_L - \beta_c E_L)$$

But the total c.m. momentum is zero by definition so that

$$\beta_c = p_L/E_L = \beta \gamma m/(M + \gamma m)$$

Similarly, by transforming energies

$$E_L = \gamma_c(E_c + \beta_c p_c)$$

and with $p_c = 0$ we find

$$\gamma_c = E_L/E_c = (M + \gamma m)/E_c$$

To find E_c take the squared four-vector:

$$E_c^2 = (p_\mu^c)^2 = (p_\mu^1 + p_\mu^2)^2 = M^2 + m^2 + 2p_\mu^1 \cdot p_\mu^2$$

$$= M^2 + m^2 + 2(E_1 E_2 - \mathbf{p}_1 \cdot \mathbf{p}_2)$$

$$= M^2 + m^2 + 2m\gamma M \quad \text{since} \quad p_2 = 0$$

1.20   (a) Use the approximate expression $q = (2E/\hbar c) \sin \frac{1}{2}\theta$.
       (b) Use equation (1.6).

1.21   By conservation of four-momentum

$$p_\mu^1 + p_\mu^2 = p_\mu^3 + p_\mu^4$$

Taking the scalar product and inserting the values of $(p_\mu)^2$

$$M^2 + M^2 + 2ME_1 = M^2 + M^2 + 2(E_1' E_2 - \mathbf{p}_1' \cdot \mathbf{p}_2)$$

where E_1' and E_2 are the two total energies after collision. This equation gives $\boldsymbol{p}_1' \cdot \boldsymbol{p}_2$ in terms of E_1, E_1' and E_2 and by converting to kinetic energies using $E_1' = M + T_1'$, $E_2 = M + T_2$, $E_1 + M = E_1' + E_2$ we obtain $\boldsymbol{p}_1' \cdot \boldsymbol{p}_2 = T_1' T_2$. The value of $\cos \psi$ follows by noting that $\boldsymbol{p}_1' \cdot \boldsymbol{p}_2 = p_1' p_2 \cos \psi$ and that $p = \sqrt{T(T + 2M)}$ for each particle.

1.25 Equation (1.26) is

$$T_1 = -Q[M_f + M_i]/2M_2 \qquad (Q \text{ negative})$$

and

$$M_i = M_f + Q, \qquad M_i = M_1 + M_2$$

It follows that

$$Q^2 - 2(M_1 + M_2)Q - 2M_2 T_1 = 0$$

whence

$$Q = (M_1 + M_2) \pm [(M_1 + M_2)^2 + 2M_2 T_1]^{1/2}$$

from which, by expanding the square root

$$Q = -M_2 T_1/(M_1 + M_2)[1 - \tfrac{1}{2}M_2 T_1/(M_1 + M_2)^2]$$

The correction is the quantity in square brackets.

1.33 In an obvious notation we have

$$m_1 v_1 = m_2 v_2, \qquad m_1 r_1 = m_2 r_2$$

The total angular momentum is

$$I = m_1 r_1^2 \omega + m_2 r_2^2 \omega$$
$$= \mu R^2 \omega \quad \text{where} \quad \mu = m_1 m_2/(m_1 + m_2) \quad \text{and} \quad R = r_1 + r_2$$

by using the above relations.

The magnetic moment is

$$\mu_{\text{mag}} = \pi r_1^2 i_1 + \pi r_2^2 i_2$$

and using $i_1 = -e\omega/2\pi$, $i_2 = +e\omega/2\pi$

$$\mu_{\text{mag}} = -e\omega(r_1^2 - r_2^2)/2$$

To relate to μ_B, suppose that $m_2 \gg m_1$. Then $I = m_1 r_1^2 \omega = [l(l+1)]^{1/2}\hbar$ if quantized and

$$\mu_{\text{mag}} = -e\hbar[l(l+1)]^{1/2}/2m = -[l(l+1)]^{1/2}\mu_B$$

1.34 For s-wave scattering $d\sigma/d\Omega = \sigma/4\pi$, and the element of solid angle is $d\Omega = 2\pi \sin \theta \, d\theta$. From (1.15) we then obtain $d\sigma = -\sigma \sin \phi_L \cos \phi_L \, d\phi_L$ and from simple kinematics the recoil energy at laboratory angle ϕ_L is $T_r = T_0 \cos^2 \phi_L$. It follows that $dT_r = -2T_0 \cos \phi_L \sin \phi_L \, d\phi_L$ so that $d\sigma/dT_r = \sigma/2T_0 = \text{constant}$.

1.36 Putting $\eta_l = 0$ in (1.79) and (1.81) for all l up to kR we obtain

$$\sigma_{\text{el}} = \sigma_{\text{inel}} = (\pi/k^2) \sum_0^{kR} (2l + 1)$$
$$= (\pi/k^2)[kR + 1]^2 = \pi(R + \lambda)^2$$

1.37 For an impenetrable sphere, $\eta_0 = 1$ and $\psi_f = 0$ at the surface $r = R$. Equation (1.72) then gives

$$\exp 2i\delta_0 \exp ikR = \exp(-ikR), \quad \text{or} \quad \delta_0 = -kR$$

Elements of nuclear physics

Equation (1.77) then gives

$$d\sigma_{el}/d\Omega = |e^{i\delta_0} \sin \delta_0/k|^2 = \sin^2 \delta_0/k^2$$

In the classical limit $kR \ll 1$ $d\sigma_{el}/d\Omega = R^2$. Equation (1.79) gives

$$\sigma_{el} = 4\pi \sin^2 \delta_0/k^2 \quad \text{and for} \quad kR \ll 1 \quad \sigma_{el} = 4\pi R^2$$

The classical calculation for a hard sphere gives $\sigma_{el} = \pi R^2$.

1.41 The energy eigenvalues of the square-well Hamiltonian are

$$E_{nl} = X_{nl}^2 \hbar^2/2MR^2$$

where

$$X_{nl} = KR \quad \text{and} \quad E = T + V$$

The eigenfunctions must vanish at the well boundaries so that X_{nl} can be found by locating zeros of $J(KR)$. From the explicit expressions, or from tables, we find that: for $l = 0$ ($J_{1/2}$) minima occur at $KR = \pi, 2\pi, 3\pi, \ldots$ for $l = 1$ ($J_{3/2}$) minima occur at $KR = 4 \cdot 49, 7 \cdot 73, 10 \cdot 90, \ldots$ so that the order of levels is

$$(n, l): (1, 0); (1, 1); (2, 0); (2, 1); (3, 0); (3, 1)$$

where n is the serial number for a given l.

1.42 Referring to equation (1.102) there are $2(2l + 1)$ degenerate levels for each l. If N is even, l is any even integer between 0 and N. The total number of possibilities is $2 \sum_0^N (2l + 1)$ in steps of 2, i.e. $(N/2 + 1)$ terms $= 2(N/2 + 1) 2N/2 + 2(N/2 + 1) = (N + 1)(N + 2)$. Similarly when N is odd; $(N + 1)/2$ terms.

1.43 The effective harmonic oscillator potential is

$$V = -U + \tfrac{1}{2}m\omega^2 r^2 + l(l + 1)\hbar^2/2mr^2$$

If we require $dV/dr = 0$ we have

$$\omega = [l(l + 1)]^{1/2} \hbar/mr^2$$

and substituting for r in the formula for V gives

$$V = -U + [l(l + 1)]^{1/2} \hbar\omega$$

d^2V/dr^2 is positive for real ω, so this is the minimum V.

1.44 From Example 1.41 the energy of a state that is just bound is equal to the well depth, i.e.

$$X_{nl}^2 \hbar^2/2MR^2 = V$$

and this will be the same in other wells with the same VR^2.

For the 3p level, from Example 1.41, $X_{nl} = 10 \cdot 90$ and with $M = 1 \cdot 67 \times 10^{-27}$ kg, $R = 1 \cdot 25 (118)^{1/3}$, we find

$$E_{3p} = 65 \cdot 2 \text{ MeV}$$

Since the scattering resonance shows this state to be $6 \cdot 2$ MeV above the well (neglecting c.m. corrections) the well must be $59 \cdot 0$ MeV deep. For the larger radius well, scaling according to R^2 gives $E_{3p} = 62 \cdot 7$ MeV, $V = 56 \cdot 7$ MeV (if VR^2 is constant). The scattering resonance then occurs at $6 \cdot 0$ MeV.

Chapter 2

2.3 (a) If the electron and positron have four-momenta p_μ^- and p_μ^+, then in both cases

$$p_\mu^\gamma = p_\mu^+ + p_\mu^-$$

Taking the scalar product of each side with itself gives

$$0 = 2m^2 + 2E^+E^- - 2p^+ \cdot p^-$$

In the c.m. (zero momentum) system p^+ and p^- must be equal and opposite so that

$$0 = 2m^2 + 2E^2 + 2p^2$$

which is not consistent with the relation $E^2 = p^2 + m^2$ so that the process is not possible.

(b) If the four-momentum of the electron is p_μ before emission and p'_μ afterwards then

$$p_\mu = p'_\mu + p_\mu^\gamma$$

whence

$$m^2 = m^2 + 2E'E^\gamma - 2p' \cdot p^\gamma$$

Again, in the c.m. system p' and p^γ must be equal and opposite so that

$$0 = 2E'E^\gamma + 2(p^\gamma)^2$$

and the only real solution is $p^\gamma = E^\gamma = 0$, i.e. no photon process is possible.

Chapter 3

3.1 Let the energy transferred to the electron be ω so that the energy and momentum of the electron after the collision are $(m + \omega)$, $(\omega^2 + 2m\omega)^{1/2}$. If the initial energy of the heavy particle is $E = \gamma M$ its final energy is $E' = E - \omega$. The initial momentum is $\beta \gamma M$. Conservation of four-momentum requires

$$p_\mu - p'_e = p'_\mu - p_e$$

where p_μ is for the heavy particle, p_e for the electron and the accent indicates the final state.

Taking the scalar product of each side by itself and using equation (1.22) gives

$$M^2 + m^2 - 2E(m + \omega) + 2p \cdot p'_e = M^2 + m^2 - 2mE'$$

For maximum energy transfer we assume that the electron is projected forward, and this equation then reduces to

$$2(\omega^2 + 2m\omega)^{1/2}\beta\gamma M = 2\omega E + 2m(E - E') = 2\omega(\gamma M + m)$$

Squaring and rearranging gives

$$\omega[\beta^2\gamma^2 - \gamma^2 + m^2/M^2 - 2m\gamma/M] = -2m\beta^2\gamma^2$$

and if $m \ll M$ and $2m\gamma/M \ll 1$ then

$$\omega = 2m\beta^2\gamma^2 = 2mv^2/(1 - \beta^2).$$

The condition is equivalent to $E \ll M^2/2m$.

3.9   If a nucleon absorbs a pion, the total energy is at least $m_N + m_\pi$. The corresponding momentum is given by $p^2 = (m_N + m_\pi)^2 - m_N^2$ which implies

$$p = 529 \text{ MeV/c}$$

The Fermi momentum is $p_F = \hbar k_F = (\hbar k_F/1 \cdot 6 \times 10^{-13}) \text{ MeV/c}$ and with the given value of k_F this is 268 MeV/c. Pions are therefore captured preferentially by a *pair* of nucleons rather than by a single nucleon.

3.15   Let the γ-ray energy be E_γ (mass zero). At threshold, the particles are at rest in the c.m. system. Conservation of four-momentum gives

$$p_\gamma + p_e = p_{3m}$$

and taking a scalar product of four-momenta

$$(p_\gamma)^2 + (p_e)^2 + 2\mathbf{p}_\gamma \cdot \mathbf{p}_e = (p_{3m})^2$$

Inserting values for an electron initially at rest $0 + m^2 + 2E_\gamma m = 9m^2$ so that $E_\gamma = 4m$.

3.19   Let E be the energy of the forward-going quantum. Then the backward going quantum has energy $T + 2m - E$. Conservation of four-momentum gives

$$p_+ + p_- = p_\gamma$$

and taking a scalar product of four-momenta

$$(p_+)^2 + (p_-)^2 + 2p_+ \cdot p_- = (p_\gamma)^2$$

Inserting values, remembering that for a γ-ray $E = p$,

$$m^2 + m^2 + 2(T + m)m = 4E(T + 2m - E)$$

Solution of this quadratic in E, for $T \gg m$, gives the result.

3.20   From the vector equation $\mathbf{F} = \mathbf{I} + \mathbf{J}$ we have

$$F^2 = I^2 + J^2 + 2\mathbf{I} \cdot \mathbf{J} \cos \theta$$

and using eigenvalues for the squared angular momenta

$$\cos \theta = [F(F+1) - I(I+1) - J(J+1)]/2[I(I+1)J(J+1)]^{1/2}$$

For $F = 2$, $\theta = 63 \cdot 4°$; for $F = 1$, $\theta = 138 \cdot 2°$.

Chapter 4

4.13   The kinetic energy available to overcome the Coulomb barrier is the kinetic energy in the c.m. system, which is $\frac{1}{2}\mu v^2 = A_2 T/(A_1 + A_2)$ where T is the laboratory kinetic energy. Putting this equal to the potential energy when the nuclei are in contact gives T, which is the Coulomb barrier V_C required.

4.14   Since $E = \gamma m$, uniform increments of energy mean that γ increases uniformly with distance along the accelerator, so that

$$\gamma = \gamma_0 + kl \quad \text{or} \quad dl = k^{-1} d\gamma$$

From the point of view of the electron, a Lorentz transformation gives $dl_e = dl/\gamma = d\gamma/k\gamma$ so that $l_e = k^{-1} \ln \gamma_f/\gamma_0$ where γ_f refers to the final energy. Substituting the values given $l_e = 0 \cdot 57$ m.

4.15   If we use $\mathbf{B} = \text{grad } V$ we immediately obtain

$$B_x = Gy \quad \text{and} \quad B_y = Gx$$

The force on a charged particle is given by

$$F = ev \times B$$

so the x- and y-components are

$$F_x = -evGy \quad \text{and} \quad F_y = evGx$$

i.e. proportional to the field gradient and to the component displacement.

Chapter 5

5.2 In the c.m. system, both nucleons have momentum $\beta_c \gamma_c m$ where $\beta_c = (\text{total lab. momentum/total lab. energy}) = p/(E+m)$ where $p^2 = E^2 - m^2$. For $E = (250+m)$ MeV and $m = 938$ MeV it follows that $p_c = 342$ MeV/c and by the uncertainty principle the separation is $\approx \hbar/p_c = 0.58$ fm.

5.5 The normalization constant is given by

$$4\pi \int_0^\infty |\psi|^2 \, r^2 \, dr = 1$$

and is found to be $c = (\alpha/2\pi)^{1/2}$. The probability of a separation greater than R is

$$4\pi c^2 \int_R^\infty e^{-2\alpha r} \, dr = e^{-2\alpha R} = 0.395$$

The expectation value of r is given by

$$\langle r \rangle = \int r \, |\psi|^2 \, r^2 \, dr \Big/ \int |\psi|^2 \, r^2 \, dr = 1/2\alpha = 2.2 \text{ fm}$$

5.12 For elastic scattering $\eta_l = 1$ so for s- and p-waves only we have

$$f(\theta) = 1/k[(e^{2i\delta_0} - 1)/2i + 3(e^{2i\delta_1} - 1)\cos\theta/2i]$$

The first term can be put in the form

$$\cos\delta_0 \sin\delta_0 + i \sin^2\delta_0$$

and the second can be approximated to $3\delta_1 \cos\theta$. From this we find

$$|f(\theta)|^2 = 1/k^2[\sin^2\delta_0 + 3\delta_1 \sin 2\delta_0 \cos\theta]$$

from which the result follows.

5.13 See Lock, W. O. and Measday, D. F., *Intermediate Energy Nuclear Physics*, Methuen 1970, pp. 184 and 93.

Chapter 6

6.1 Some formulae relating to a Fermi gas are given in Section 6.4.1. For the proton gas we have $\rho \propto Z/R^3$ and the Fermi wavenumber $k_F \propto Z^{1/3}/R$. The corresponding kinetic energy is $\langle T_p \rangle \propto Z^{2/3}/R^2$ per proton, i.e. $\propto Z^{5/3}/R^2$ for the nucleus. Similarly the neutron kinetic energy is $\propto N^{5/3}/R^2$ and the asymmetry energy is therefore proportional to

$$[Z^{5/3} + N^{5/3} - 2(A/2)^{5/3}]/A^{2/3}$$

If this is now expressed in terms of the quantity $(A - 2Z)/A$ and expanded to second order it is found that the asymmetry energy varies as $(A - 2Z)^2/A$.

6.2 If a particle of charge Z_1e and energy T is Coulomb-scattered through an angle θ by a much heavier particle of charge Z_2e the distance of closest approach is

$$D = (Z_1 Z_2 e^2 / 8\pi\varepsilon_0 T)(1 + \mathrm{cosec}\, \theta/2)$$

For the given Z_1, Z_2, T and θ we find $D = 1\cdot3 \times 10^{-14}$ m which gives the radius estimate.

6.5 If we assume that the proton is a uniformly charged sphere of radius R then the potential is

$$\left(\frac{1}{4\pi\varepsilon_0}\right)\frac{Ze}{r} \quad \text{for} \quad r > R,$$

and

$$\left(\frac{1}{4\pi\varepsilon_0}\right)\left(-\frac{1}{2}\frac{Zer^2}{R^3} + \frac{3}{2}\frac{Ze}{R}\right) \quad \text{for} \quad r \leqslant R$$

Inside the proton the perturbing Hamiltonian is therefore

$$\left(\frac{1}{4\pi\varepsilon_0}\right)\left(-\frac{1}{2}\frac{Zer^2}{R^3} + \frac{3}{2}\frac{Ze}{R} - \frac{Ze}{r}\right)(-e)$$

and the energy shift given by first-order perturbation theory is

$$\langle\psi|\,H\,|\psi\rangle = \left(\frac{1}{4\pi\varepsilon_0}\right)\left(\frac{4\pi}{\pi a_0^3}\right)\int_0^R \left(\frac{Ze^2 r^2}{2R^3} - \frac{3Ze^2}{2R} + \frac{Ze^2}{r}\right) r^2 \exp\left(-\frac{2r}{a_0}\right) dr$$

using the given wavefunction.

As the integral need only extend to R ($\ll a_0$) the exponential may be approximated to unity and the integration gives the final result

$$\left(\frac{1}{4\pi\varepsilon_0}\right)\frac{2}{5}\frac{e^2 R^2}{a_0^3}$$

corresponding with an energy shift of about 10^{-9} eV.

6.18 Equation (6.7b) gives $F(q)$. If the nuclear charge distribution is uniform with a sharp cut-off at $r = R$ this expression becomes

$$F(q) = \frac{4\pi\rho}{qZe}\int_0^R r \sin(\boldsymbol{q} \cdot \boldsymbol{r})\, dr$$

and integration by parts gives the required result.

Chapter 7

7.4 From $\boldsymbol{j} = \boldsymbol{l} + \boldsymbol{s}$, we obtain by taking the scalar product of each side with itself,

$$j^2 = l^2 + s^2 + 2\boldsymbol{l} \cdot \boldsymbol{s}$$

Inserting the eigenvalues for j^2, l^2, s^2 operators divided by $\hbar^2$

$$2\boldsymbol{l} \cdot \boldsymbol{s} = j(j+1) - l(l+1) - s(s+1)$$

For $j = l \pm \frac{1}{2}$, $2\boldsymbol{l} \cdot \boldsymbol{s}$ then has the values l and $-(l+1)$ so that the energy difference is proportional to $(2l+1)$.

7.6 From Reference 1.3b, p. 288 the electromagnetic spin-orbit energy may be put in the form

$$\frac{Ze^2\hbar^2}{8\pi\varepsilon_0 m^2 c^2}\frac{1}{\langle r^3\rangle}(\boldsymbol{l} \cdot \boldsymbol{s})$$

Setting $l.s \approx 1$, $Z = 10$ and $r \approx 1 \cdot 2(20)^{1/3}$ fm we obtain an energy of about 12 keV. Expression (7.6) for $l = 1$ and $A = 20$ gives an energy of $4 \cdot 1$ MeV which is very much greater than the magnetic energy and comparable with the shell spacing.

7.7   The expectation value of the potential energy for a harmonic oscillator is

$$\langle V \rangle = \langle \tfrac{1}{2}m\omega^2 r^2 \rangle = \tfrac{1}{2}(N + \tfrac{3}{2})\hbar\omega$$

so that

$$\langle r^2 \rangle = (\hbar/m\omega)(N + \tfrac{3}{2})$$

From Example 1.42 the number of neutrons or protons for a given N cannot exceed $(N+1)(N+2)$. Hence the total number of nucleons in the specified nucleus is

$$A = 2 \sum_0^N (N' + 1)(N' + 2) \rightarrow 2N^3/3 \quad \text{as} \quad N \rightarrow \infty$$

It follows that

$$\sum \langle r^2 \rangle = 2 \sum_0^N (\hbar/m\omega)(N' + \tfrac{3}{2})(N' + 1)(N' + 2) \rightarrow (\hbar/2m\omega)N^4 \quad \text{as} \quad N \rightarrow \infty$$

The mean-square radius for a nucleus of radius R is $\tfrac{3}{5}R^2$ so that for A nucleons we also have $\Sigma\langle r^2 \rangle = \tfrac{3}{5}AR^2$ and equating this to the previous value and using $R = r_0 A^{1/3}$

$$\hbar\omega = (\tfrac{3}{5}A^{-5/3})(\hbar^2/2mr_0^2)N^4$$

Substituting for N in terms of A then gives

$$\hbar\omega = (\tfrac{5}{3}A^{-1/3})(\tfrac{3}{2})^{4/3}(\hbar^2/2mr_0^2)$$

or

$$\hbar\omega \approx 40A^{-1/3} \text{ MeV}$$

7.8   In a nucleus containing a single Λ-particle

$$\langle r^2 \rangle_\Lambda = (\hbar/m\omega)(N + \tfrac{3}{2}) = \tfrac{3}{2}(\hbar/m\omega) \quad \text{as} \quad N = 0$$

From Example 7.7

$$\langle r^2 \rangle_\Lambda = \tfrac{3}{5}r_0^2 A^{2/3}$$

so that

$$\hbar\omega = (\tfrac{5}{2}A^{-2/3})(\hbar^2/m_\Lambda r_0^2) \approx 60A^{-2/3} \text{ MeV}$$

7.11   Using equations (7.10) and (7.11) with $g_l = 1$ and $g_s = 5 \cdot 586$ since the nuclei specified all have an odd proton we obtain

$$^7\text{Li} \qquad l = 1, \qquad I = \tfrac{3}{2}, \qquad \mu_I = 3 \cdot 79\mu_N$$

^{23}Na is an anomalous case because the SPSM indicates $I = \tfrac{5}{2}$ but the observed spin is $\tfrac{3}{2}$. For $l = 2$, $I = \tfrac{3}{2}$ we would have $\mu_I = 0 \cdot 12\mu_N$ in violent disagreement with observation.

$$^{39}\text{K} \qquad l = 2, \qquad I = \tfrac{3}{2}, \qquad \mu_I = 0 \cdot 12\mu_N$$

$$^{45}\text{Sc} \qquad l = 3, \qquad I = \tfrac{7}{2}, \qquad \mu_I = 5 \cdot 79\mu_N$$

Chapter 8

8.3   Let the elliptical cross-section of the nucleus be defined by the equation $x^2/a^2 + y^2/b^2 = 1$ and let the charge density be ρ. Then

$$eQ = \int (3x^2 - r^2)\rho \, dV = (8\pi\rho/15)ab^2(a^2 - b^2)$$

But $Ze = \int_{-a}^{a} \pi y^2 \rho \, dV = \tfrac{4}{3}\pi\rho ab^2$ so that

$$Q = \tfrac{2}{5}Z(a^2 - b^2)$$

The average (radius)^2 of the protons is

$$\left(\frac{1}{Z}\right)\left\langle \sum r^2 \right\rangle = \int \rho r^2 \, dV \bigg/ \int \rho \, dV = \tfrac{1}{5}(a^2 + 2b^2)$$

Using the deformation parameter as defined

$$Q = \frac{4}{3}\left\langle \sum r^2 \right\rangle \delta$$

For a spherical distribution

$$\left(\frac{1}{Z}\right)\left\langle \sum r^2 \right\rangle = \int \rho r^2 \, dV \bigg/ \int \rho \, dV = \tfrac{3}{5}R^2$$

from which

$$Q = \tfrac{4}{5}ZR^2\delta$$

8.5   Using (8.12) for the 8^+ level, $\hbar^2/2\mathscr{I} = 6\cdot94$ keV and the 10^+ level is then at 764 keV. The rigid-body moment of inertia is given by $\tfrac{2}{5}MR^2$ with $R = 1\cdot25A^{1/3}$ fm and $M = 1\cdot66 \times 10^{-7}A$ kg. If we express it as $2\mathscr{I}_{rig}/\hbar^2$ we obtain 269 MeV^{-1} whereas the observed value found above is 144 MeV^{-1}.

8.6 $E_I - E_{I-2}$ follows directly from (8.12). For a $K = 0$ band $E_I = \tfrac{1}{2}\mathscr{I}\omega^2 = \hbar^2/2\mathscr{I}[I(I+1)]$ so that $\hbar^2\omega^2 = \hbar^4/\mathscr{I}^2[I(I+1)]$ whence using (8.15) for $\mathscr{I}^2$

$$\hbar^2\omega^2 = (E_I - E_{I-2})^2 \cdot 4I(I+1)/(4I-2)^2$$

For large I, the second term on the right-hand side $\to \tfrac{1}{4}$.

8.9   The number of two-phonon states is limited by the requirement of symmetry under exchange (bosons). For $\lambda = 2$, $|\mu| \leq \lambda$ there are 25 states, which may be grouped together to give an equal number of states with definite symmetry. To see this, list the individual and resultant μ-values as follows:

$\mu_1\mu_2$		$\mu_1\mu_2$	μ	symmetry
22	$+$	22	4	s
21	$\pm$	12	3	s, as
20	$\pm$	02	2	s, as

. . . .

Completion of the table shows that there are 15 symmetrical states (and 10 antisymmetrical) with μ-values appropriate to $I = 4^+$, 2^+ and 0^+.

8.10   From (8.38) the two-nucleon state may be written

$$|IM\rangle = \sum_m (jjmm' \,|\, IM) \,|jm\rangle \,|jm'\rangle$$

If the nucleons are exchanged

$$|IM\rangle = \sum_m (jjm'm \mid IM) |jm'\rangle |jm\rangle$$

$$= \sum_m (jjmm' \mid IM) |jm\rangle |jm'\rangle \times (-1)^{2j-I}$$

by the Clebsch–Gordan property. The state function is thus symmetric or antisymmetric under exchange according as $2j - I$ is even or odd, i.e. as I is odd or even, since j is half-integral.

Since $T = 0$ is an antisymmetric isospin state and $T = 1$ is symmetric, the fermion statistics require I odd and I even in these cases respectively.

8.11   In a pure simple harmonic oscillator well the 2s and 1d levels are degenerate, with energy E, say. Spin-orbit coupling raises the $d_{3/2}$ level by an energy $\propto (l+1)$ (Ex. 7.4) and lowers the $d_{5/2}$ level by an energy $\propto l$, leaving the $s_{1/2}$ level unaltered. We write $E_{3/2} - E_{5/2} = E_{so} \propto 2l + 1$. The 4^+ *level* energy may be written $E_4 = E + \text{spin-orbit} + \text{residual interaction (R.I.)}$ and since it can only result from the $(d_{5/2})^2$ configuration we have, using figures given and $l = 2$,

$$-0.35 = E + (-2 \times \tfrac{2}{5} \times E_{so}) + 0.14 \text{ MeV}$$

$$= E - 4.0 + 0.14 \text{ MeV}$$

so that

$$E = 3.5 \text{ MeV}$$

The 2^+ *level* is formed from $(d_{5/2})^2$ and $(d_{5/2}s_{1/2})$ and the diagonal terms of the Hamiltonian are

$$\tfrac{5}{2}\tfrac{5}{2} |H| \tfrac{5}{2}\tfrac{5}{2} = E + \text{spin-orbit} + R.I.$$

$$= 3.5 - 4.0 - 0.94 = -1.44$$

$$\tfrac{5}{2}\tfrac{1}{2} |H| \tfrac{5}{2}\tfrac{1}{2} = 3.5 - 2.0 - 1.09 = 0.41$$

The $R.I.$ contributes off-diagonal terms and the full 2^+ Hamiltonian is then

$$\begin{pmatrix} -1.44 & -0.81 \\ -0.81 & 0.41 \end{pmatrix}$$

which has eigenvalues $\lambda = -1.85$ and $+0.82$ MeV.

The 0^+ *level*, by similar construction, has the Hamiltonian

$$\begin{pmatrix} -3.03 & -1.09 \\ -1.09 & 1.15 \end{pmatrix}$$

which has eigenvalues $\lambda = -3.30$ and $+1.42$. The states are therefore 0^+, 2^+, $4^+ \equiv -3.30, -1.85 \ -0.35$ MeV. The eigenvector of the 2^+ state is given by

$$\begin{pmatrix} -1.44 & -0.81 \\ -0.81 & 0.41 \end{pmatrix} \begin{pmatrix} a \\ b \end{pmatrix} = -1.85 \begin{pmatrix} a \\ b \end{pmatrix}$$

with $a^2 + b^2 = 1$. Solution of this equation gives $a = 0.9$, $b = 0.4$ so

$$|2^+\rangle = 0.9 |d_{5/2}^2\rangle + 0.4 |d_{5/2}s_{1/2}\rangle$$

Similarly

$$|0^+\rangle = 0.97 |d_{5/2}^2\rangle + 0.24 |s_{1/2}^2\rangle$$
$$|4^+\rangle = |d_{5/2}^2\rangle$$

Chapter 9

9.2   From (7.10) and (7.11) we have for an odd-neutron nucleus ($I = j$)

$$\mu_I = \mu_n = -1 \cdot 91 \mu_N \quad \text{when} \quad j = l + \tfrac{1}{2}$$
$$\mu_I = (-j/j + 1)\mu_n \quad \text{when} \quad j = l - \tfrac{1}{2}$$

The experimental μ_I thus indicates $j = l - \tfrac{1}{2} = \tfrac{1}{2}$. Evaluating expressions (9.18) for an energy of $0 \cdot 57$ MeV gives

$$\Gamma(\text{E1}) = 0 \cdot 45 \text{ eV} \qquad \Gamma(\text{M1}) = 3 \cdot 9 \times 10^{-3} \text{ eV}$$
$$\Gamma(\text{E2}) = 3 \cdot 6 \times 10^{-6} \text{ eV}$$

A transition with a *halflife* of 130 ps has a width Γ of $3 \cdot 5 \times 10^{-6}$ eV, using $\Gamma\tau = \hbar$ which agrees well with the E2 width. We therefore identify the ground state as $p_{1/2}$ (and it is a *hole* state because the closed shell is at ^{208}Pb). The excited state must have $j = \tfrac{3}{2}$ or $\tfrac{5}{2}$ and $f_{5/2}$ (hole) would agree with the shell model, and would provide 2 units of orbital angular momentum for the transition.

9.3   From Fig. 9.4 we find that suitable halflives are produced by M4 and E4 transitions of 662 keV. The corrections for A-value are given by equations (9.17) and the predicted halflives are still in the right region, with M4 closer than E4. From Fig. 9.8, $\alpha_K < 10\%$ so that internal conversion can be neglected for the present purpose.

With a $\tfrac{3}{2}^+$ ground state, an M4 transition can couple to an $\tfrac{11}{2}^-$ excited state.

9.4   The transitions are M4 and E3 and there is a 10 keV M1 transition from the first excited state to ground.

The relative intensity is given by $\Gamma(\text{M4})/\Gamma(\text{M3})$ and using equations (9.15) and (9.17) we obtain a ratio of 3×10^{-9}.

9.11   The DSAM lifetime implies a total width $\Gamma = 0 \cdot 039$ eV using $\Gamma\tau = \hbar$. Since this is much greater than the proton width we have $\Gamma_\gamma \gg \Gamma_p$ (in analogy with the slow-neutron case discussed in the text) and the cross-section at resonance is

$$\sigma = 4\pi\lambdabar^2 g\Gamma_p/\Gamma \quad \text{from (9.32)}$$

The factor g is given by equation (11.16) with $s_1 = \tfrac{1}{2}$, $s_2 = 0$ (even–even target) and $I = \tfrac{9}{2}$, i.e. $g = 5$.

The reduced de Broglie wavelength is $\lambda = \hbar/\sqrt{2m_p T_p} = 3 \cdot 35 \times 10^{-15}$ m. Substitution gives $\sigma = 0 \cdot 9$ b.

Chapter 10

10.3   The electron energy is a maximum when the neutrino energy is zero and the energy and momentum equations are then

$$Q = T_e + T_R \qquad \text{(note } Q \text{ is the } \textit{kinetic} \text{ energy available)}$$
$$\boldsymbol{p}_e = \boldsymbol{p}_R$$

Using the relativistic expression for p^2 we have

$$E_e^2 - m_e^2 = E_R^2 - M_R^2.$$

But for both particles $E = T + M$ so that

$$T_e^2 - T_R^2 = 2M_R T_R - 2m_e T_e$$

Substituting for $(T_e + T_R)$ from the first equation and rearranging gives the result.

10.8   From equation (10.16)

$$d\lambda \propto p_e p_\nu E_e E_\nu \, dE_e$$

and $p_\nu = E_\nu = (E_0 - E_e)$, putting $c = 1$. Also, at high energies $p_e = E_e$ so that

$$d\lambda \propto E_e^2 (E_0 - E_e)^2 \, dE_e$$

By integrating we find $\lambda \propto E_0^5$.

10.10 The electron momentum is a maximum when the neutrino momenta are collinear and then the energy and momentum equations are, for decay of a muon at rest

$$m_\mu = E_e + E_{\nu\bar{\nu}}$$

$$p_e = p_{\nu\bar{\nu}} = E_{\nu\bar{\nu}}$$

These give

$$m_\mu = E_e + (E_e^2 - m_e^2)^{1/2}$$

so that

$$(m_\mu - E_e)^2 = E_e^2 - m_e^2$$

from which the result follows.

10.11 From equation (10.21a) the integrated Fermi function is

$$f(Z, p) = \int_0^{p_0} p^2 (W_0 - W)^2 F(Z, p) \, dp$$

and we assume that $F(Z, p) = 0\cdot71 =$ constant. For the energy $4\cdot47$ MeV we have

$$W_0 = (4\cdot47/0\cdot511) + 1 = 9\cdot75$$

and

$$p_0 = (W_0^2 - 1)^{1/2} \qquad = 9\cdot7$$

Take $p = 1, 2, \ldots, 9$, $9\cdot7$ and tabulate $(W_0 - W)$ where $W^2 = (1 + p^2)$. The integral can then be approximated by a summation and is found to be equal to 1954. With the given $t_{1/2}$ we then have $ft_{1/2} = 1954 \times 1\cdot57 = 3068 \; s$ (omitting errors).

10.12 From equation (10.29b) the coupling constant g_β is obtained by use of the $ft_{1/2}$ value of Example 10.11 as

$$g_\beta = g_V = 1\cdot416 \times 10^{-62} \; \text{J m}^3.$$

Chapter 11

11.5   Equation (1.79) gives

$$\sigma_{\text{el}}^0 = \frac{4\pi}{k^2} \left| \frac{\eta_0 \exp(2i\delta_0) - 1}{2i} \right|^2$$

$$= \frac{4\pi}{k^2} \left| \frac{\exp(-2ikR)}{2i} \frac{\rho_0 + ikR}{\rho_0 - ikR} - \frac{1}{2i} \right|^2$$

using (11.12). This can be put in the form

$$\sigma_{\text{el}}^0 = \frac{\pi}{k^2} \left| \frac{2ikR}{\rho_0 - ikR} - (\exp(2ikR) - 1) \right|^2$$

$$= \frac{4\pi}{k^2} \left| \frac{\exp(2ikR) - 1}{2i} - \frac{1}{2i} \frac{2ikR}{(\text{Re}\,\rho_0 + i(\text{Im}\,\rho_0 - kR))} \right|^2$$

in which the real and imaginary parts of ρ_0 have been separated.

If this is compared with (1.78) we see that the first term (squared) gives the same scattering as a hard sphere ($\delta_0 = -kR$). The second term is identified with compound scattering.

Equation (1.81) gives

$$\sigma^{0}_{\text{inel}} = (\pi/k^2)(1 - |\eta_0 \exp(2i\delta_0)|^2)$$

and by use of (11,12) and again separating the parts of ρ_0 this may be reduced to

$$\sigma^0_{\text{inel}} = \frac{\pi}{k^2} \frac{-4kR\,\text{Im}\,\rho_0}{\text{Re}^2\,\rho_0 + (\text{Im}\,\rho_0 - kR)^2}$$

We now assume that near the resonance $T = T_0$ we may write (Ref. 11.1a)

$$\rho_0 = -\alpha(T - T_0) - i\beta$$

and the compound elastic cross-section then becomes

$$\sigma^0_{\text{el}} = \frac{\pi}{k^2} \frac{4k^2R^2}{\alpha^2(T-T_0)^2 + (\beta + kR)^2}$$

which is of the Breit–Wigner form (11.13)

$$\sigma^0_{\text{el}} = \frac{\pi}{k^2} \frac{\Gamma_a^2}{(T-T_0)^2 + \Gamma^2/4}$$

Similarly the inelastic cross-section σ^0_{ab} has the form (11.14).

11.6 Example 11.5 shows that the hard-sphere amplitude can be written in the form

$$(e^{2i\phi} - 1)/2i$$

with $\phi = kR$. If we define β by the equation

$$\tan\beta = \Gamma/2(T - T_0)$$

then the compound scattering amplitude $(\Gamma/2)/(T - T_0 + i\Gamma/2)$ can be expressed as

$$-(e^{-2i\beta} - 1)/2i$$

If the two amplitudes are combined in the expression for σ^0_{el} we obtain

$$\sigma^0_{\text{el}} = \frac{4\pi}{k^2} \left| \frac{e^{2i(\phi+\beta)} - 1}{2i} \right|^2$$

$$= \frac{4\pi}{k^2} |e^{i\delta_0} \sin\delta_0|^2$$

where $\delta_0 = \beta + \phi$.

11.8 The entropy of a gas may be written

$$\int_{S(0)}^{S(E)} dS = \int_0^E dE/T$$

Integrating, the left-hand side becomes

$$k[\ln\omega(E) - \ln\omega(0)]$$

and using $dE = 2ak^2T\,dT$, the right-hand side is $2ak^2T$. It follows that

$$\omega(E) = \omega(0)e^{2akT} = \omega(0)\exp[2(aE)^{1/2}]$$

11.10 From equation (5.12) by expanding the tangent term we get

$$k \tan KR = K \left[\frac{\tan kR + \tan \delta_0}{1 - \tan kR \tan \delta_0} \right]$$

which may be rearranged to give

$$\tan \delta_0 = \frac{k \tan KR - K \tan kR}{K + k \tan kR \tan KR}$$

The k-value in m^{-1} is given by $k^2 = 2m_n T/\hbar^2 = (2 \cdot 2 \times 10^{11})^2$ for 1-eV neutrons neglecting c.m. corrections. For $R = 4 \cdot 55$ fm

$$\tan kR = kR = 10^{-3}$$

and then

$$\tan \delta_0 = (\tan KR - 5)/[5 \times 10^3 + 10^{-3} \times \tan KR]$$

This shows resonances as K varies at $KR = (n + \frac{1}{2})\pi$ as successive s-levels are bound by the increasing potential well depth ($\approx \hbar^2 K^2/2m_n$). For neutrons of near-zero energy the scattering anomalies occur for radii given by $KR = (n + \frac{1}{2})\pi$. With the fixed K given and $R = 1 \cdot 2A^{1/3}$ fm successive A-values are $0 \cdot 5$, 12, 59, 162. Actual anomalies are seen at $A \approx 20$, 60, 150.

11.11 Using the notation of Chapter 5 we find that continuity of ψ and $d\psi/dr$ for the $l = 1$ wavefunction at the potential radius R leads to the expression

$$\tan (kR + \delta_1) = \frac{kK^2R^3 \tan KR}{(K^2 - k^2)R^2 \tan KR - k^2KR^3}$$

11.13 The s-state wavefunctions have no angular dependence, so the angular integral of the matrix element (11.41) reduces to

$$\int_0^\pi \exp (iqR \cos \theta) \sin \theta \, d\theta$$

$$= [\exp (iq R \cos \theta)/iqR]_0^\pi$$

$$= [\exp (-iqR) - \exp (iqR)]/iqR$$

which is proportional to $(\sin qR)/qR$. The intensity is proportional to the square of the quantity and peaks when $q = 0$ (i.e. $k_i = k_f$, $\theta = 0$).

11.19 For s-wave emission the Schrödinger equation is

$$\frac{d^2\psi}{dr^2} + \frac{2\mu}{\hbar^2} \left(T_\alpha - \frac{Z_1 Z_2 e^2}{4\pi\varepsilon_0 r} \right) \psi = 0$$

The total energy T_α of the α-particle is equal to the Coulomb energy at a distance b from the nucleus, i.e.

$$T_\alpha = Z_1 Z_2 e^2/4\pi\varepsilon_0 b$$

and substituting in the equation we have

$$\frac{d^2\psi}{dr^2} + \frac{2\mu}{\hbar^2} \frac{Z_1 Z_2 e^2}{4\pi\varepsilon_0} \left(\frac{1}{b} - \frac{1}{r} \right) \psi = 0$$

Now taking $\psi = \exp [-\gamma(r)]$ and neglecting $d^2\gamma/dr^2$ we obtain

$$\frac{d\gamma}{dr} = \left[\frac{2\mu}{\hbar^2} \frac{Z_1 Z_2 e^2}{4\pi\varepsilon_0} \left(\frac{1}{r} - \frac{1}{b} \right) \right]^{1/2}$$

and γ is then as given in equation (11.48). The integration from R to b is performed using the substitution $r = b \cos^2 \theta$, and equation (11.50) follows using the expression for T_α and the barrier height

$$B = Z_1 Z_2 e^2 / 4 \pi \varepsilon_0 R$$

11.20 Write the final bracket in (11.50) in the form

$$\cos^{-1} x^{1/2} - x^{1/2}(1-x)^{1/2}$$

where $x = T_\alpha / B = R/b$ is small. If this is so then $\cos^{-1} x^{1/2} = \pi/2 - \varepsilon$ with $\varepsilon = x^{1/2}$.

Also the second term is $\approx x^{1/2}$ so that altogether the bracket becomes

$$\pi/2 - 2x^{1/2}$$

and for the quantity γ

$$\gamma = \frac{Z_1 Z_2 e^2}{4 \varepsilon_0 \hbar v} - \frac{2e}{\hbar} \left(\frac{Z_1 Z_2 \mu R}{2 \pi \varepsilon_0} \right)^{1/2}$$

11.23 Excluding the asymmetry and pairing terms the semi-empirical mass formula gives

$$-B = -\alpha A + \beta A^{2/3} + \varepsilon Z^2 / A^{1/3}$$

If a nucleus of mass number A splits into $A_1 + A_2$ the energy released is given by the mass difference

$$\beta(A^{2/3} - A_1^{2/3} - A_2^{2/3}) + \varepsilon(Z^2/A^{1/3} - Z_1^2/A_1^{1/3} - Z_2^2/A_2^{1/3})$$

Write $A_2 = A - A_1$ and $Z_2 = Z - Z_1$ and differentiate each term with respect to A_1. The first term gives a maximum for $A_1 = A/2$ and the second has turning-points at $A_1 = A$ and $A_1 = A/2$. Differentiation with respect to Z_1 gives turning-points at $Z_1 = Z$ and $Z_1 = Z/2$. Of these possibilities, one describes no fission at all and the other, $A_1 = A_2 = A/2$ and $Z_1 = Z_2 = Z/2$, is the required condition. It may be checked to be a maximum by a second differentiation.

By setting the energy release to zero for the case of equal division and using $\beta = 17 \cdot 8$ MeV and $\varepsilon = 0 \cdot 71$ MeV we find

$$Z^2/A \geqslant 17 \cdot 6$$

11.24 The work done in bringing the two identical nuclei from infinity to contact is easily found to be

$$\frac{1}{4} \frac{Z^2 e^2 2^{-2/3}}{4 \pi \varepsilon_0 r_0 A^{1/3}}$$

and this is the corresponding energy release in fission. Setting it equal to the energy available from the mass change, as calculated in Example 11.23, we find

$$Z^2/A \geqslant 56 \cdot 4$$

11.27 The formula is proved in the author's *Nuclear Physics, an Introduction*, Longman 1973, Sect. 5.1.4. See also Ref. 1.3b, p. 36, and Burge, E. J., *Atomic Nuclei and their Particles*, Oxford 1977, Ch. 3.

11.28 Formula 11.62 may be written in terms of the impact parameter b as

$$R_{\min} = b(\tan(\theta/2) + \sec(\theta/2)) = R$$

say. To prove this, let the tangential velocity at the closest point be v. Then conservation of angular momentum and energy give

$$Rv = bv_1$$

$$\tfrac{1}{2}\mu(v_1^2 - v^2) = Z_1 Z_2 e^2 / 4\pi\varepsilon_0 R$$

Eliminating v

$$\tfrac{1}{2}\mu v_1^2 \left(1 - \frac{b^2}{R^2}\right) = \frac{Z_1 Z_2 e^2}{4\pi\varepsilon_0 R} = \mu v_1^2 \tan\frac{\theta}{2} \cdot \frac{b}{R}$$

from the formula in Example 11.27. Writing $x = R/b$ we obtain the quadratic

$$x^2 - 2x \tan(\theta/2) - 1 = 0$$

whose positive solution is $x = (\tan(\theta/2) + \sec(\theta/2))$ as required.

11.29 (a) The laboratory velocity of the Fe ion is $v_1 = 4\cdot3 \times 10^7$ ms^{-1}. Using this we obtain $\eta = 122$.

(b) To find the angular momentum number L we use (11.63) but we require the c.m. angle of deflection. From equation (1.16) and the laboratory angle of 39° this is found to be 47° and L is then $122/\tan 23\cdot5 = 280$.

(c) For the interaction distance R_{min} we use (11.65) but the c.m. wave-number k is required. This is given by

$$k = \mu v_1/\hbar \quad \text{where} \quad \mu = (56 \times 238)/(56 + 238)\,\text{a.m.u.}$$

$$= 29\cdot7\,\text{fm}^{-1}$$

and R_{min} is then 14·4 fm.

(d) The reaction cross-section is πb^2 where b is the impact parameter for the grazing orbit, for which absorption begins. From the calculated R_{min} and equation (11.62) $b = 0\cdot66 R_{min}$, whence $\sigma = 2\cdot84$ b.

Appendix 6 Fundamental constants

The constants and conversion factors are taken from E. R. Cohen, 'The 1973 Table of the Fundamental Physical Constants', *Atomic Data and Nuclear Data Tables* **18,** 587, 1976. Errors in the constants, usually not more than a few parts per million, have been omitted. Symbols are generally in accord with the recommendations of the report 'Quantities, Units and Symbols', The Royal Society, London 1975.

The permeability of the vacuum μ_0 $(=1/c^2\varepsilon_0)$ is taken to be $4\pi \times 10^{-7}$ H m^{-1} exactly and the permittivity of the vacuum is then $8\cdot854\,19 \times 10^{-12}$ F m^{-1}. The atomic mass scale is based on the mass of the isotope ^{12}C (Sect. 6.1).

Quantity	Symbol	Value
General		
Gravitational constant	G	$6\cdot672 \pm 0\cdot004 \times 10^{-11}$ N m^2 kg^{-2}
Velocity of light	c	$2\cdot997\,92 \times 10^8$ m s^{-1}
Avogadro's number	N_A	$6\cdot022\,05 \times 10^{23}$ mol^{-1}
Elementary charge	e	$1\cdot602\,19 \times 10^{-19}$ C

Elements of nuclear physics

Quantity	Symbol	Value
Faraday constant	$F = N_A e$	$9.648\,46 \times 10^4$ C mol^{-1}
Electron mass	m_e	$9.109\,53 \times 10^{-31}$ kg
Electron charge to mass ratio	e/m_e	$1.758\,80 \times 10^{11}$ C kg^{-1}
Classical electron radius	$r_0 = \mu_0 e^2/4\pi m_e$	$2.817\,94 \times 10^{-15}$ m
Thomson cross-section	$\frac{8}{3}(\pi r_0^2)$	$6.652\,45 \times 10^{-29}$ m^2
Gas constant	R	$8.314\,41$ J mol^{-1} K^{-1}
Boltzmann constant	$k = R/N_A$	$1.380\,66 \times 10^{-23}$ J K^{-1}
Planck constant	h	$6.626\,18 \times 10^{-34}$ J s
	$\hbar$	$1.054\,59 \times 10^{-34}$ J s
Compton wavelength:		
of electron	$h/m_e c$	$2.426\,31 \times 10^{-12}$ m
of proton	$h/m_p c$	$1.321\,44 \times 10^{-15}$ m

Atomic and nuclear masses

Quantity	Symbol	Value
Atomic mass unit	a.m.u. or u	$1.660\,57 \times 10^{-27}$ kg
Atomic mass:		
of electron	—	$5.485\,80 \times 10^{-4}$ u
of proton	M_p	$1.007\,276$ u
of hydrogen atom	M_H	$1.007\,825$ u
of neutron	M_n	$1.008\,665$ u
Ratio of proton to electron mass	—	1836.15
Mass:		
of proton	m_p	$1.672\,65 \times 10^{-27}$ kg
of neutron	m_n	$1.674\,92 \times 10^{-27}$ kg
Neutron–H atom mass difference	$M_n - M_H$	0.782 MeV/c^2

Spectroscopic constants

Quantity	Symbol	Value
Rydberg constant	$R_\infty = m_e e^4/8h^3 \varepsilon_0^2 c$	$1.097\,37 \times 10^7$ m^{-1}
Bohr radius	$a_0 = h^2 \varepsilon_0/\pi m_e e^2$	$5.291\,77 \times 10^{-11}$ m
Fine-structure constant	$\alpha = \mu_0 e^2 c/2h$	$7.297\,35 \times 10^{-3}$

Magnetic quantities

Quantity	Symbol	Value
Bohr magneton	$\mu_B = eh/4\pi m_e$	$9.274\,08 \times 10^{-24}$ J T^{-1}
Nuclear magneton	$\mu_N = eh/4\pi m_p$	$5.050\,82 \times 10^{-27}$ J T^{-1}
Magnetic moment:		
of electron	μ_e	$9.284\,83 \times 10^{-24}$ J T^{-1}
of muon	μ_μ	$4.490\,47 \times 10^{-26}$ J T^{-1}
of proton	μ_p	$\begin{cases} 1.410\,62 \times 10^{-26} \text{ J T}^{-1} \\ 2.793 \mu_N \end{cases}$
of neutron	μ_n	$\begin{cases} -0.966\,30 \times 10^{-26} \text{ J T}^{-1} \\ -1.913 \mu_N \end{cases}$
Gyromagnetic ratio of proton (corrected for diamagnetism of water)	γ_p	$2.675\,13 \times 10^8$ rad s^{-1} T^{-1}
g-factors:		
of proton	$\begin{cases} g_l(p) \\ g_s(p) \end{cases}$	$\begin{aligned} &1 \\ &5.586 \ (=2\mu_p/\mu_N) \end{aligned}$
of neutron	$\begin{cases} g_l(n) \\ g_s(n) \end{cases}$	$\begin{aligned} &0 \\ &-3.826 \ (=2\mu_n/\mu_N) \end{aligned}$

Quantity	Symbol	Value
Conversion factors		
1 a.m.u. (u)		$\begin{cases} 1{\cdot}660\,57 \times 10^{-27}\ \text{kg} \\ 931{\cdot}502\ \text{MeV}/c^2 \end{cases}$
1 electron mass (m_e)		$0{\cdot}511\,003\ \text{MeV}/c^2$
1 proton mass (m_p)		$938{\cdot}280\ \text{MeV}/c^2$
1 eV		$\begin{cases} 1{\cdot}602\,19 \times 10^{-19}\ \text{J} \\ 2{\cdot}417\,97 \times 10^{14}\ \text{Hz} \\ 8{\cdot}065\,48 \times 10^{5}\ \text{m}^{-1} \\ 1{\cdot}160\,45 \times 10^{4}\ \text{K} \end{cases}$
Energy–wavelength product		$1{\cdot}239\,85 \times 10^{-6}\ \text{eV} \times \text{m}$
Width–lifetime product ($\hbar$)		$6{\cdot}582\,18 \times 10^{-16}\ \text{eV} \times \text{s}$

Index

Index

D

data analysis (track chamber), 139
de Broglie waves, 19, 21, 27
decay constant, 59
decay law, 59
deep inelastic scattering, 367
deformation, nuclear, 220–42, 247–51
delayed coincidence method, 271
delayed neutrons, 351
delta particle (Δ), 32, 63–64, 171, 178, 317–21
delta rays, 73
density of states, 296–301
depletion layer, 135
detectors, general, 128–42
  junction, 134–137
  scintillation, 128–33
  semiconductor, 133–8
  surface barrier, 135–7
deuteron, 147, 155–8
  binding energy of, 155–8, 171
  electromagnetic moments of, 156, 171–3
  photodisintegration of, 171–2
  radius of, 156
  stripping reactions of, 209
  virtual level of, 153–60, 172
differential cross-section, 18
diffused junction detector, 135
dipole radiation, 262–4
direct interaction processes, 334–2
  of heavy ions, 361–5
direct inelastic scattering, 337–40
disintegration, of beryllium, 55–8
  of deuterium, 171–2
  of lithium, 7, 107
  of nitrogen, 7, 55
  of nucleus, 6–7, 107
dispersion (spectrometer), 126
displacement laws, 49–50
distorted waves, 339
doorway state, 332
Doppler broadening of resonances, 280–1
Doppler shift attenuation method (DSAM), 273–4
double β-decay, 306–7
double-closed shell, 207, 243
double-focusing spectrometer, 126
double hump, 356–8
doublet structure, 206
drift chamber, 141
drift tube accelerator, 112–15

E

effective interaction, 189, 244
effective potential, 25, 36–7
  in nuclear matter, 189
effective range, 155, 378–80
elastic collisions, 4–6
elastic scattering, 4–6, 30, 182–5, 214–15, 332–3
  of α-particles, 3, 31–2, 60, 67, 147, 178
  of electrons, 182–4
  of heavy ions, 361–2
  of neutrons, by protons, 149–55
  of protons, 160–3, 184, 214–5, 325
  of radiation, 85
elasticity (reactions), 326
electric dipole radiation, 262–4
electric quadrupole moment, 100–1, 210–3, 226, 240–1
electromagnetic interaction, 69–71, 255–88, 374
electromagnetic moments, 99–101, 172–4, 209–13, 240–2
electromagnetic radiation, absorption of, 81–7
electron, 46–9
electron capture, 195, 300–1
electron scattering, elastic, 182–4
  inelastic, 67, 276–8
electron synchrotron, 124–5
electrostatic generator, 108–11
elementary particles, table, 374, 376–7
elements, abundance of, 195–7
endothermic reaction, 8
energy gap, 193
energy levels, 35–40, 205–9, 247–9, 317–30
energy loss, by collisions, 76–8
  by radiation, 78–81
energy-mass relation, 12–15, 179
energy of mirror nuclei, 184
evaporation spectra, 330
exchange forces, 175, 177
excitation function, 329
exclusion principle, 160, 161, 166, 188–90, 194, 205, 208, 332
exothermic reaction, 8
exotic atoms, 97–8

F

Fano factor, 90, 135
Feather relation, β-rays, 90
Fermi function, 299–300
Fermi gas model, 186, 188, 198
Fermi-Kurie plot, 292–3, 299, 315
Fermi theory of β-decay, 295–301
Fermi transitions, 302–3
Feynman diagram, 61

402